# Theorizing Digital Cultural Heritage

**Media in Transition**

David Thorburn, *series editor*

Edward Barrett, Henry Jenkins, *associate editors*

*New Media, 1740–1915*, edited by Lisa Gitelman and Geoffrey B. Pingree, 2003

*Democracy and New Media*, edited by Henry Jenkins and David Thorburn, 2003

*Rethinking Media Change: The Aesthetics of Transition*, edited by David Thorburn and Henry Jenkins, 2003

*Neo-Baroque Aesthetics and Contemporary Entertainment*, Angela Ndalianis, 2004

*Theorizing Digital Cultural Heritage: A Critical Discourse*, edited by Fiona Cameron and Sarah Kenderdine, 2007

# Theorizing Digital Cultural Heritage

A Critical Discourse

*Edited by Fiona Cameron and Sarah Kenderdine*

The MIT Press

Cambridge, Massachusetts

London, England

MIT Press books may be purchased at special quantity discounts for business or sales promotional use. For information, please e-mail ⟨special_sales@mitpress.mit.edu⟩ or write to Special Sales Department, The MIT Press, 55 Hayward Street, Cambridge, MA 02142.

This book was set in Perpetua on 3B2 by Asco Typesetters, Hong Kong.
Printed and bound in the United States of America.

Library of Congress Cataloging-in-Publication Data

Theorizing digital cultural heritage : a critical discourse / edited by Fiona Cameron and Sarah Kenderdine.
    p.   cm. — (Media in transition)
Includes bibliographical references and index.
ISBN 978-0-262-03353-4 (alk. paper)
1. Cultural property—Digitization. 2. Digital art. 3. Library materials—Digitization. 4. Virtual museums. 5. Museums—Collection management. I. Cameron, Fiona. II. Kenderdine, Sarah. III. Series.

CC135.T47   2007
363.6'90285—dc22                                                          2005056754

10   9   8   7   6   5   4   3   2   1

# Contents

## Series Foreword

David Thorburn, *editor*
Edward Barrett, Henry Jenkins, *associate editors*

New media technologies and new linkages and alliances across older media are generating profound changes in our political, social, and aesthetic experience. But the media systems of our own era are unique neither in their instability nor in their complex, ongoing transformations. The Media in Transition series will explore older periods of media change as well as our own digital age. The series hopes to nourish a pragmatic, historically informed discourse that maps a middle ground between the extremes of euphoria and panic that define so much current discussion about emerging media—a discourse that recognizes the place of economic, political, legal, social, and cultural institutions in mediating and partly shaping technological change.

Though it will be open to many theories and methods, three principles will define the series:

- It will be historical—grounded in an awareness of the past, of continuities and discontinuities among contemporary media and their ancestors.
- It will be comparative—open especially to studies that juxtapose older and contemporary media, or that examine continuities across different media and historical eras, or that compare the media systems of different societies.
- It will be accessible—suspicious of specialized terminologies, a forum for humanists and social scientists who wish to speak not only across academic disciplines but also to policymakers, to media and corporate practitioners, and to their fellow citizens.

### *Theorizing Digital Cultural Heritage*
### edited by Fiona Cameron and Sarah Kenderdine

These essays identify some of the ways in which digital technologies have transformed the traditional museum, altering our understanding of such fundamental words as *indigenous*, *artifact*, *heritage*, *space*, *ecology*, *the past*. If the new technologies promise to enhance and valuably complicate the archiving, preservation, and display of cultures, they also may strengthen and extend embedded forms of ideological appropriation and control. The emergence of these powerful technologies has created an extended moment of transition and reevaluation, as these writers argue, in which the ground assumptions of the museum and of the knowledge communities devoted to preserving and representing the cultural heritage must be reconsidered, newly theorized, re-imagined.

# *Acknowledgments*

*Digital Cultural Heritage: A Critical Discourse* has been a thrilling and demanding mission, and its foundation is the professional and personal generosity of the thirty authors whose words make up this volume. This willingness to contribute their research to such a tome made it possible for the editors to confidently approach their task, and to provide a work of significance to the sector. For this, we thank you all. Most importantly, we have esteemed regard for the commissioning editors of the work at MIT Press, David Thorburn and Doug Sery, who allowed this volume of relatively young authors to be compiled, and who have helped steer it to a successful conclusion. Also, thanks to Valerie Geary at MIT Press for her administrative support.

The successful completion of the book has been made possible by the affirmation provided by Museum Victoria, under the direction of Tim Hart, Director of Information, Multimedia, and Technology. The Museum's timely financial contribution endorsed the belief that this work is of great importance to the development of technologies in cultural organizations.

The work was mentored by Professor Stephen Garton, Dean of Arts at Sydney University. Thanks also to Merilee Robb, Director, Research Development at the University of Sydney, and Professor Ien Ang at the Centre for Cultural Research, University of Western Sydney, for their research support and advice.

The scholastic rigor of the contributors was honed with the help of a significant number of peer reviewers and we extend our gratitude to: Ross Parry, Andrea Witcomb, Angelina Russo, Helena Robinson, Beryl Graham, Harald Kraemer, Klaus Muller, Rose Bollen, Martin Doerr, Gerhard Budin, Kevin Sumption, Lynda Kelly, Anita Kocsis, Erik Champion, Norbert Kanter, Franca Garzotto, Leonard Steinbach, Paolo Paolini, Hayley Townsend, Alonzo Addison, David Butts, and Lynne Teather.

We would like to thank Fiona Hile and Mary Watson for their keen proofing and editorial eye, and Gus Gollings of Common Ground Publishing for taking on the volume in all its variation.

# Theorizing Digital Cultural Heritage

# Introduction

Museums and heritage organizations have institutionalized authority to act as custodians of the past in Western societies. As such, they hold a significant part of the "intellectual capital" of our information society. The use of emerging digital technologies to activate, engage, and transform this "capital" is paralleled by shifts in the organizational and practice culture of the institutions entrusted with its care.[1] In a symbiotic relationship, cultural heritage "ecologies" also appropriate, adapt, incorporate, and transform the digital technologies they adopt. Why and how this transformation occurs in our cultural organizations requires serious investigation and is the subject of this compilation.

The renowned and much quoted media theorist Marshall McLuhan (1964) stated "we become what we behold that we shape our tools and thereafter our tools shape us." Simply put, McLuhan argued that new ways of perceiving the world, embedded in knowledge structures and societal transformations, enable the development of tools that emulate new social and theoretical ideas. Thereafter these tools, through technological innovation, have the ability to offer a range of possibilities beyond those originally imagined. Looking at the relationship between technology, cultural theory, and society from another angle, Lev Manovich (2001) rightly argues that new media is culture "encoded" in a digital form.

This anthology, international in scope, draws together thirty authors to critically analyze and theorize on themes of museums and heritage in relation to "digital culture." It is concerned not simply with the implementation of digital technologies themselves but with the relationships created within cultural complexes such as the philosophical, historical, social, artistic, biological, geographic, and the linguistic.

This book represents the first comprehensive theoretical discourse on cultural heritage and digital media since 1997. That edited collection, *The Wired Museum: Emerging Technology and Changing Paradigms*, by Katherine Jones-Garmil, followed the advent of the museum Internet revolution and discusses the future of these institutions through a technological lens. Today, nine years since this ground-breaking publication (a period of unprecedented technical development within the field), it is timely that a more critical and theoretical appraisal—of the specific roles that museums and cultural heritage institutions play in interpretation and representation using new and emerging digital media—should be realized.

An abundance of publications have emerged in recent years that investigate critical cultural theory and global digital cultures at a meta-level. These texts[2] discuss issues regarding "virtuality," the relationships between the virtual and real, the body and machine, place, space, hypertext, cyberspace, and interactivity. A number of the works are directed specifically toward cinema and the visual arts sector and Internet cultures. Yet other issues specific to the museums and heritage sector remain, that have yet to receive significant exposure in the marketplace outside of conference circuits.[3] These include: digital cultural heritage as a political concept and practice; the representation and interpretation of cultural heritage such as digital objects (including questions of aura and debates of "virtual" versus "real"); issues of mobility and interactivity both for objects and for consumers of digital heritage; the relations between communities and heritage institutions as mediated through technologies; the reshaping of social, cultural, and political power in relation to cultural organizations made possible through communication technologies; and the visualization and interpretation of archaeological sites and historic environments, embracing both digitally based knowledge management tools and virtual reality. Necessarily, the discussion of these issues embraces other concerns common to critical digital theory such as virtual reality, cognitive science, artificial intelligence, visual art history and theory, cultural communication and learning theory, social research, information management and indigenous knowledge, cultural studies, communications, history, anthropology, museum studies, film studies, and information management.

The suite of essays in this publication is intended to serve a broad international audience. As a reference work, it provides a resource for professionals, academics, and students working in all fields of cultural heritage (including museums, libraries, galleries, archives, and archaeology) as well as education and information technology. It is envisioned that the publication will be used as a primary and a secondary text at both undergraduate and postgraduate levels. This rich compendium will also be of value to non-specialists who are interested in cultural heritage and its future interpretation.

In the last few years much of the discourse about the relation between cultural heritage and digital technology has been descriptive and introspective, focusing on projects and their technical considerations. These discussions are appearing mostly in journals arising from conferences or on respective project-based Web sites. These fora continue to make valuable contributions to the ongoing development of the sector, although they have yet to foster a body of sustained critical thinking about the meanings and implications of the apparent transformations, challenges, and possibilities posed by communications technologies. Moreover, digital heritage's standing as heritage has been a source of considerable debate over recent years. It is only recently that digital heritage has accorded status as an entity in its own right. The *UNESCO Charter on the Preservation of Digital Heritage* articulates this turn by creating a new legacy—the digital heritage: "resources of information and creative expression are increasingly produced, distributed, accessed and maintained in digital form, creating a new legacy—the digital heritage."[4] As a result, digital technology has been largely unmapped in terms of a critical theory for cultural heritage per se, and for heritage institutions.

*Digital Cultural Heritage: A Critical Discourse* departs from the volumes preceding it[5] with the range of timely theoretical contributions it brings to an analysis of cultural heritage and technology. The choice of authors, combining both theorists and practitioners in various disciplines pertaining to cultural heritage, moves the discussion beyond purely theory described by Peter Lunenfeld as "science fictionalized discourse" to ground these discourses in practice (1999, ix). This combination of theory and praxis lies at the foundation of this book.

Collecting organizations are vehicles for the enduring concerns of public spectacle, object preservation, shifting paradigms of knowledge and power. Digital technologies are implicated with historical transformations in language, society, and culture, and with shifting definitions of the museum. To speak of the digital is to engage simultaneously with the impressive array of virtual simulacra, instantaneous communication, ubiquitous media, global interconnectivity, and all their multifarious applications.[6] Given how important digital technologies have become in our lives, this collection of essays arose as the result of a perceived need for a sustained interchange between digital cultural theory and heritage practices. Although the cultural heritage sector acknowledges that digital technology requires institutions to face new challenges, many of these issues have not yet been fully imagined, understood, or critically explored outside of conference roundtables. The authors here consider the extent to which digital technologies are a cultural construct and how they might be used purposefully to transform institutional cultures, methods, and, most importantly, relationships with audiences—into the future.

This book demonstrates that the idea of a technologically determined future for museums is not a debate unique to our time; it is one that has a long historical trajectory. Thoughts of technological determinism have always taken center stage when apparent ruptures and dramatic shifts in the *modus operandi* occur—thereby enabling us periods of acute observation. As Fredric Jameson argues, "radical breaks between periods do not generally involve complete changes but rather the restructuring of a certain number of elements already given" (1983, 123). Identifying and predicting the parameters for future exchange between cultural organizations and audiences, together with the operational paradigm shifts required in the face of dynamic technological change this process demands, is the task of the authors in this volume.

Given the breadth of discussion, the book has been broken into three parts. Part I: *Replicants/Object Morphologies*; Part II: *Knowledge Systems and Management—Shifting Paradigms and Models*; and Part III, which focuses upon *Virtual Cultural Heritage*. The series of chapters in part I: *Replicants/Object Morphologies* is primarily concerned with the confluence of technology and culture in the representation of art and heritage collections for both Western and Indigenous communities. As Christiane Paul argues, in her book *Digital Art*, "Technologies often tend to develop faster than the rhetoric evaluating them, and we are still in the process of developing descriptions for art using digital technologies as a medium—in social, economic, and aesthetic respects" (2003, 67). The authors in part I seek to engage with this debate in relation to the wider applications of cultural heritage.

The reproducibility and so called "immaterial" nature of digital objects and assets has induced some anxiety within the museum sector. The idea that "real" objects and works of art are under threat, exacerbated by theories of mechanical reproduction and simulation by proponents such as Walter Benjamin and more recently Jean Baudrillard, has had a persuasive effect on the way museum collections and digital objects have been viewed, used, and assigned meaning.[7] To this end these contributors critically examine and re-theorize prevailing arguments about the relationships between material and digital objects. They examine definitions and morphologies assigned to objects, art works and other cultural objects, and multimedia, in both exhibitions and digital space and speculate on how these have impacted on the role, function, and imaginative uses of digital media and digital art. The theses of these arguments rest on one overarching theme, that is, what new understandings can be brought to bear on the relationship between digital and physical collections, art works, and on the digital object? Moreover, these commentaries challenge conventional understandings of museum representation, art, history, and culture, and their application to digital objects and Indigenous collections.

From this starting point the authors explore a number of avenues around this theme. Peter Walsh looks to older visual technologies such as photography, and the historical foundations of cultural heritage institutions, to illuminate possible meanings for digital objects. Andrea Witcomb offers new perspectives on the display of objects alongside digital media in exhibitions through an analysis of interpretive approaches and forms of narrative construction. Fiona Cameron examines the poetics and politics of the "digital" historical object, the relationship between virtual and material objects and more abstract concepts of materiality, aura and authenticity, authority, interpretation, representation, knowledge, and affect. The potential uses of augmented and virtual reality in managing, researching, and sustaining Indigenous collections is a major theme of Deidre Brown's contribution.

Authors Beryl Graham and Sarah Cook investigate the histories of digital art works and critically examine their place within physical spaces and the Internet as a means of offering new perspectives on their meaning and the challenges posed for conventional practices of curatorship and documentation. A discussion of the history and politics of art communities in networked and real-world locations forms a basis for narratives on collaborative art creation, creative content creation, and education, accessibility, and usability.

Most significantly, these chapters offer new discourses of meaning for digital objects, art works, and media within cultural heritage institutions. Discussions illuminate the political and economic implications of the use of digital media in the production of art works and in the acquisition, representation, and conservation of collections. Furthermore, they foreground transformations required in institutional mission and practice, and in relationships with audiences.

In chapter 1, *The Rise and Fall of the Post-Photographic Museum*, author Peter Walsh gives his attention to the history of the development of technology as a cultural practice. He looks to the history of the development of technology, notably photography, art, and the modern art museum in the nineteenth century to find a parallel revolution to the one currently being experienced with the introduction of new media. Walsh examines the way photography changed the manner in which art was presented, reproduced, analyzed, and evaluated and the profound impact the introduction of this technology had on the everyday practices of museums other than the range of its collecting activities. In his celebrated essay, *The Work of Art in the Age of Mechanical Reproduction*, Walter Benjamin writes, "that which withers in the age of mechanical reproduction is the aura of the work of art." Here Walsh makes the opposite claim that the "aura" of original art works was, in fact, created by the invention and distribution of photographic technologies.

Andrea Witcomb, in *The Materiality of Virtual Technologies: A New Approach to Thinking about the Impact of Multimedia in Museums* (chapter 2), discusses the impact of digital media technologies in museum exhibitions. Witcomb argues that existing discussions tend to be based on an opposition between the virtual and the material world; the virtual is thus interpreted either as a threat or as a radical process of democratization. Here, Witcomb adopts a different viewpoint, one that begins by taking digital media technologies as material objects in their own right, thus enabling the emergence of new perceptions on the relationship between the display of objects alongside digital media elements.

Following on from this theme, *Beyond the Cult of the Replicant: Museums, Objects—Traditional Concerns, New Discourses* (chapter 3), Fiona Cameron examines prevailing debates and bifurcations used to describe and define historical collections and virtual/digital "historical" objects. Cameron illustrates how digital historical collections have been bounded by an object-centered museum culture and material culture paradigms. The author argues that the roles and uses of the digital object must also be understood as part of the broader heritage complex—an institutionalized culture of practices and ideas that is inherently political, socially and culturally circumscribed. Cameron concurs that there is a need to move away from the formalist notions of technology and materiality that make digital objects fit into the specific rubric of "replicant" that have constrained their value, meaning, and imaginative uses. This chapter rethinks the relationships between historical collections and their digital counterparts while offering new understandings of digital "historical" objects as material, representation, presence, affect, experience, and value.

*Te Ahu Hiko: Digital Cultural Heritage and Indigenous Objects, People and Environments* by Deidre Brown (chapter 4), examines the possible heritage applications of three-dimensional augmented and virtual reality (AR and VR) to Indigenous New Zealand Maori treasures, bodies, and landscapes. Here Brown argues that the digital landscape presents itself to Developing and Fourth World people as a new frontier ready for settlement. While generally the product of First World ideologies, Brown demonstrates how digital media offer non-Western people the opportunity to challenge the authority of such structures and reinterpret the way in which cultural arts, artifacts, practices, and environments are managed and presented.

In *Redefining Digital Art: Disrupting Borders* (chapter 5), Beryl Graham examines current working categories for classifying digital art in physical installations and looks at how these definitions are affecting how art is shown. Graham goes on to argue that digital art as a relatively new medium is disrupting "safe" categories and shows how its particularly fluid characteristics are inherently worrisome to arts institutions. Graham uses her own working definition of digital art to examine the particular challenges of collecting, documenting, and conserving new media art in institutions.

Beginning with the working definitions of digital art first proposed by Beryl Graham in chapter 5, Sarah Cook, in *Online Activity and Offline Community: Cultural Institutions and New Media Art* (chapter 6) further refines the taxonomy of digital art with a discussion of strictly online practices—net.art, net-art, and art on the Internet. Definitions and examples, as well as established criteria, are provided. In particular the question of the necessity for collaborative art creation in an online environment is addressed through an examination of the inherent multiplicity of agendas and structures prevalent in the online world. The role that the shifting politics and economics of the Internet play in the creative process of content-generation online is the focus of Cook's discussion.

The series of chapters in part II: *Knowledge Systems, Management and Users—Shifting Paradigms and Models*, critically investigates the confluence of knowledge, learning, information management, digital technologies, and user research in the cultural heritage sector. These discussions address current imperatives and those likely to be operative in the future, pertaining to cultural heritage institutions as knowledge providers, the subject of knowledge and learning brought about by current cultural complexes, technologic possibilities, and users. As Lev Manovich argues, the shift of culture to computer-mediated forms of production, distribution, and communication "affects all stages of communication, acquisition, manipulation, storage, and distribution" (2001, 19). In the context of this discourse, prevailing themes in part II, by authors Fiona Cameron and Helena Robinson, and Harald Kraemer, address shifting epistemological paradigms, digital landscapes, and the implications for knowledge creation and the documentation of collections in museums. The histories of taxonomies, methodologies, and information standards for creation and access to knowledge are addressed by Ingrid Mason and Gavan McCarthy. Of primary concern are the polemics and politics of digital technologies, documentation, and information standards. The authors also address notions of documentation: as discursive, as a navigational system, and as a conduit for new theoretical, conceptual, access, and learning paradigms.

Other contributors, in particular Susan Hazan, Angelina Russo and Jerry Watkins, and Ross Parry and Nadia Arbach, address more abstract issues around power and knowledge, expertise, authority, objectivity, learning and narrative construction, user/audience relationships, and notions of social inclusion and exclusion in knowledge access.

From these starting points, contributors address the topics within the various disciplines of art, history, museology, the domain of information management, and Web design. Most significantly, they offer fresh insights into traditional concerns around knowledge and cultural heritage, and challenge the authority of such structures. Furthermore they propose ways in which the aforementioned imperatives might reshape institutions, foreground new knowledge models and practices in knowledge creation, in narrative formation, in documentation, learning, and in exhibition curation.

In chapter 7, *A Crisis of Authority—New Lamps for Old*, Susan Hazan questions the role of digital media in museums and how it modifies the relationship between museums and their audiences in terms of knowledge. As traditional owners of an ideological expert system that was once perceived as the primary authority on the knowledge systems they manage, Hazan argues that museums must now face the realization that their audiences may avail themselves of such knowledge through the media and other online resources. At the same time, the educational goals and social responsibilities that have traditionally fallen within the parameters of the insular institution of the museum are being challenged from many different directions. This chapter maps out the various adjustments that museums are making, and are required to make, as they reach a critical moment in a crisis of authority.

Following on from this argument, Angelina Russo and Jerry Watkins (chapter 8) *Digital Cultural Communication: Audience and Remediation* discuss the potential for convergent new media technologies to connect cultural institutions to new audiences through community cocreation programs using a framework termed "Digital Cultural Communication." The authors argue that this connection requires more than the provision of convergent technology infrastructure: the cultural institution must also consider the audience's familiarity with new literacies, and supply and demand within the target cultural market. To establish this framework successfully, the authors pose ways in which cultural institutions can seek to expand curatorial missions from exhibitions of collections to the remediation of cultural narratives and experiences.

In *Digital Knowledgescapes—Cultural, Theoretical, Practical, and Usage Issues Facing Museum Collection Databases in a Digital Epoch*, Fiona Cameron and Helena Robinson offer new insights into ways that collections documentation and interpretation within collections databases can be reconceptualized to form new knowledge models in line with contemporary theoretical and pedagogical public access concerns, and the capabilities of emerging technological innovation. The discussion illuminates how the transformative process may occur, the nature and form of new knowledge models, how these relate to user needs, and the way they may be applied to collections documentation and information capture. To conclude, the discussion poses a tangible discourse on how these new paradigms may radically reform museum practice, for example museum policy, institutional knowledge creation and management, and staff roles and tasks.

The theme of collections documentation and emerging digital technologies is further discussed in chapter 10, *Art Is Redeemed, Mystery Is Gone: The Documentation of Contemporary Art* by Harald Kraemer. Here Kraemer examines the specific traits of contemporary art and the necessity for new documentation procedures in the context of the possibilities offered by digital technologies. He argues that the diverse prerequisites and specific demands of contemporary art involve a changed methodology of analytic access, and suggests methods for significantly extending traditional static methods of aesthetic documentation. He examines this by addressing two questions. In what respect could a renewed methodology of documentation question contemporary art meaningfully? And, what roles might the interactive possibilities of digital and multimedia technology play here?

Integral to accessing cultural heritage in a digital environment, the argument for cultural information standards and frameworks becomes a persuasive one. In chapter 11, *Cultural Information Standards—Political Territory and Rich Rewards*, Ingrid Mason examines the role of cultural information standards in providing the infrastructure for the collection, preservation, and accessing of digital cultural heritage. The significance of critically examined standards, Mason argues, is because of their role in supporting the collection and preservation of digital cultural heritage. That is, they determine what is collected (selection policies and guidelines) and kept (preservation policies and methods). To do this, Mason explores the relationships between cultural information standards and digital cultural heritage and what drives their development—the sociopolitical forces that influence cultural information standards and the benefits of virtual access to digital cultural heritage.

Moreover, Gavan McCarthy, in *Finding a Future for Digital Cultural Heritage Resources Using Contextual Information Frameworks* (chapter 12), explores the use of contextual information frameworks as a means by which knowledge can be passed from generation to generation. Following the emergence of the Internet, the creation and curation of digital cultural heritage resources has become a major cultural force, enabling the linking and sharing of information. McCarthy argues that one of the major challenges for all those creating and curating digital cultural heritage resources is in finding a path that will enable the resources to be utilized by future generations, one that can mitigate both media redundancy and epistemic failure. This chapter extends current thinking by conceptualizing how the elements of our cultural heritage knowledge base might be managed in the future.

The following two chapters engage new models for audience research and learning in an online environment. First, *Engaged Dialogism in Virtual Space: An Exploration of Research Strategies for Virtual Museums* by Suhas Deshpande, Kati Geber, and Corey Timpson (chapter 13), proposes a new theoretical framework based on Appraisal theory and classical rhetoric for formulating an audience-centered strategy in researching the optimal performance of virtual museums. Knowledge of how varied audiences use and interpret the content contained in a virtual museum is essential for designers to plan, create, and monitor the performance in virtual space. To this end, this model provides a unique way to assess discourse and the persuasive use of language to understand audience behavior. And chapter 14, *Localized, Personalized, and Constructivist: A Space for Online Museum Learning* by Ross Parry and Nadia Arbach, examines the confluence of user-driven software, learner-centered education, and visitor-led museum provisions and their relevance to online museum learning. From the assessment of these variables, the authors argue that what emerges is a paradigm of increased personalization, localization, and constructivism characterized by a greater awareness of and responsiveness to the experiences, preferences, and contexts of the distant museum learner. To this end Parry and Arbach suggest how these types of experiences are impacting on the status, value, and role of new media within organizations, the capacity of museums to use technological innovation, and ways institutions might conceive and articulate their involvement with Web-based learning.

Part III of this anthology, *Cultural Heritage & Virtual Systems*, examines the intersection of cultural heritage research, documentation, and interpretation—as it is mediated through the techniques and modalities of virtual reality. The term "virtual cultural heritage" is generally accepted to mean: virtual reality (3D and 4D computational and computer graphics systems that support real-time, immersive, and interactive operations), employed specifically for the presentation, preservation, conservation, and documentation of natural and cultural heritage. As Bernard Fischer and his colleagues note, in reference to the possibilities available to us through the development of techniques in virtual reality, ". . . cultural virtual reality . . . has now come of age" (2000, 7–18). While most of the preconditions for virtual reality have been available since the 1990s, the cultural, aesthetic, sociological, and scientific implications of the use of these technologies for heritage are still being formulated. Those engaged in the practice of virtual cultural heritage (hereafter called virtual heritage) are particularly concerned with the lack of translation of the scientific principles of the discipline of archaeology to the modalities of virtual reality. Does virtual reality offer the potential to make significant contributions to "spatial understandings, human-landscape interaction patterns, temporal ordering of material remains and fragments assembly" (Sideris and Roussou 2002, 32)? Virtual heritage is often labeled "edutainment," and even the "Disnification" of culture, and as Juan Barceló states ". . . in most cases the use of virtual reality in archeology seems more an artistic task than an inferential process" (2000, 9–36). Further, as Athanasios Sideris and Maria Roussou point out, "we observe a growing number of VR projects related to Archaeology evolving independently while any consistency to the methodological and conceptual logic remains specific to each project and is usually self justified" (2002, 31). Indeed, these researchers go on to point out that virtual heritage, as it has developed so far, may be best suited for interpretive products for the general public rather than for valid scientific inquiry.

The authors in part III take up the challenge of articulating what it means to develop both scientific objects and interpretive products for cultural heritage using the modalities offered by virtual reality. Each are engaged in the practical business of fieldwork and data capture using the most advanced scientific tools available. They *are* concerned about the issues of registration and capture technologies, storage, longevity, transparency, and the metadata of the models they create. In keeping with the purpose of this book, the writing below reflects the theoretical issues confronting virtual heritage. These include the described "newness"[8] of technologies engaged by users in virtual travel and education, navigation as paradigm in virtual space—and the requirements for "presence" and immersion combined with cognitive science and semantics for learning in virtual environments. The creation of spatial archives with the use of mobile and augmented technologies and the affordances of hyperdocuments for defining the multifarious knowledge contained in architectural heritage spatial representations are addressed by the authors below. The use of complex systems and artificial intelligence as an aid to problem solving in archaeology demonstrates a visionary future.

Sarah Kenderdine, in *Speaking in Rama: Panoramic Vision in Cultural Heritage Visualization* (chapter 15), examines the chronological use of the panorama—to decode its sociocultural implications as it reemerges in virtual heritage. In this way the work traces the panoramic scheme "in transition" to reach an applied virtual heritage. The argument develops to demonstrate that it is the provocative tension that exists for virtual heritage as a tool for scientific and cultural visualization, the gap between the scientific requirement to reproduce rational material reality and those "sensations" produced by the illusion inherent to panoramic immersion—that intrigue visitors to the (virtual) space.

How immersive virtual spaces can be articulated as modalities for virtual heritage are the subjects of the next two chapters. Erik Champion and Bharat Dave, in *Dialing Up the Past* (chapter 16), investigate the centrality of "place" as a structuring concept for virtual heritage environments. Both "place" and "placeness" in virtual heritage require more than just a high degree of visual realism. The authors draw upon concepts from architecture, cultural geography, and theory in virtual environments (especially "presence" research), to identify a number of specific features of place-making. It is the argument of the authors that virtual heritage requires the same hermeneutic features of place.

Bernadette Flynn in *The Morphology of Space in Virtual Heritage* (chapter 17), identifies one of the core disappointments in virtual heritage, that is, that an algorithmically accurate large-scale 3D model of a cathedral or castle is taken as the hallmark of authenticity, and that the reduction of the monument or artifact to visual simulation disrupts its connection to material evidence, and thus to history. As Walter Benjamin observed in discussing the replacement of the original by the copy, art is severed from its relationship to ritual and magic—it loses its aura. Evoking the presence of the past relies on a different treatment of the space that creates cultural and social presence. Applying methods of ludology, the constraints, affordances, and challenges within the architecture of virtual reconstruction can be used to convey a direct negotiation with a specific cultural context. This can lead to the imagined "presence" of the organically social experience. In this way, Flynn demonstrates that heritage is restored not only as a spatial representation onscreen, but also as a place for habitation—as a kinaesthetic sense of presence in the past.

In chapter 18, *Toward Tangible Virtualities: Tangialities*, Slavko Milekic focuses on two characteristics of virtual environments: the absence of support for meaningful experiential interactions with virtual information, and the fact that currently the emphasis in virtual environments is placed on the quantity of information rather than its quality.

He draws on cognitive science and aspects of learning theory to propose how the designers of virtual environments can meet the challenges of supporting user interactions that contribute to information transfer and retention, and make the quality of virtually presented information meet or exceed a real life experience.

In *Ecological Cybernetics, Virtual Reality, and Virtual Heritage* (chapter 19), Maurizio Forte also investigates the cognitive attributes of virtual systems as created in virtual heritage. He maintains that in virtual heritage the risk is to enhance the amazing aesthetic features despite the lack of informative/narrative feedback and cognitive research within the virtual worlds themselves. Forte reiterates that the importance of the virtual reality systems in the applications of cultural heritage should be oriented towards the capacity to change ways and approaches of learning, and he makes the case for ecological and cybernetic approaches to investigating virtual worlds for heritage.

In chapter 20, *Geo-Storytelling: A Living Archive of Spatial Culture*, Scot Refsland, Marc Tuters, and Jim Cooley propose applications for mobile computing and virtual heritage. What is interesting about locative media in the context of virtual heritage is that it makes possible the notion of a collaborative mapping of space, and the intelligent social filtering or "narrowcasting" of that space. This has the potential to significantly impact upon the dominant modes of representation, most notably that of the linear, expository narrative.

Rodrigo Paraizo and José Ripper Kós (chapter 21), drawing on their experience of architectural heritage in Brazil, discuss different methodologies for the use of "hyperdocuments" as a basis for revealing urban heritages through electronic tools. *Urban Heritage Representations in Hyperdocuments* considers spaces that are also stages for social events; and that hyperdocuments can be powerful tools to display not only the physical urban structure but also to demonstrate the connections that create the urban spaces people dwell in.

Finally, in chapter 22, Juan Antonio Barceló, in a visionary essay presents a theory for the use of artificial intelligence in virtual heritage as a specific tool for archaeological inquiry. *Automatic Archaeology: Bridging the Gap between Virtual Reality, Artificial Intelligence, and Archaeology* moves the discussion away from interpretive virtual heritage to interdisciplinary scientific applications. In general, archaeologists are looking at exploring *how* perceptual properties (shape, size, composition, texture, location) allow us to indirectly solve the archaeological problem, and that finding the social *cause* (production, use, distribution) of what we "see" creates meaningful scientific outcomes. Artificial intelligence can make this process of "reverse engineering" easier—and an automatic method of analysis.

This anthology gathers in one place the issues pertinent to practitioners charged with the study and interpretation of cultural heritage. The volume necessarily must accommodate an expansive collection of viewpoints to reflect the multidisciplinary developments being undertaken across the world. The essays act as introductory statements, and we hope that the issues presented find resonance with readers of varying interests, to invigorate lively debates in cultural institutions and to posit new ideas for future research.

## Notes

1. For example, the Internet, environments of virtual and augmented or mixed reality, mobile computers, wearable technologies, automatons, artificial intelligence, intelligent agents. For a discussion of culture and digital technologies, see Gere 2002.

2. For example, see Lunenfeld 1999, Cubitt 1998, Manovich 2001, Barrett and Redmond 1997, Herman and Swiss 2001, Miller and Slater 2000, Hillis 1999, Allen and Hecht 2001, Vince and Earnshaw 2000, and Graham 2001.

3. Museums and the Web, International Cultural Heritage Informatics Meeting, International Symposium on Virtual Reality, Archaeology and Intelligent Cultural Heritage, Computing Archaeology, Virtual Systems and Multimedia, Electronic Visualization and the Arts, to name a few of the established conferences.

4. See UNESCO 2003, n.p.

5. For examples, see Thomas and Mintz 1998 and Keene 1998. Both examples have a technical/practical orientation based on case examples and do not deal with the wider heritage field. The former deals primarily with multimedia applications in museums rather than a theoretical analysis of the Internet or other emerging digital media. The latter provides a manual for the digitization of museum collections, raising issues and challenges while positioning the discussion within the broader information economy.

6. See Gere 2002, 11.

7. See Benjamin 1968 and Baudrillard 1982.

8. Refer to MIT Press series *Media in Transition*, ed. David Thorburn, in particular Gitelman and Pingree 2003.

## References

Barceló, Juan Antonio. 2000. "Visualizing What Might Be: An Introduction to Virtual Reality Techniques in Archaeology." In *Virtual Reality in Archaeology*, ed. J. A. Barceló, M. Forte, and D. H. Sanders, 9–36. Oxford: Archaeopress.

Baudrillard, Jean. 1982. *Simulations*. Trans. P. Foss, P. Patton, and P. Beitch. New York: Semiotext(e).

Baudrillard, Jean. 1994. *Simulacra and Simulation*. Trans. Sheila Faria Glaser. Ann Arbor: The University of Michigan Press.

Benjamin, Walter. 1968. *Illuminations: The Work of Art in the Age of Mechanical Reproduction*. London: Fontana Press.

Frischer, Bernard, Franco Niccolucci, Nick Ryan, and Juan Barceló. 2000. "From CVR to CVRO: The Past, Present, and Future of Cultural Virtual Reality." In *Virtual Archaeology between Scientific Research and Territorial Marketing, Proceedings of the VAST Euro Conference, November, Arezzo, Italy*, ed. Franco Niccolucci, 7–18. Oxford: Archaeopress.

Gere, Charlie. 2002. *Digital Culture*. London: Reaktion Books.

Gitelman, Lisa, and Geoffrey B. Pingree, eds. 2003. *New Media 1740–1915*. Cambridge, Mass.: MIT Press.

Jameson, Fredric. 1983. "Postmodernism and Consumer Society." In *The Anti-Aesthetic: Essays on Postmodern Culture*, ed. H. Foster, 123–132. Seattle: Seattle Bay Press.

Jones-Garmil, Katherine. 1997. *The Wired Museum: Emerging Technology and Changing Paradigms*. Washington, D.C.: American Association of Museums.

Keene, Suzanne. 1998. *Digital Collections, Museums and the Information Age*. London: Butterworth-Heinemann.

Lunenfeld, Peter, ed. 1999. *The Digital Dialectic: New Essays on New Media*. Cambridge, Mass.: MIT Press.

Manovich, Lev. 2001. *The Language of New Media*. Cambridge, Mass.: MIT Press.

McLuhan, Marshall. 1964. *Understanding Media: The Extensions of Man*. New York: McGraw-Hill.

Paul, Christiane. 2003. *Digital Art*. London: Thames and Hudson.

Thomas, Thomas, and Ann Mintz. 1998. *The Virtual and the Real: Media in Museums*. Washington, D.C.: American Association of Museums.

UNESCO. 2003. *UNESCO Charter on the Preservation of Digital Heritage*. Available at ⟨http://portal.unesco.org/ci/en/ev.php-URL_ID=8967&URL_DO=DO_TOPIC&URL_SECTION=201.html⟩ (accessed 1 November 2005).

# I  *Replicants / Object Morphologies*

# 1 Rise and Fall of the Post-Photographic Museum: Technology and the Transformation of Art

Peter Walsh

## Haskell's Hierarchies

Two Parnassian gatherings of great artists open Francis Haskell's book, *Rediscoveries in Art: Some Aspects of Taste, Fashion, and Collecting in England and France* (1976). The first is Paul Delaroche's huge, semicircular mural painted for the apse of the Ecole des Beaux Arts in Paris, begun in 1837. The second is the congress of great painters in the sculpted podium of the Albert Memorial in London. Henry Hugh Armstead executed the sculpture and the Memorial itself was designed around 1864.

Haskell makes the point that, in the roughly seventeen years between the starting dates of these two works, the roster of important artists had changed. What seemed in 1837 like a gathering of immortals was thoroughly revised and edited by the time Armstead set to work.

The point Haskell does not make, but I wish to underline, is that one of these schemes, the one Delaroche painted in Paris, was begun just *before* the introduction of photography. The second, the one sculpted by Armstead, was begun about fifteen years *after* the effects of photography had had a chance to percolate through Western culture. In the meantime, photography had changed the way art was presented, reproduced, analyzed, and evaluated.

## The Artist-Inventors

Photography is usually considered a technological development. In fact, artists were very intimately involved in the development of photography. The inventors who produced the first photographic processes—Jacques Mandé Daguerre in France and William Henry Fox Talbot in England—were both artists, though Talbot's early career was as a scientist, linguist, and mathematician. Ironically, both Talbot's photographic invention, the "calotype," and his considerable artistic talents, developed entirely in his pursuit of the new medium, are today more highly regarded by historians of photography than those of the far more famous Daguerre.

One of the first to introduce the daguerreotype into the United States was Samuel F. B. Morse, also a formally trained and prominent American painter. He is also credited, of course, with the invention of the telegraph and Morse Code.

The invention of photography created an enormous cultural reaction. Photography was a much bigger and more immediate sensation in its time than the advent of the World Wide Web in our own. Daguerre's achievements were announced at the Institut de France on August 19, 1839, with speeches, pomp, and ceremony intended to etch the date as an historic day.[1] In the next couple of years, hundreds of "first photographs" of, among other things, the moon, followed.

Within a decade, photographic technology had spread around the globe. A variety of competing methods for taking, developing, and reproducing photographs sprang up overnight, each providing its own special qualities and advantages, including long and short exposure times. These included Talbot's calotype (later talbotype) method, and collodion glass negative systems, as well as the daguerreotype.

### Myths and Realities of Photographic Culture

The effects of photography on art and on culture have been enormously complex and are still incompletely understood. A great deal of what we think we know about photography, especially about the relationship between art and photography, is largely mythological.

One myth of photography is of a poor, misunderstood medium that struggled for decades for recognition as art form, finally achieving that high status in the late twentieth century. In fact, the status of photography—nearly always the focus of one aesthetic controversy or another—varied throughout the nineteenth and early twentieth century. At different times and locations, photography's prestige both increased and declined, depending in part on other streams of the culture.

Early in the nineteenth century, critics and connoisseurs were more likely to attach high value to photographs than they were later. Early photographic images—especially daguerreotypes—were difficult to reproduce or make in multiple copies. Thus many were as unique and even rarer than images made by traditional methods. Photographic prints were often considered a kind of automated drawing.[2]

Nevertheless, in 1855, despite heavy lobbying, officials excluded photography from the fine arts exhibitions at the Exposition Universelle in Paris. Instead, it was relegated to the Industry Pavilion. But three years later, when Camille Silvy exhibited his *River Scene* at the annual exhibition of the Photographic Society of Scotland, *The Edinburgh Evening Courant* commented, "We cannot pass this picture of a new artist without notice. The water, the foliage, the figures, the distance, the sky are all perfect." Another critic wrote in the *Daily Scotsman*, "If this photograph is untouched, and taken from nature, it is a triumph of the art, and equal to any picture by Van der Neer or other famous delineator of similar scenery."[3]

Another set of assumptions, frequently taught in the history of nineteenth-century art, is that photography exerted certain formal influences on how art was executed, composed, and framed. Some scholars have pointed out, however, that Western painting had begun to look like photography some years before photography existed. In fact, much of the refinement of photographic equipment consisted of making sure that camera lenses would capture photographs that fit into established Renaissance ideas of linear perspective.[4]

### Powers of the Photograph

Two things particularly impressed the first viewers of photographs. One was the virtually microscopic detail they contained.

Morse visited Daguerre in 1839, shortly before the Institut de France announcement. Astonished by Daguerre's achievement, Morse reported to his family:

> You cannot imagine how exquisite is the fine detail portrayed. No painting or engraving could ever hope to touch it. For example, when looking over a street one might notice a distant advertisement hoarding and be aware of the existence of lines or letters, without being able to read these tiny signs with the naked eye. With the help of a hand-lens, pointed at this detail, each letter became perfectly and clearly visible, and it was the same thing for the tiny cracks on the walls of buildings or the pavements of the streets.[5]

The renowned scientist, Alexander von Humboldt, was also struck by the inhuman precision of Daguerre's images, which omitted nothing, not even the most insignificant detail. "One could see in the image," von Humboldt exclaimed, "that in one skylight—and what a trifle—a pane had been broken and mended using gummed paper."[6]

Beside such tiny details, the other thing that photographs seemed to capture—almost miraculously—was truth itself. Apparently free of any subjective manipulation, the photograph seemed capable of recording any moment perfectly, preserving it forever.

Art, by contrast, seemed a kind of fiction to photography's unvarnished truth. Confronted by the historical photograph, the grand allegories and events of formal history painting—in the French academic hierarchy, the highest form of art—began to look silly and old-fashioned.

This idea of the reality of the photographic image persisted, somewhat anachronistically, well into the twentieth century. Susan Sontag wrote that "while a painting, even one that meets photographic standards of resemblance, is never more than the stating of an interpretation, a photograph is never less than the registering of an emanation (light waves reflected by objects)—a material vestige of its subject in a way that no painting can be" (1982, 350). Earlier, Walter Benjamin had noted, "photographs become standard evidence for historical occurrences" (1969, 226).

These two impressive and unique qualities of photography—infinite detail and unassailable truthfulness—had an almost immediate impact on two forms of art, both of which fell so quickly in status and popularity that they all but dropped out of the history of art.

Miniature portrait painting, although it never ranked especially high in official estimation, was a popular and fashionable form of personal art from the Renaissance to the early nineteenth century, especially in large, wealthy cities like Paris. The daguerreotype killed it almost overnight, not so much because the photographic portraits were cheaper—they weren't at first—but because they usurped the miniature's three main claims to fame: fine detail, truthfulness to the original, and social status.

Fueled by the society business of photographers like Nadar, the photographic portrait became the rage of the fashion-set by the 1850s. Such associations did much to raise the general status of photography. As Rosen and Zerner have noted, in the mid-nineteenth century, "Many artists and critics affected to despise the new invention until . . . they decided they wanted photographs of their mothers" (1984, 99).

The other kind of art to fall off the map after the invention of photography was the reproductive engraving. Handmade engravings reproducing paintings and other works of art had been produced since the Renaissance—almost, in fact, since the invention of printing itself. Apart from direct copies, prints were the principle means of spreading artistic compositions and ideas from one place to another.

Fine reproductive engravings were traditionally highly regarded as art in their own right. The plates typically noted the names of both the original artist and the engraver. Connoisseurs eagerly collected the prints, which commanded high prices. By the early nineteenth century many print collectors imagined that, as the originals were clearly deteriorating, reproductive engravings would eventually replace them.[7]

With the advent of photography, however, the bottom dropped out of the market for reproductive engravings. They have never recovered either their value or their status. The reason, of course, was that works of art could now be photographed. From this point forward, photographs would play a central and defining role in the study, dissemination, and appreciation of art.

### Photography and the New Art Museum

Within a generation of its invention, photography had become a leading avenue for distributing images—overwhelming, in quantity and breadth of distribution, all images made in the history of humanity. Famous monuments and works of art were among the most popular subjects for photographers.[8]

In England, pioneering photographers and publishers of photography, including Roger Fenton, Philip Henry Delamotte, Joseph Cundall, the Arundel Society, and William Talbot himself, devoted many of their experiments to photographing works of art and promoting photography as a means of reproducing art images and making major works of art better known. The Italian firm of Fratelli Alinari, founded in Florence in 1852, rapidly gained an international reputation through its contacts with major art museums and art historians like John Ruskin and Bernard Berenson, both themselves avid users of photography in their own research.

The Alinari firm was the first to photograph Piero della Francesca's frescoes in Arezzo and their images of works of art, along with those produced by their competitors, were distributed throughout Europe and North America.[9]

Finally, photography helped give rise to an entirely new kind of art museum, the *post-photographic museum*. This model became so dominant in modern culture that it eventually absorbed and transformed the older idea of an art museum. Nowadays, we tend to forget there was ever a *pre-photographic museum*. But it was through the post-photographic museum that photography not only influenced art that came after its invention but also the entire history of art that came before it.

Pre-photographic museums—the Ashmolean at Cambridge, the Louvre, the Uffizi, London's National Gallery, the Prado, the Vatican Galleries, the Hermitage, the

British Museum, the Accademia in Venice, the Pinacoteca di Brera in Milan, and similar institutions—almost all of them in Europe—were founded around existing collections of originals, previously assembled by individual collectors, learned societies, or ruling dynasties.

Often housed in former palaces, religious structures, or converted government buildings, these "cabinets of curiosities" and "princely collections" typically reflected a regional school or taste and were not intended to represent art as a whole.

During the Napoleonic era, many of these collections were reorganized more systematically into public national museums, like Amsterdam's Rijksmuseum, which brought together art works from churches and monasteries and collections confiscated from pre-Napoleonic monarchies and aristocratic families. As the nineteenth century progressed, art works from European colonies and archaeological expeditions joined these national collections, making them primarily showcases for imperial power and national prestige.[10]

Like photography itself, however, the post-photographic museum set out to capture everything. They were founded to create a collection, not to house an existing one, and their avowed function was "educational."

### The Example of the South Kensington Museum

By the second half of the nineteenth century, the photographic approach to art had become the norm. Vast numbers of photographs of works of art spread throughout the world—in individual prints, in stereopticon views, in inexpensive chromolithographic reproductions, framed and hung in thousands of homes, public buildings, and classrooms, and, as printing technologies improved, increasingly in printed books and periodicals. Moreover, photography helped inspire a new kind of specialized art museum, which was itself to transform the way art was displayed and interpreted for the public.

The prototype of this post-photographic art museum, and the direct inspiration for many that came later, especially in the United States, was the South Kensington Museum, now known as the Victoria and Albert. Founded at the urging of Albert, the Prince Consort, the South Kensington Museum's inspiration was the large collection of objects, brought from around the world, for London's Crystal Palace exposition in 1851—itself the prototype of the "world's fair" approach of representing all the cultures and industries of the globe in one place and making them into popular entertainment.

The South Kensington Museum had two main goals in mind: to elevate public taste, especially the taste of British manufacturers, and to elevate society through the morally beneficial influence of great art. Collecting valuable original art works was not, at least initially, especially important. When originals were unavailable or too costly, photographs, plaster casts, and other reproductions made entirely satisfactory substitutes. Typically, the South Kensington Museum exhibited all four together—photographs, casts, copies, and originals—in the public galleries.

In the educational approach of the South Kensington Museum, and at its offspring elsewhere, the photographic reproduction was only slightly less valuable than an original. In some ways—especially in its lower cost—the photograph was superior.

The South Kensington Museum's first superintendent, Henry Cole, was very much committed to what we would now call "new imaging technologies" and their educational potential. By 1856, Cole had established a collection of photographs, including work by Julia Margaret Cameron, and many examples of different photographic printing methods.[11] Cole's understanding and support of photography, and its use in museum administration, was shared by his influential lieutenant at the museum, the artist Richard Redgrave, and by John Charles Robinson, another artist, who, as Superintendent of Art Collections at South Kensington, played a major role in the museum's acquisition of original works of art.

Cole appointed Charles Thurston Thompson to serve as the museum's photographer. Thompson's duties were complex. When the famous Raphael Cartoons were loaned to the museum in 1864, Thompson photographed them. The Arundel Society, organized to make known major works of art, published and distributed Thompson's photographs.

Thompson also photographed works being cast for installation in the museum. In 1866, for example, he was sent to Santiago, Spain, to photograph the late twelfth-century portal of Santiago Cathedral. These photographs were used to complement the exhibition of the casts in the museum. Like the photographs of the Raphael Cartoons, they were also published and distributed by the Arundel Society. As the medieval portal was concealed behind the later west front of the cathedral, it was virtually unknown before Thompson photographed it. The Arundel Society publication secured, in fact, its place in medieval art history.[12]

But the influence of photography was even more involved than that. In 1855, Thompson traveled to Toulouse to photograph the Jules Soulage collection of seven hundred and forty-nine objects, rich in French Renaissance material. This was the first known use of photography to reinforce a museum acquisition proposal. The collection was cataloged and shown at Marlboro house, raising support for purchasing the collection with government support.[13] The practice of evaluating potential acquisitions first from photographs, then from originals, has persisted in the post-photographic museum ever since.

Once the original objects were acquired, the South Kensington Museum used photography and printed museum catalogs strategically to establish such names as Andrea and Luca della Robbia and Giovanni Pisano in the art history canon. The museum sold photographs of the collections to the public and circulated selections to provincial art colleges.[14] These efforts were so successful that, by the early 1860s, public demand for these photographs had outgrown the supply.[15]

## Triumph of the Post-Photographic Museum

Significantly, pre-photographic museums, at least in London, had much more difficulty adapting to the new medium. By the 1840s, the British Museum, in cramped and antiquated quarters, was trying both to rid itself of its image as "an elitist cabinet of curiosities"[16] and to publicize the importance of its growing collections. Photography—as the South Kensington Museum was to demonstrate—was potentially an important tool for both ends.

Although photography, already the rage of London as elsewhere, was much discussed at the British Museum and experiments—by Talbot himself—were carried out there as early as 1843, the trustees did not establish a photographic studio until the early 1850s. Even then, they may have been responding to South Kensington's already well-established lead in the still-new technology.

To manage the new studio, the British Museum trustees appointed Roger Fenton the museum's first official photographer. Originally a barrister, Fenton was one of the most distinguished and influential photographers of his day. Sponsored by the royal family, he was commissioned in 1855 to document the Crimean War. The combat pictures he made there—now his most famous works—are among the first battlefield photographs ever made.

Fenton was also one of photography's most active early promoters and experimenters. He served, among other activities, as the first honorary secretary of the Photographic Society of London, now the Royal Photographic Society, and as its agent in research-similar organizations in France.

But Fenton's pioneering role at the British Museum ended early and unhappily by 1859. The trustees, unable to understand the importance of photography and reluctant to invest in it, regarded Fenton as merely "a temporary technician, fulfilling a specific yet limited function."[17] Museum photography in London was eventually centralized in South Kensington's studios and Fenton was not replaced at the British Museum until 1927.[18] Similarly, London's National Gallery did not have an official photographer until well into the twentieth century.[19]

In North America, where the vast majority of art museums were founded well after the invention of photography, a very different attitude prevailed. The majority of museums established in the United States and Canada in the late nineteenth century and early twentieth century—including the Museum of Fine Arts, Boston, The Metropolitan Museum of Art in New York City, the Philadelphia Museum of Art, the Saint Louis Art Museum, the Royal Ontario Museum in Toronto, and the Corcoran Gallery in Washington, DC—deliberately modeled themselves on the example of the Cole's South Kensington Museum. Most incorporated photography into their plans.

In 1869, for example, Charles Callahan Perkins proposed "the establishment of a Museum of Art of the character of that at South Kensington" in Boston, Massachusetts. The new institution became the Museum of Fine Arts, Boston.

"As its aims are educational," wrote Perkins, "and its funds are likely to be for some time limited" the new museum's collections must be those that could be assembled quickly and cheaply. "Original works of art being out of our reach on account of their rarity and excessive costliness, and satisfactory copies of paintings being nearly as rare and costly as originals, we are limited to the acquisition of reproductions in plaster and other analogous materials of architectural fragments, statues, coins, gems, medals, and inscriptions, and of photographs of drawings by the old masters, which are nearly as perfect as the originals from which they are taken, and quite as useful for our purposes."[20]

In a less overt way, photography began to influence the way museums treated art and displayed it to the public. Up until the early nineteenth century, art works were rarely shown to the public in an obviously damaged or incomplete form. Damaged Roman and Greek sculptures were provided with replacement parts, sometimes created by prominent neoclassical sculptors like Albert Thorvaldsen, sometimes taken from large inventories of body parts from other sculptures. Paintings were typically retouched to hide damage or to reflect changing taste, cut down or enlarged to fit new displays.

By the early twentieth century, these practices, at least in art museums, had all but disappeared. A new scientific discipline of art "conservation" was rapidly replacing art restoration. Among the most important technical tools of the new, specially trained art conservator was photography—not just conventional photography but also X-ray images and infrared and ultraviolet photographs—that revealed how art works had changed or been altered over time.

The post-photographic museum aesthetic—reinforced and aided by these new photographic tools—was to remove as much of the earlier restorations as practically and aesthetically possible, replacing only as much as necessary to understand the work as it was originally created and to protect it from further deterioration. Museums displayed art objects with obvious damage and other visible effects of time—now considered important signs of their age, originality, and authenticity.

Thus, even as the post-photographic museum replaced photographs and other reproductions with originals in their galleries, the direct influence of photography lingered in the displays. Photographs had, ironically, made the *originals themselves* so important and valuable that curators could no longer tolerate anything less. From the galleries of "progressive" institutions like the Museum of Fine Arts, Boston, and the Fogg Art Museum nearby on the Harvard University campus, the old "educational" photographs and plaster casts slowly disappeared.

As we have seen, this complex system of photography, acquisition, conservation, publication, and circulation became the established system for the post-photographic museum, and it became the model for all art museums. The educational and evangelical system developed by Cole in the 1850s and 60s has itself become the basic paradigm of the modern educational art museum. Coupled with the development of academic art history, itself a discipline developed and taught largely with photographs and printed books, photography became the main means by which art is understood, interpreted, and evaluated.

## A New Hierarchy

Let me return, at last, to Haskell's comparison of the Delaroche and Armstead canons of art. It is tempting to note that the artists Armstead dropped from Delaroche's pantheon—including Giorgione, Van Dyke, and all the Dutch masters except Rembrandt—tend not to come across especially well in black and white photographs. On the other hand, the new arrivals on the Albert Memorial, including Ingres and the Carracci, fare better in that medium.

The real change, however, was the new emphasis, on the Albert Memorial, on the great Italian masters—precisely those painters, in other words, whose work was being photographed, collected, and published by the post-photographic museum and its branches and allies.[21] Thus those painters and art works that rise to the top of the post-photographic hierarchy of art are those best known through photographs.

Benjamin, in his essay "The Work of Art in the Age of Mechanical Reproduction"—so famous now that it has almost become a sacred text—complains that photographic reproductions lack the "aura" of the original: "its presence in time and space, its unique existence at the place where it happens to be." He concludes, "that which withers in the age of mechanical reproduction is the aura of the work of art" (1969, 220–221).

By Benjamin's youth in the 1890s, the status of photography had changed. The Arts and Crafts movement was in full sway. Machine production, once admired for its god-like perfection, had given way to a new taste for the handmade. The cultural value of mechanically produced objects, like the photograph, had fallen. Factory-made was bad; craftsman-made was good.

In fact, Benjamin has the aura of art exactly the wrong way around. It is the mechanical reproduction—the photograph—that created the aura of the original, much as it was the machine that created the "handmade," the negative that created the "positive," and the digital that gave retroactive birth to its latent opposite, the "analog."

Before photography, there were no "handmade originals" because there were no factory-produced photographs of the originals. It is, as we have already seen, the reproduction that confers status and importance on the original. The more reproduced an artwork is—and the more mechanical and impersonal the reproductions—the more important the original becomes.

Artists themselves were among the first to notice the change photography made to works it blessed with frequent reproduction. Paintings, like Millet's *Sower* and *Angelus*, that were mass-produced as Goupil photolithographs, quickly attained the status of religious relics.

"I went to the Millet exhibition yesterday with Amelia," Camille Pissaro wrote to his son, Lucien, in 1887:

> A dense crowd. . . . There I ran into Hyacinthe Pozier, he greeted me with the announcement that he had just received a great shock, he was all in tears, we thought someone in his family had died. . . . Not at all, it was *The Angelus*, Millet's painting which had provoked his emotion. This canvas, one of the painter's poorest, a canvas for which in these times 500,000 francs were refused, has just this moral effect on the vulgarians who crowd around it: they trample one another before it! This is literally true . . . and makes one take a sad view of humanity; idiotic sentimentality which recalls the effect Greuze had in the eighteenth century. . . . These people see only the trivial side in art.

They do not realize that certain of Millet's drawings are a hundred times better than his paintings which are now dated. . . . What animals! It is heartbreaking.[22]

Conversely, less reproduced art is less significant. The unphotographed, unpublished work of art exists in a kind of limbo. In fact, under the aura of the post-photographic museum, the unphotographed work can hardly be said to exist at all. In post-photographic art history, discovering and publishing such a work is almost a second act of creation.

In his biography of the legendary art dealer, Lord Joseph Duveen, S. N. Behrman notes how Duveen flattered his millionaire art collector clients with lavishly produced and illustrated catalogs of their collections. "One Christmas," Behrman writes, "Duveen got up for Kress a sumptuous book called *The Collection of Paintings, Sculptures, etc., of Samuel H. Kress*. It was an enormous weight, not easy to lift. Nevertheless, its title was something of an exaggeration, for the book contained histories and reproductions of only what Duveen had sold to Kress; it ignored the vast reaches of Kress's other purchases. But then those came from other dealers, so for Duveen they were nonexistent."[23]

In the post-photographic world, printed art books have replaced monumental displays like the base of the Albert Memorial. These books present the works and artists annointed "important" by experts to the world, and their illustrations essentially constitute the official canon of art. Since the late nineteenth century, virtually all arguments about the nature and significance of art have been made directly via the medium of photography or with its influence.

Far from diminishing the "aura" of works of art, these endless photographic reproductions have vastly added to their significance, ultimately converting the museums that hold them from the Imperial warehouses and curio cabinets of pre-photography to the vast, echoing temples of post-photography. Even the massive, symbol-laden architecture of the post-photographic museum reflects this transformation of art from precious objects into sacred icons of deep quasi-religious power—all thanks to this power of photographic reproduction.

## Fall of the Post-Photographic Hegemony

Today we are in the era of the post-Internet art museum. Once, art historians said "it's not in my books or slide library, thus it must not be important." Now, their students tend to say "It's not on the Internet. It must not exist."

With relatively little fanfare, the advent of inexpensive digital imaging, the spread of the World Wide Web, and the ability of these two technologies to quickly and easily alter, publish, and distribute photographic images, has radically changed the old, post-photographic hegemony. It has changed the perceived nature of photography as well. When image-altering software is cheap and easy to use, and manipulated images are commonplace, photographs no longer set the standard for visual truth.

Museums now recognize that photographs of their own collections have never been neutral representations of works of art, but aesthetic objects in themselves, carefully adjusted to fit the tastes and ideas of the photographers and their employers. Museum photographs are, in fact, interpretations of art objects, which, like all art historical documents, change over time. Through exhibitions like "Camera Obscured: Photographic Documentation and the Public Museum," curators even made these points to the public.[24]

Once, the expense and complexity of making quality art museums and publishing them helped confine the discourse on art to a privileged circle of museums, art scholars, and elite publishers. Now that discourse is available to anyone with access to modern computer networks.[25] On the World Wide Web, anyone can be a museum curator or art historian, promoting his or her own view of the meaning of art, borrowing images from museum Web sites, from hundreds of other sources on the Web, or scanning them on cheap equipment attached to their computers.

Art museums, now major presences on the Web themselves, have yet to respond directly to these technological changes with shifts in their basic philosophy towards the display and interpretation of works of art. For the most part, despite the transfer of masses of art-related material to electronic media, the uses museums make of digital photography to interpret and assess art remain similar if not identical to those made of "analog" photographs in the late nineteenth century. Online material and other digital publications produced by art museums still closely resemble printed catalogs and exhibition brochures, with a few technological flourishes rather than a fundamental change in approach.

But the new museum Web sites are only the first manifestation of the post-Internet museums. Now fully engaged in a new electronic world, art museums are being pressured to move in new, yet-to-be-defined directions, as they were by photography one hundred and fifty years ago. Like photography when it appeared on the scene, digital technology is, simultaneously, a new art medium, a new way of interpreting and publicizing art, and a distinct challenge to art itself. A new South Kensington Museum may already be in operation, either in the physical world or in virtual space, for one era in museum history has surely ended while another has only begun.

*Notes*

Portions of this paper were first presented, in a different form, in "The Painters of the Albert Memorial: Canons, Taste, and Technology," a keynote address at ICHIM 2001, Instituto Politechnico di Milano, September 2001.

1. See Frizot 1998, 23.

2. Ibid.

3. Both quoted in Baker and Richardson 1997, 139.

4. See Rosen and Zerner 1984, 99–110.

5. Frizot 1998, 28.

6. Ibid.

7. For a fuller discussion of the original status, decline, and eventual disappearance of reproductive engravings, see Cohn 1986.

8. For an early history of photographic images of works of art, see Hamber 1989, parts I, II, III, and IV.

9. See Heilbrun, "Around the World: Explorers, Travelers, and Tourists," in Frizot 1998, 156.

10. In post-imperial Europe, such collections have posed political problems. The Netherlands has returned to Indonesia most of the artwork it collected there during its colonial days, but Greece still presses for the return of the "Elgin Marbles" long housed in the British Museum. The imperialist nature of such collections has been much discussed in critical studies of art museums.

11. See Baker and Richardson 1997, 137–138, and Physick 1975.

12. Baker and Richardson 1997, 129–130.

13. Ibid., 153.

14. Ibid., 139.

15. Date and Hamber 1990, 309.

16. Ibid.

17. Ibid., 322.

18. Ibid.

19. See Hamber 1989.

20. Whitehill 1970, 9–10.

21. Haskell 1976, 17–18.

22. Pissarro 1972, 110–111. Of course, reproductions of the *Angelus* were Goupil's best-sellers while Millet's drawings were little known to the general public.
    In lectures, I have illustrated this process of aura creation with a photograph I took in the Louvre years ago. It shows a vast crowd trying to peer into the murky glass case holding the *Mona Lisa* while completely ignoring the *Madonna of the Rocks* and other major Renaissance masterpieces hanging around it. I have also made the comment that, although the Louvre is one of the best attended of major museums in the world it is also one of the least crowded—the reason being that these vast crowds concentrate their attentions on the three works that have been endlessly reproduced and pass by the rest.

23. Behrman 1952, 99–100.

24. See, for example, Born 1998, which discusses this exhibition held in London in 1997. Similar exhibitions have been held at American museums since about 1990, including the Museum of Fine Arts, Boston.

25. For more details on the effects of new technology on breaking down hierarchies of intellectual authority, see my essay, "That Withered Paradigm: The Web, the Expert, and the Information Hegemony," in Jenkins and Thorburn 2003.

## References

Baker, Malcolm and Brenda Richardson, eds. 1997. *A Grand Design: The Art of the Victoria and Albert Museum*. New York: Harry N. Abrams and The Baltimore Museum of Art.

Behrman, S. N. 1952. *Duveen*. New York: Random House.

Benjamin, Walter. 1969. "The Work of Art in the Age of Mechanical Reproduction." In *Illuminations*, ed. H. Arendt. New York: Schocken Books.

Born, Georgia. 1998. "Public Museums, Museum Photography, and the Limits of Reflexivity." *Journal of Material Culture* 3 (July): 223–254.

Cohn, Marjorie B. 1986. *Francis Calley Gray and Art Collecting for America*. Cambridge, Mass.: Harvard University Press.

Date, Christopher, and Anthony Hamber. 1990. "The Origins of Photography at the British Museum, 1839–1860." *History of Photography* 14, no. 4 (October–December): 309.

Frizot, Michel. 1998. "Photographic Developments." In *A New History of Photography*, ed. M. Frizot. Cologne: Köneman.

Hamber, Anthony. 1989. "The Photography of the Visual Arts, 1839–1880." Part I: *Visual Resources V*, 280–310; Part II: *Visual Resources VI*, 19–41; Part III: *Visual Resources VI*, 219–241.

Haskell, Francis. 1976. *Rediscoveries in Art: Some Aspects of Taste, Fashion, and Collecting in England and France*. Ithaca, N.Y.: Cornell University Press.

Heilbrun, Françoise. 1998. "Around the World: Explorers, Travelers, and Tourists." In *A New History of Photography*, ed. M. Frizot. Cologne: Köneman.

Jenkins, Henry, and David Thorburn, eds. 2003. *Democracy and New Media*. Cambridge, Mass.: MIT Press.

Physick, John. 1975. *Photography and the South Kensington Museum*. London: Victoria and Albert Museum.

Pissarro, Camille. 1972. *Letters to His Son Lucien*. Ed. John Rewald with assistance of Lucien Pissarro, 3d ed., revised. Mamaroneck, N.Y.: Paul P. Appel.

Rosen, Charles, and Henri Zerner. 1984. *Romanticism and Realism: The Mythology of Nineteenth-Century Art*. New York: Viking Press.

Sontag, Susan. 1977. *On Photography*. New York: Farrar, Straus, and Giroux.

Walsh, Peter. 2003. "That Withered Paradigm: The Web, The Expert, and the Information Hegemony." In *Democracy and New Media*, ed. H. Jenkins and D. Thorburn, with B. Seawell. Cambridge, Mass.: MIT Press.

Whitehill, Walter Muir. 1970. *Museum of Fine Arts, Boston: A Centennial History*. Cambridge, Mass.: Harvard University Press.

# 2 The Materiality of Virtual Technologies: A New Approach to Thinking about the Impact of Multimedia in Museums

Andrea Witcomb

*Modern multimedia exhibitions reflect not the international world of museums as repositories, but the external world in which museums now find themselves. This is the world of our post-industrial society—dominated by technology, with pervasive media and advertising industries, and instantaneous electronic communications; a society with a pluralistic culture in which the boundaries between high art and mass culture have broken down.*
—Roger Miles, "Exhibiting Learning"

Contemporary discussions on the impact of multimedia technologies on museums tend to assume a radical difference between the virtual and the material world, a difference that is conceived in terms of a series of oppositions. The material word carries weight —aura, evidence, the passage of time, the signs of power through accumulation, authority, knowledge, and privilege. Multimedia, on the other hand, is perceived as "the other" of all of these—immediate, surface, temporary, modern, popular, and democratic. It is an opposition which is captured by the quotation that begins this essay. The character of the opposition is rarely disputed. What is disputed is its significance. The introduction of multimedia is either a threat to the established culture and practices of the museum complex or its opportunity to reinvent itself and ensure its own survival into the twenty-first century. For those who see it as a threat, the implications are a loss of aura and institutional authority, the loss of the ability to distinguish between the real and the copy, the death of the object, and a reduction of knowledge to information. For those who interpret it as a positive move, such losses are precisely what enable new democratic associations to emerge around museums. For them, the loss of institutional authority equates with the need for curators to become facilitators rather than figures of authority, an openness to popular culture, the recognition of multiple meanings, and the extension of the media sphere into the space of the museum. These are all interpreted as positive steps.

### Developing a Materialist Approach to Understanding Multimedia in Museums

This chapter will suggest that the development of this opposition in relation to the introduction of multimedia into museums is sometimes unhelpful. There is a need to begin from a different starting point, one that does not presume such a radical and politicized opposition between the material and the virtual. One of the reasons for this is that positioning the multimedia exhibit as "the other" to objects by focusing on its immaterial qualities has, in the main, resulted in limited applications which do little more than supplement conventional forms of textual interpretation. Despite claims to the contrary, many multimedia stations continue to operate within traditional didactic frameworks.[1] The problem is a similar one to that identified by Paul Carter in relation to the use of sound in museums. As he says, "Conventionally, the spoken voice in museums and galleries is confined to the informative monologue heard by pushing a button or hiring a headset" (Carter 1992, 15).

Part of the problem is that multimedia, in the exhibition context, is understood as a tool for interpretation and rarely as a material expression in its own right. This tends to limit its more imaginative uses. By thinking of multimedia as a material form of expression, however, it might be possible to open up a space for thinking about multimedia displays in ways that go beyond the offer of a further information point, such as a touchscreen interactive or a Web site. The advantage is that we can begin to recognize the ways in which multimedia installations in museums can enhance what a number of writers are beginning to refer to as the "affective" possibilities of objects.[2] This entails a recognition of the way in which objects, and in my argument multimedia installations, are able to engage emotions and in the process produce a different kind of knowledge—one that embodies in a very material way, shared experiences, empathy, and memory. Using the work of Dipesh Chakrabarty (2002), I will argue that these forms of knowledge, and consequently multimedia installations, have important parts to play in contemporary democracies. This chapter, then, will focus on particular types of multimedia displays—those that act as installations or objects in their own right rather than as an interpretive layer that is added to the display of objects from the collection.

I am particularly interested in establishing the function of these multimedia objects within the context of their display in an exhibition. How do they affect the display of objects from the collection? And what is their broader message within the museum context? Here I want to suggest the existence of at least two functions for multimedia installations in exhibitions. The first is their ability to privilege the role of interpretation itself. It is the possibilities of the medium as an art object, as installation, that provides one of the most powerful means of making explicit the nature of the contemporary museological revolution. This is a revolution which has revealed the process of making meaning in exhibitions, a process which previously was made to appear neutral through a focus on the conventional world of objects (Witcomb 2003). As Chakrabarty (2002) points out, objects in museums were always interpreted within broader analytical frameworks which served specific pedagogical interests—for example, the education of the citizen to serve the interests of the modern nation state. These analytical frameworks have always been understood as objective and rational precisely because they eschewed more experiential forms of knowledge production. Multimedia installations, however, in playing to notions of experience, are one of the primary means through which the open politicization of contemporary museum narratives has occurred.

In so doing, however, their second function is also made manifest. Some multimedia installations in museums act as releasers of memory in much the same way as objects can make unconscious memories conscious. This they achieve through their power to affect us by "touching" us or "moving" us. When this happens the visitor undergoes an experience in which their sense of self is altered. Until now, such encounters with objects in museums, while recognized, have tended to be haphazard as they were not part of the formal strategies of museological interpretation. The explicit attempt to produce such embodied experiences through multimedia, however, strains against the more rational practices of interpretation, opening up new interpretive possibilities for museum collections.

### Background: The Virtuality of Multimedia as the Symbol of the New Museum

Since the introduction of digital multimedia into the world of museums through such things as Web sites, computer interactives, holograms, and digitized film and sound, museum practitioners and critics have either lamented or celebrated these developments. For its proponents, these developments have become emblematic of the emergence of a new museum—one which is focused on providing both intellectual and physical access, democratizing the interpretation of its holdings by using a mode of communication associated with contemporary popular culture.[3] The museum they describe is an open, flexible institution, attentive to the needs of its audiences, rather than the remote, elite institution of old. It is an institution which reflects the processes of globalization and the emergence of multiculturalism, and which welcomes polysemy. The introduction of digital media into the museum space is seen as the means to reflect these wider developments within the museum, bringing audiences together, representing difference and even contestation.

An excellent example of this way of thinking can be found in the writings of George MacDonald, especially during the time in which he presided over the reopening of the Museum of Civilization in Canada. At the time, MacDonald claimed that the

> values, attitudes, and perceptions that accompany the technological transition from industrial to information society can make it possible for museums to achieve their full potential as places for learning in and about a world in which the globetrotting mass media, international tourism, migration, and instant satellite links between cultures are sculpting a new global awareness and helping give shape to what Marshall McLuhan characterized as the global village. (1992, 161)

The tool that enabled this transition to take place was of course the introduction of multimedia into the space of the museum.

It strikes me that one of the most remarkable aspects of comments like these is not whether or not such claims can be supported empirically but that they substantiate an opposition between objects and multimedia technologies. The latter are extras that support and extend the possibilities created by the existence of collections in museums. This view is, of course, supported by many applications of these technologies in the museum context. It is there in the creation of Web sites that construct a virtual version of the museum, in the touch-screen interactives that provide an extension to the information available in the exhibition itself, and in the use of these technologies for collection management purposes.

## Multimedia as Object

Increasingly, however, there are examples of the use of multimedia in the museum context that are not extras. Instead, they have come to stand as displays in their own right, almost replacing the historical role of objects in exhibition making. In a sense they have become objects. This is not only because they share many of their characteristics with the material world—they are three or two dimensional, can be touched, looked at, in some cases viewed from all angles, exist in real time, and are the results of a human production process. It is also because they offer meaning in their own right, not as interpretive aids but as creative art objects. As a result, like other objects in exhibitions, they play a structural role in the production of a meaningful text. Such installations demand our interpretive reading of them just like the traditional object on display does. In what follows, then, I am interested in interpreting multimedia installations in exhibitions that stand on their own and which are not an interactive accompanying a display of objects from the collection. By this I do not mean multimedia installations in a contemporary art museum context,[4] but installations in social history and ethnographic exhibitions. I am interested in two things—the meanings produced through the interaction of viewers with such installations, and their effect on the interpretation of traditional, object-based displays.

## Case Studies: The Bunjilaka Gallery

My case studies come from the Museum of Melbourne in Victoria, Australia. As the main campus of the Museum of Victoria, the Museum of Melbourne represents the recent reinvention of the MoV. Opening its doors in October 2000, the museum has been described by its critics as both politically correct and an expression of infotainment. For others, however, it is a positive expression of the new museology. One of these expressions is the Bunjilaka Gallery, which, unlike other galleries at the museum, is not just an exhibition gallery. It is also a community center for Victorian Aboriginal communities who use its community access gallery, meeting rooms, and education center for their own purposes. As two of the gallery's former curators, Tony Birch and Gaye Sculthorpe, recently put it, the gallery is focused not on critique but on conciliation between the museum and Aboriginal communities and between Aboriginal people and the wider Australian society (2004). The exhibition itself, I would argue, achieves these political effects by devoting approximately half of its two introductory exhibitions to multimedia installations that act as objects in their own right, and serve to position the audience in a receptive frame of mind.

## Koori Voices: Establishing Dialogue

The first of these installations is a display of 280 photographs together with video interviews with indigenous people as part of the *Koori Voices* exhibition. The first hint that this installation is to be treated as an object rather than as an interpretation of other objects comes with the use of an introductory label, which reads:

> I think at every place where there are people now, there is some kind of story.
> Iris Lovett-Gardiner, 1999.
> The stories and voices of aboriginal people of Victoria speak with dignity.
> We ask you to listen to these voices.

There are two points to be made in regard to this label. The first is that the installation has a label in its own right rather than being an extension of the label function for a group of objects. In other words, it is accorded the status of object within the exhibition system. The other point has to do with the effect of the label's content. The effect of these two statements, one from a named Aboriginal person, the other from the anonymous curators, is to position the visitor to the exhibition as a civil person, capable of showing respect to those whom they may see as different and whom they know to have suffered oppression. It is ethically impossible therefore, to comfortably ignore the invitation to look into the eyes of the Aboriginal people captured for the most part by the colonial photographic gaze.

The impact, however, is not one of a depiction of an abject people but of an amazing ability to survive. The installation is an articulation of the importance of family, community, and just simple survival, spanning, as it does, a time frame from the 1800s to the present. The point is not made simply through the use of mute photographs. As one walks from photo to photo, looking up, down, and across the rows of photographs, an occasional face responds to your gaze and becomes "alive," engaging you in a direct dialogue. Interspersed amongst the photographs, and framed in exactly the same manner, are a number of interviews with a range of Aboriginal people.[5] Filmed with their eyes directly facing the camera, they look like just another photograph when the video is not activated. When the video is activated by the viewer passing by, it feels as if they are engaging you in a dialogue. As the viewer you become a captive of their gaze in a strange inversion of the colonial experience. Paying one's respect by listening to these voices seems to be the least one can do. The impact is both ethical and physical. Just as you would not turn your back on someone who is talking directly to you, so it is hard to turn away from their gaze and ignore what they have to say. Their messages are mostly about their families, their communities, and the places where they grew up. While some of the people do provide accounts of the brutality of colonization and its effects on them or their families, the intent is not one of open hostility or critique. The message is indeed one of conciliation—"we share in your humanity," they seem to say.

The interesting thing here is that there is a choice—no doubt some people do choose to ignore the request or perhaps don't even notice it. But even without the label, the power of this installation is an affective one—the object demands your attention, stops you in your tracks, and invites you to listen. It has the power to touch the viewer at an emotional level and thus to have a material effect.

## Multimedia and Affect

The installation works through what Ross Gibson (2004) has described as the museum's power to affect alteration. This is a process whereby the museum visitor undergoes a change from unknowing to knowing, from partial to holistic comprehension. Importantly, this process occurs when there is an opportunity to experience what it is to be "other," as Gibson puts it, "alteration occurs in and through otherness." This requires both imagination and the ability to empathize. One of the ways in which alteration can happen through imagination and empathy is in the process of affect. This occurs when a physical reaction to an object involves an emotional response that leads to a greater degree of understanding. Museum exhibitions are prime sites for such affective responses. It is the basis, for example, of the narrative in the film *The Russian Ark* in which the two protagonists both experience and witness a series of affective responses to the objects on display at the Hermitage Museum. As Charles Saumarez Smith has put it, the knowledge of objects in museums is "more involved with the process of obtaining some form of felt relationship to the past, intuitive, sensual, three-dimensional" (quoted in Stewart 1999, 30). I would suggest that installations such as the ones described above offer precisely such an opportunity. By engaging in a very direct and physical way with the viewer, they are able to activate an emotional response based, in part, on partial knowledge of what has occurred in the past and, in part, on the opportunity the installation provides to extend that partial knowledge through a simulation of dialogue with those who experienced that past.

The process of alteration that occurs as a result is not necessarily based on factual knowledge. Rather, it is a knowledge that is felt rather than rationally understood. The effect is similar to that described by Messham-Muir (2006) in his discussion of the object-based installations in the Holocaust exhibition at the Imperial War Museum in London. While there is plenty of factual information about the Holocaust to give visitors an empirical knowledge of what happened, the most powerful exhibits are those that elicit a physical and emotional response. Messham-Muir discusses a display of masses of shoes that were confiscated by the Nazis from prisoners at the Majdanek and Aktion Reinhard concentration camps by describing their emotional impact on him:

> Standing next to these shoes, sharing their space, I found myself looking closely at a particular shoe amongst the innumerable browning mass. It was an intensely affecting experience. Looking at that one shoe, in its raw stark materiality, I could easily imagine slipping it onto my own foot. . . . At that moment, I shared with its anonymous

murdered owner, from 60 or so years earlier, a common and very every-day habitual relationship to that shoe. And it is through these kind of shared relationships with objects that we can enter into powerful, empathetic relationships that seem to transcend place and time. (2006, n.p.)

The point here is that the object offers an entry point for imagination to play a role in the process of coming to know. The process is exactly the same with the multimedia installation in the Banjiluka Gallery. The experience offered to the viewer is both physical and emotional. Moreover, the two cannot be distinguished from one another as the emotional, empathetic response is a reaction to the physical power of hundreds of eyes making contact with the viewer via the media of photography and video, together with the content provided by human narratives told in the first person and delivered by those who experienced the past or its effects themselves. The fact that this contact is activated by the physical presence of the viewer walking around the installation only makes the materiality of the installation all the more important.

## Our Grief: *Recovering Erased Memories*

The shared base of understanding produced by the activity of interacting with the photographic/video installation is built upon by another multimedia installation in the same gallery. Called *Our Grief*, this installation works by using visual and documentary fragments to build a picture of a shared but forgotten history. Set out as a small theatre, the installation works by juxtaposing a single object—a gun—alongside a historical quotation and a multimedia "slide show." The slide show itself, while clearly shown on a digital screen, is framed by a black border making it into a picture on the wall. This is effective because much of the content works through a combination of aesthetic appreciation and cultural memories of landscape. The installation on the wall is basically composed of a sound track containing an ominous musical background interrupted by readings from a report by G. A. Robinson, Chief Protector of Aborigines in the 1840s. The direct quotations are of graphic descriptions of massacres of Aboriginal people conducted by the white settlers. Accompanying this sound track are idyllic images of the Australian landscape—images that remind us of quiet pastoral paintings with no hint of the dark deeds that supported the settlement and development of these landscapes. Images of rivers framed by trees on their banks, images of valleys made picturesque with the ruins of a pioneer's barn.

The sensory contrast between the auditory track and the visual record are made powerful not just because of the contrast. Their power to affect also comes by the way in which the visual record works as a memento mori for two different cultures. For Aboriginal people the landscape once cradled the bodies of those who were killed there. It offers therefore a connection to the past, a way to remember and memorialize. For the white settlers, however, the landscape that is depicted conjures up another memory, this time a nostalgic one. For the landscape that is pictured is also the landscape of an Australian romantic tradition in the visual arts. Brought forward into our consciousness are our memories of the paintings by the Heidelberg artists, the pioneer legends built up in Australian literature and in films such as *The Man from Snowy River*. These are idyllic landscapes in which Australians, we are told by historians, finally depicted themselves as belonging in the landscape. Based on a forgetting of historical experience, these visual images are torn asunder by the installation. The landscape is not idyllic, silent, or evidence of progress and development. Instead it becomes a witness to a terrible shared history, one that whites have tried to forget but which is powerfully brought alive in this installation.

In this way, documentary and visual fragments rub against our collective memories, jolting them and producing a new understanding. The process that occurs here is akin to Susan Stewart's argument that "visual perception becomes a mode of touching when comparisons are made and the eye is 'placed upon' or 'falls upon' relations between phenomenon" (1999, 32). *Our Grief* is an installation that reaches out to touch us in much the same way as objects can become vehicles for the process of remembering and making connections between different phenomena. Interestingly, the museum object—the gun on the wall—becomes the extra layer of interpretation, extending to the multimedia presentation an irrefutable materiality to the evidence presented. The gun is used as an aid to materialize the documentary evidence into a clear message—whites killed blacks. As a package, the installation thus works in the same way that an artwork does—making you look anew at your surroundings and in this case reassessing received narratives about the Australian past. The installation is powerful because it makes conscious that which is unconscious in Australian society—the memory, as well as the forgetting, of the processes involved in the colonization of Aboriginal people.

This installation's explicit manipulation of dominant cultural narratives by creating new contexts for received understandings of the Australian landscape points to the way in which the past is made anew in many of these installations. In providing a contrasting interpretation of the past the installation points not only to differences in how the past is represented by different interests but to the process of interpretation itself.

In a sense history becomes interpretation. It also becomes a form of both memory and memory making.

## Two Laws: *Highlighting the Role of Interpretation in Museums*

This point is made explicit in the final installation I wish to discuss. Part of the *Two Laws* exhibition, the installation consists of a filmic presentation of an imaginary dialogue between the anthropologist and former Director of the Museum of Victoria, Baldwin Spencer, and Arunta elder, "Charlie," known as Irrapmwe by his people. In this dialogue the viewer sees two actors dressing up for their respective roles and moving back in time. It is very clear that this is not a reconstruction but a creative piece with a particular aim in mind—to make explicit the different cultural values between the two men and to bring about a process of "alteration" in Baldwin Spencer and, by extension, the museum itself. The dialogue between the two men carries the narrative. It centers on Baldwin's gradual perception that his culture is different, rather than superior, to that of his Arunta informant. The exchange is done sensitively with care taken not to ridicule the character of Spencer, but to place him in the context of his times. In the process, the role of interpretation in defining cultures and history is again highlighted. Western knowledge about indigenous cultures is revealed as a process of interpretation, often misguided. While Western culture is shown up as arrogant it is also depicted as capable of learning and changing. So while there is critique, there is also redemption. This is achieved by the way in which the piece looks at the past through the eyes of the present, enabling a process of negotiation over the status of each other's culture.

The character of Irrapmwe shows Spencer how he was culturally insensitive, upsetting indigenous belief systems by displaying secret sacred material in public spaces, as well as to the wrong gender, and how a lack of understanding led to misinterpretations of the culture of the Arunta people. For example, Irrapmwe shows Spencer how he denied women a cultural role by assuming that because no women showed him cultural rites and objects these did not, therefore, exist. In fact, women were simply keeping such objects and rites away from the eyes of those for whom they were not intended. In the process of explanation, differences between oral and written cultures are explored, as is the anthropologist's role in defining indigenous culture as inferior to Western culture. Complex issues around who has the right to communicate and display indigenous knowledge, who "owns" that knowledge, are touched upon, as are issues of repatriation and access to material culture. Spencer comes to understand that he does not have the right to make pronouncements on these things without first entering into a dialogue with indigenous elders and gaining their consent. By extension, the present day museum is placed in a position where dialogue is the first prerequisite for any relations with indigenous cultures. As Irrapmwe says to Spencer, indigenous people "trust museums that talk to our people." All is not lost and museums can make amends for past deeds provided they form partnerships and respect and recognize difference.

In the course of the exchange, reference is made to particular objects in the museum's collection. These were collected by Spencer in his expeditions to Central Australia. They "emerge" out of the darkness when the exhibition cases below the digital screen light up to show the object that is the subject of the discussion between the two men. In a similar manner to the gun in the *Our Grief* installation, the objects become the props rather than the other way round.

The stage is thus set for the display of traditional ethnographic material. After viewing all of these multimedia installations, traditional Aboriginal culture can only be read against a history of dispossession. In the process it becomes not only a celebration of continuity, but also a testament to the existence of a complex culture, despite the threats posed by colonization. The ethnographic material on display cannot be interpreted as part of a static culture. Thus a cabinet displaying objects in nineteenth century museum style is revealed as a museological interpretation, which had more to do with the values of Spencer's culture than it did with traditional Aboriginal culture. And in the final exhibition, *Belonging to Country*, the continued maintenance of indigenous cultures gains center stage through the display of contemporary and historical indigenous material culture.

## Affect and Democracy

The political implications of the way in which multimedia installations can work through affect are significant for museums. According to Dipesh Chakrabarty (2002), museums, in their recognition of the importance of experiential forms of knowledge, are ideally positioned to participate in what he calls an "experiential" form of democracy. For Chakrabarty, there are two models of democracy. The first and oldest model is one he names a pedagogical democracy. This is a form of democracy, which is based on the notion that to be a citizen, to be political, a person first needs to be educated in the values of democracy. Such an education took place through modern educational institutions—the school, the university, the museum, and the library—because such places privileged a form of knowledge based on analytical skills or the power to abstract from the immediate example. It was a form of education which prioritized the written word above all other means of communication because of its capacity to develop abstract reasoning. As Chakrabarty puts it, "Abstract reasoning made it possible for the citizen to conceptualize such imaginary entities as 'class,' 'public,' or 'national' interests, and adjudicate between competing claims" (2002, 5). While this model of democracy still exists, Chakrabarty suggests that another model is emerging and becoming prevalent. This he names an experiential model of democracy. Born as a result of the new social movements of the 1960s and 1970s and the privileging of a politics of identity—a politics, which privileges the experienced, rather than the analytic—this form of democracy "orients us to the realms of the senses and the embodied" (2005, 9). It is the knowledge embedded in the senses, Chakrabarty argues, which provides an increasingly important information base to citizens of contemporary mass and media democracies.

Thus, in responding to political demands to be more representative, museums have had to move from displays based on abstract forms of reasoning to displays which privilege experience. One of the main ways in which experience can be communicated is through affect. While there is an increasing interest in the way in which objects can produce affective responses,[6] multimedia installations also offer the opportunity to explicitly explore the possibilities of affective responses for the production of cultural narratives which seek to work across cultural divides. The benefits of this work in a society that sometimes finds it difficult to acknowledge the repercussions of the colonial experience, as well as the implications of multiculturalism, are obvious. In reaching out to the senses, in seeking affective responses, multimedia installations in museums may well be more politically effective in achieving alteration in the mind of the citizen than the more traditional use of objects in didactic displays intent on *reforming* the minds of the citizen. The lessons learned here could well be used to inform the ways in which museums display their collections.

## Conclusion

What then do multimedia installations, like the ones in this study, enable in the exhibition context? First, they make the act of interpretation into an object of attention. It is almost as if the medium becomes the object. Second, these installations facilitate an affective response both to history and to the display of objects. They enable the creation of a physical, mental, and emotional space which prepares the audience for a sensitive re-reading of those objects that are on display. It is a space that can work in nondidactic ways by avoiding the practice of framing objects through abstract concepts. The result is the ability to empower the museum visitor to undergo a process of alteration, just as much as it is an attempt to tell history from a different point of view. However, because affective responses are embodied and not abstract, the process of coming to know is not framed by a didactic top-down approach characteristic of conventional museological interpretations of the material world, but rather is more open to the process of dialogue and interactivity. In a sense, multimedia installations, if seen alongside objects and not as their other, can bring a new understanding of the way in which virtual media can have both physical expressions and material effects.

*Notes*

1. See Witcomb 2003 and 2006.

2. For examples, see Hooper-Greenhill 2000, Kwint 1999, and Messham-Muir 2005.

3. For examples, see Noack 1995, Miles 1993, and MacDonald 1987, 1991, and 1992.

4. Multimedia installations in art museums are treated as objects in their own right since they are part of the collection. There is, for example, a growing literature on issues to do with their conservation. For an example, see Hunter and Choudhury 2003.

5. See ⟨http://melbourne.museum.vic.gov.au/exhibitions/exh_bunjilaka.asp?ID=560416⟩ for an image of this installation.

6. See, for example, Clifford 1985, Csikszentmihalyi 1993, and Pearce 1992.

*References*

Birch, T., and G. Sculthorpe. 2004. "Marking the Historical Moment in a Museum Context." Paper presented at The Rebirth of the Museum? An International Symposium, July 8–9, University of Melbourne.

Carter, Paul. 1992. "Performing History: Hyde Park Barracks." *Transition* 36/37: 15.

Chakrabarty, D. 2002. "Museums in Late Democracies." *Humanities Research* IX, no. 1: 5–12.

Clifford, J. 1985. "Objects and Selves: An Afterword." In *Objects and Others: Essays on Museums and Material Culture*, ed. G. W. Stocking, 236–246. Madison: University of Wisconsin Press.

Csikszentmihalyi, M. 1993. "Why We Need Things." In *History from Things: Essays on Material Culture*, ed. S. Lubar and W. D. Kingery, 20–29. Washington, D.C.: Smithsonian Institution Press.

Gibson, R. 2004. "The Museum as Cultural Laboratory." Paper presented at The Rebirth of the Museum? An International Symposium, July 8–9, University of Melbourne.

Hooper-Greenhill, E. 2000. *Museums and the Interpretation of Visual Culture*. New York: Routledge.

Hunter, Jane, and Sharmin Choudhury. 2003. "Implementing Preservation Strategies for Complex Multimedia Objects." Paper presented at The Seventh European Conference on Research and Advanced Technology for Digital Libraries, August, Trondheim, Norway. Available at ⟨http://metadata.net/newmedia/Papers/ECDL2003_paper.pdf⟩ (accessed April 18, 2005).

Kwint, M. 1999. "Introduction: The Physical Past." In *Material Memories*, ed. M. Kwint, C. Breward, and J. Aynsley, 1–16. Oxford: Berg.

MacDonald, G. F. 1987. "The Future of Museums in the Global Village." *Museum* 155: 212–213.

MacDonald, G. F. 1991. "The Museum as Information Utility." *Museum Management and Curatorship* 10: 305–311.

MacDonald, G. F. 1992. "Change and Challenge: Museums in the Information Society." In *Museums and Communities: The Politics of Public Culture*, ed. I. Karp, C. Mullen Kreamer, and S. D. Lavine, 158–181. Washington, D.C.: Smithsonian Institution Press.

Messham-Muir, K. 2006. "Affect, Interpretation and Technology." *Open Museum Journal* 7. Available at ⟨http://amol.org.au/omj/journal_index.asp⟩.

Miles, R. 1993. "Exhibiting Learning." *Museums Journal* 93, no. 5: 27–28.

Noack, D. R. "Visiting Museums Virtually." *Internet World* October (1995): 86–91.

Pearce, S. M. *Museums, Objects and Collections*. Leicester: Leicester University Press, 1992.

Stewart, S. 1999. "Prologue: From the Museum of Touch." In *Material Memories*, ed. M. Kwint, C. Breward, and J. Aynsley, 17–36. Oxford: Berg.

Witcomb, A. 2003. *Reimagining the Museum: Beyond the Mausoleum*. New York: Routledge.

Witcomb, A. 2006. "Interactivity: Thinking Beyond." In *Blackwell Companion to Museum Studies*, ed. S. J. Macdonald, 353–361. London: Blackwell.

# 3

## Beyond the Cult of the Replicant: Museums and Historical Digital Objects—Traditional Concerns, New Discourses

Fiona Cameron

---

Prevailing debates on the meaning and relationships between historical collections and virtual/digital "historical" objects are bounded by established heritage discourses, material culture paradigms, and the object-centeredness of museum culture. Digital objects, their value, meaning, and presence, have been informed by these conventions and subsequently judged from the standpoint of the "superior" physical counterpart. Furthermore, the use and intention of digital "historical" objects are rooted in older conventions of representation, that is, as historical documents, functioning materials, and as semiotic texts.

In referring to established theories of the material and immaterial, theorist Katherine Hayles (1999, 94) argues that technologies are cultural constructs, and therefore their subsequent meanings and imaginative uses are dependent on the cultural values and meanings attributed to them.

> If we articulate interpretations that contest the illusion of disembodiment and these interpretations begin to circulate through the culture, they can affect how the technologies are understood and consequently how they can be developed and used. Technologies do not develop on their own. People develop them and people are sensitive to cultural beliefs about what technologies can and should mean.

Put simply, challenging the illusion of the immaterial can be a liberating force, opening up new discursive possibilities. Therefore, in this chapter I critically examine the debates around the original-material/copy-immaterial divide and map new definitions and terminologies for digital historical collections. Most research undertaken on destabilizing the material/immaterial binary and the mapping of new definitions and terminologies has been done in the field of digital art and the uses of multimedia in exhibitions.[1] This chapter takes these discourses in a new direction by rethinking definitions of, and relationships between, analog and digital historical objects drawing on specific case examples. I argue that the roles and uses of the digital object must be understood as part of the broader heritage complex—an institutionalized culture of practices and ideas that is inherently political, socially and culturally circumscribed, and as such implicated in the cycle of heritage value and consumption.

## Positing the Digital Object as Terrorist

Discourses have centered around the status of the digital copy as inferior to its nondigital original, and the potential of the former to subvert the foundational values and meanings attributed to the original. Western concepts of object-centeredness, historical material authenticity, and aura play a central role in upholding this differential relationship.

Historian Graeme Davison (2000) explores this further and argues that a preoccupation with material remains, elevated materiality along with the unique and handmade, and references a distaste for mass production.

> In its preoccupation with the material remains of the past—"the things" you "keep"—
> it endorses our own materialism; yet in its reverence for what is durable, handmade or
> unique it also reinforces our underlying distaste for a culture of mass production and
> planned obsolescence. (Davison 2000, 115)

Theorists Walter Benjamin and more recently Jean Baudrillard embrace these ideas—the value of material authenticity and a repugnance for reproductions both arguing that mechanical reproduction and more recently simulations pose a threat to the "real" object and works of art leading to the loss of their auraic, iconic, and ritualistic qualities. Benjamin, in his much-quoted essay "The Work of Art in the Age of Mechanical Reproduction," claims that original art works have an auraic presence defined as the "essence of all that is transmutable from its beginning, ranging from its substantive duration to its testimony to the history which it has experienced" (1968, 3). He also maintains "that which withers in the age of mechanical reproduction is in the aura of the work of art" (1968, 3).

Similarly, theorist Jean Baudrillard claims that virtual reproductions will be sold as a perfect analogon of the "real," and as such will merge with the real in meaning, carry the rhetoric of the "real" object, and act as a continuous, faithful, and objective reproduction. Viewing media as an instrument for destabilizing the real and true, Baudrillard concurs that all historical and political truth will be reduced to information—a semiotically self-referring existence. This apocalyptic view of the material/immaterial relationship is based on a fear that as 3D simulations become more convincing, surrogates will merge in "form" and affectual tome (the ability for objects to engage the senses and trigger emotional responses and memory) with the physical object, and viewers will be unable to perceptually distinguish the replica from the real. Collections could then become obsolete, thus undermining museum culture and practice. Here the digital is posed as a terrorist.

By contrast and adding credence to these fears, then director of the Canadian Museum of Civilization George MacDonald proposed an antimaterialist museological epistemology. Engaging Marshall McLuhan's argument that the information society and new media has the potential to facilitate social interaction and accessibility, MacDonald (1992) reframes museums primarily as places for the dissemination of information rather than a central repository for objects. Here objects, according to MacDonald, are important only in that they contain information that can be communicated through a variety of media (Witcomb 2007).

Like earlier reproductive technologies such as line engravings and photography, assigning an individual creativity, a visible materiality, and authorship to digital historical objects has been an area of contestation (Fyfe 2004, 51). In discussing the cultural status and claims of photographs to be admitted to twentieth-century art museums, media theorist Douglas Crimp (1993, 56) contends that photography has the potential to disrupt the homogeneity of art.

> So long as photography was merely a vehicle by which art works entered the imaginary museum, a certain coherence obtained. But once photography itself enters, an object among others, heterogeneity is re-established at the heart of the museum; its pretensions to knowledge are doomed. For even photography cannot hypostatize style from the photograph.

The severing of the photograph from its referent and hence its role as an analogon, according to Crimp, by its admittance to the museum, had the potential to undermine the idea of creativity, artistic merit, the original, and authentic as key artistic and museological concepts of value while opening a universe of new values.

In summary these arguments variously raise questions about museum culture as a series of practices for defining object value and meaning, in particular the concepts material authenticity, originality, and aura. Those of Benjamin and Baudrillard aim to hold established object-centered values in place by posing the copy as a dissent force, whereas MacDonald seeks to redefine object-centered museum culture and posit museums primarily as information sources rather than repositories of "authentic" objects. On the other hand, Crimp's argument around photography challenges the original/copy divide and opens up the possibility for exploring what new values might emerge for digital historical objects. Given that one of the functions of visual representations is historically to deceive, it is not surprising that the reproducibility and "immaterial" nature of digital objects and assets induced some anxiety within the museum sector.

### Object-Centeredness and Material Authenticity as a Cultural Concept

Digital historical objects and their redefinition therefore are difficult in a context where reified concepts of representation, interpretation, perception, aura, affect, authority, authenticity, material, provenance, and aesthetics are used to inscribe value and meaning from the traditional conventions of an object-centered museum culture. The basis of museological integrity and mission is founded on the formation of cultural capital and the selection of and attribution of value that upholds and sustains the meaningful deployment and circulation of collections as material evidence, original, authentic, as knowledge, as representative of the passage of time, and as signs of power and privilege (Witcomb 2003). Moreover, this focus on materiality theorist Jonathan Crary argues is inseparable from the deployment of vision as an objective source of knowledge and rational thought in the nineteenth century thus leading to the subsequent repression of other more experiential forms of knowledge production.

The culture of the modern museum is one of strong classifications between originals and reproductions, where the reproduction is back-staged through a concealed labor of production, authorship and materiality (Fyfe 2004, 48). Put simply, originals and copies are discursive objects and realizations of implicit western cultural rules where copies and originals are kept at a distance to support particular forms of knowledge and institutional claims of authority (Fyfe 2004, 61). This idea and process of distancing is most closely expressed in Benjamin's concept of aura and mechanical reproduction is a way of ensuring that the established order of object value is not subverted and conception of reality bounded within certain frames.

Epistemologically speaking, the current obsession with the bifurcation of intelligibility in which real historical objects as "original," "material," and "objective" evidence take hegemonic precedence over digital historical objects attributed as copies, immaterial, inferior, surrogate, and temporary has its roots in nineteenth-century empiricist and evolutionist ways of thinking. Removed from their cultural and social setting, historical objects were selected and valued based on a series of assumptions as evidence of deep history where temporality is reified, and variously by a process of authentication based on their materiality and method of manufacture as "original" and "authentic" (Cameron 2000). Deemed to embody the essence and moral state of a people who made them, history could be visually read through the physical characteristics of authenticated collections (Cameron 2000). Objects were thought to communicate perfectly by being what they are.

Far from being natural or innocuous, the preservation of an authentic domain of identity as exemplified in the collection and exhibition of Maori material culture at the Auckland and Canterbury Museums, New Zealand, was tied up with national politics and the need to understand and control history and the place of the indigenous people in that scheme. In this sense the reproduction had no value or "materiality" except as a servant to the real.

The cult of the real-material, attributions of original and authentic, where vision operates as the interpretive frame and physicality as a stable, truthful, and objective marker of culture, held weight well into the late twentieth century. This materialist epistemology is evident in historian Thomas Schlereth's definition of historical and technological objects as makers of cultural expression:

> Material culture properly connotes physical manifestations of culture and therefore embraces those segments of human learning and behavior which provide a person with plans, methods, and reasons for producing things that can be seen and touched. (1994, 2)

### Liberating the Digital from a Materialist Epistemology

The influences that have potentially liberated historical objects from the strict limitations of a materialist epistemology take the form of social and new history, products of the cultural turn in critical theory—the advent of poststructuralism and postmodernism and the rise of identity politics and social movements in the 1960s and 1970s. Theoretically speaking, historian Alun Munslow (2001) terms this new position as "epistemic relativism," one that views knowledge of the "real" as derived through our ideas and concepts, including linguistic, spatial, cultural, and ideological compulsions.

In this discursive framework, museological scholar Eilean Hooper-Greenhill (2000) argues that far from speaking for themselves in absolute terms according to a hegemonic discourse, objects and their meanings are now seen as contingent, fluid, and poly-semic, that is, they have the ability to be interpreted in a variety of ways. Here Hooper-Greenhill suggests that the material properties of objects such as wood and stone are no longer taken as irreducible facts, but rather disparate and discursive meanings are given equal weight (2000). Objects become the materials at hand and act as sites where discursive formation, polysemy, and affect intersect with their mate-rial properties read as signs, symbols, and texts.[2] Put simply the concept of real, orig-inal authenticity becomes a social construct, as exemplifiers of western knowledge and taste, and systems of object value become mutable.

For digital historical objects this "cultural turn" means that the strict limitations imposed by materiality, optical technologies as a research and descriptive tool, and the definitions and values accorded the real and replicant are no longer seen as abso-lute. Rather the interpretive potential of digital objects becomes legitimized, open to a variety of interpretations using a range of senses beyond the visual as well as individ-ual affectual knowledge. Likewise, the advent of digital media as a cultural construct, from which digital objects resonate is viewed in a new way. Media theorist Lev Manovich (2001a) argues that we are now "post-media" and the categories which best describe the characteristics of post-digital art concern user behavior and data organiza-tion rather than the medium per se. He goes on to argue that "synthetic computer-generated imagery is not an inferior representation of our reality, but a realistic representation of a different reality" (2001a). Therefore, digital historical objects can potentially be seen as objects in their own right, can play to notions of polysemy, the experiential, and the sensual. The denigration of vision as an objective interpretive tool, along with an interest in the haptic and interaction also enables a different rela-tionship between subject and object to emerge, thus enabling greater democratized access to collections. Therefore, to what extent do current museological practices en-trench or liberate digital historical objects from materialist sensibilities and categories of value accorded to them and engage new roles? And what epistemological considera-tions are brought to bear on them?

### Cycles of Consumption: The Politics of the Original/Copy

In the current discursive context, historical objects tend to be "museumified" as aes-thetic witnesses to their period, to historical events or people.[3] Significantly, "real" objects are deemed to have a historical actuality while acting as a visible sign of the past. They act as fragments of information, having a special place in time and space as survivors of the past ensconced in the museum (Kavanaugh 1996, 3). For example a convict love token owned by Thomas Tilley acts as testimony to a reality in the past, the arrival of the First Fleet in 1778, and the beginnings of European settlement in Australia.[4]

Objects act as a physical qualification for a sign while on the other hand they refer to clear familiar signifieds. Their functional roles are also overridden within the museum context by these metaphorical journeys (Baudrillard 1996, 81). For example, Ned Kelly's armor (a notorious Australian bushranger) no longer has a practical application as armor but signifies the romance of the bush and the heroic myth of the Kelly gang, while transporting the viewer back to the fateful day in June 1880 when Ned was captured during a standoff with police at Glenrowan, Victoria (2005). Objects that are not part of our active memory and recollection draw on our own more recent experience. For example, visions of Ned Kelly the person and the Glenrowan episode stirred by the Kelly armor are shaped by our own experiences, learned histories of this event, and popular mythology.

Drawing on sociologist Pierre Bourdieu's concept of cultural capital formation (Bourdieu and Darbel 1991), it can be argued that throughout their history reproductions such as line engravings of famous paintings, plaster casts of well-known Greek sculptures, and photographs have had a significant role in its creation (Fyfe 2004, 48). That is because reproductions are the means by which cultural capital is spread, and the rules and habits of looking are developed. Moreover, as reproductive technologies improved, as with photography, a more acute connection could be made between the medium and the creative intention of the maker.

Like earlier reproductions digital objects bring the historical object and the signs of its making into the presence of the viewer, while suppressing its own craft that made that presence possible by carrying information about the "original" object's details, its form, fabric, shape, aesthetics, and history through interpretation. In creating a surrogate, the gestures, memories, customs and intentions, and scars of their life histories are faithfully replicated in virtual space taking on the solidity, surfaces, edges, and texture of the real to ensure a more certain recovery of history, time, or aesthetic experience.[5] For example the infantry officer's red jacket or "coatee" worn by Lieutenant Henry Anderson at the Battle of Waterloo, now Belgium, on Sunday June 18, 1815, in the collection of the National Army Museum, London bears evidence of a blood stain, the trace of a wound inflicted by a musket ball (Pearce 1996, 19). The historical circumstances of time, place, and action, that the coatee and stain as an authenticated factual trace, refers contributes to the charisma, value, emotional and affectual tone of the original—its potential to move and excite the viewer (Sontag 1982, 63). Because the jacket survives physically it retains its metonymic relationship to the battle (Pearce 1996, 25).

Digital and physical collections can function as interactive conduits in engaging emotional experiences, and in extending memory, recall, and identification (Manovich 2001a, 57). The digital surrogate's similarity to the real thing allows its creators to invoke emotions in a similar way to that of earlier simulations such as panoramas and cinemagraphic techniques. The stain, for example, inscribed into the coding of the digital form, becomes one of its message-bearing qualities, the *langue* or memory for the observer when the observer examines certain features of the object, and interprets them within the context of museum information, learned experiences and a myriad of chosen meanings based on imagination and other metaphoric relationships (Leyton 1992, 1).

As part of this mourning and historical recovery process the replicant is sealed off conceptually. The virtual object is not read as having other messages, its own style, materiality, or acting as a sign substantially different from the physical object. Take, for example, the suit of Australia's first rock and roll star, Johnny O'Keefe. As Baudrillard predicted, in this instance the digital surrogate of O'Keefe's jacket is sold as a perfect analogon and rhetorical equivalent of the "real" as representative of the garment's history and the rise of rock music in Australia. Descriptive text assigns the "real" object's attributes, its media or materials, cotton, velvet, and silk, as well as its maker, tailor Len Taylor and the date of its manufacture (between 1957 and 1959), to the digital surrogate. Its own materiality, processes of production, and the author of the surrogate are excluded and concealed from the viewer. The value of the digital heritage object is derived directly from the viewer's acceptance of the real object as authentic. Its presence on a museum's Web site endorses the authority and integrity of these inscribed meanings accorded to the real.

### The Mutability of Material Authenticity and Aura

The object as a visible sign of authenticity and aura or provenance reassures museums of their mission. Benjamin maintains that the aura of an object has a stable, unchanging quality representing "its presence in time, its unique existence at the place where it happens to be." Far from being absolute and stable qualities, the physical nature of an object is subject to change through time, either deliberately or accidentally (Hooper-Greenhill 2000). Moreover, authenticity and hence the authority of the "real" is often museologically determined and contingent on curatorial evaluation of not only its physical qualities, but also its life history.

The "real" object's enchantment, its aura, for example, is its physical presence, but most important, it derives from ascribed social meanings. Its message-bearing abilities and the persuasiveness of its origin through associated stories are important ingredients in invoking its awe. Auraic or magical potency is heightened if its transformations of ownership, value, and meaning can be traced. The value of Lieutenant Henry Anderson's jacket worn at the Battle of Waterloo is its well-documented provenance and use as a souvenir of the event by a known owner (Pearce 1996, 126–127). If an object is dislocated from its systems of meaning its aura is diminished. In the case of the coatee, it becomes merely a stained jacket.

### Aura: The Original/Copy Dilemma?

"Aura" according to Benjamin originates from the object's association with ritual, one which is broken by mechanical reproduction. By separating the defining nature of an object, its accumulated history as a unique place in time, through reproduction, its aura, and hence its authority is threatened.[6]

In terms of digital copies, I would argue the opposite. Deciding what to digitize and render in 3D—and what not to—involves an active process of value and meaning-making equivalent to that of the physical object. It enacts the curatorial process of selection of what is significant, what should be remembered and forgotten, and what categories of meaning such as classification, cultural values, or aesthetic attributes are given pre-eminence. And the value of the "real" increases when digitized, enhancing its social, historical, and aesthetic importance, owing to the resources required in the compilation of a 3D rendering, and through distribution. For example, the Powerhouse Museum, Sydney, Australia chose 200 of its most significant objects from its decorative arts, science, and technology, and social history collections to be transformed into 2D and 3D visualizations.[7] By virtue of its selection, the digital surrogate illustrates, reiterates, and passes on a set of social relations constructed for the "real," while endorsing their ascribed value as the best aesthetic, historically, and technologically significant items in their collection. Here the past is at the service of the object reclaimed through its digital counterpart. As a result, these edited statements and silences and their embedded subjectivity are enhanced through selections for digitization, and where the digital surrogate also services classification and cultural meanings.

In this instance, the copy also serves to form and validate a collective social memory for the nation (Hall 2005). They operate to capture designated high points—social, cultural, technological, and aesthetic achievements while offering tangible links between past, community, and identity. This process of selective canonization confers authority and a material and institutional factuality to these frameworks making them difficult to challenge.

So has the real object lost its authority and been relegated, like its digital counterpart, to an object of signs as prophesied by Baudrillard? Surprisingly, the museum object tends to take on a new meaning of "real," a new objective stance even though we know what objects really are, embedded with cultural, disciplinary meanings, and values. This is because objects are viewed within this new space in which "real" means material—the physical presence of things as opposed to the illusion, the virtual. The materiality of an object is seen as having a more objective presence than the virtual due to its ascribed connection with a historical or aesthetic actuality. Within this context the real object is not under threat but acts as an alibi for the virtual. This poses a contradiction in meaning. Objects interpreted according to a postmodernist epistemology stand as signs, but resurrect a modernist unmediated objectivity as a physical truth when considered in relation to the digital, due to their materiality.

For example, Zeus (also known as the Poseidon of Artemision, held in the National Archaeological Museum in Athens), a fine exemplar of Early Classical bronze statuary dated to 460BC, was simulated using a laser scan by the Powerhouse Museum to allow its image and form to travel to Sydney for the exhibition *1000 Years of the Olympic Games: Treasures of Ancient Greece* (Kenderdine 2001). In this instance the value of the real sculpture was its materiality as testimony to this point in time and to techniques of early bronze casting, as well as exhibiting the tangible marks of labor and aesthetic qualities of an important sculptor.

Moreover, has a rational discourse of the real been murdered as Baudrillard concurs? I would suggest not. Strong referentials are still longed for. There is still a need to connect an object with an origin, a past, and a chain of events. It represents a social desire for rational meanings by users in order to give them a clear statement of the visual surrogate's value as a signifier, as an analogy of its physical counterpart.

Digital historical objects also maintain the referent of the subject (a disciplinary reality) while operating as the visible and intelligible mediation of the real object. For example, online collections at The Paul J. Getty Museum are interpreted within an art historical classificatory referential, categorized according to artists, subjects, and collection types.[8] Object-level information links them to anonymous authoritative narratives, placing them within the discourse of art history. Under sculpture, a Cycladic figure of a female, for example, is described in terms of stylistic and chronological metaphors with text linking the object to its cultural/geographical origin. Because they are used to aid and construct the physical object as an outside referent they are inherently biased (Manovich 2001a, 71). Hence the Cycladic figure engages discourses about power—disciplinary hegemony and classification, and knowledge of the "other."

A virtual object carries the message of the real while functioning as a sign. It takes on the physical referential of the real in which a rational disciplinary-based discourse remains a key element. For instance, with the Cycladic figure, the virtual example acts as a true simulation of the material example with the fabric, form, detail, and stylistic attributes of the physical object being attributed to its virtual counterpart.[9] A more faithful rendering in 3D will allow the observer to read its physical attributes in more detail. Information in this context has not supplanted the intrinsic material value of physical collections as suggested by George MacDonald.

In discussing photography, Roland Barthes suggested that the photograph and its referent are bonded together. They merge in meaning value, in which the former acts entirely in service of the real. In this context, without the "real" object, the virtual would have no meaning, no soul, no referential.

> The Photograph always carries its referent with itself. . . . They are glued together limb for limb. The Photograph belongs to a class of laminated objects whose leaves cannot be separated without destroying them both. . . . And this insular adherence makes it difficult to focus on photography.

In considering this in terms of digital objects, traditional systems of object value have upheld and adhered the authentic and copy binary (Barthes 1981, 5–6). But in order to reconfigure established meanings and construct a new set of relationships for them, this original copy bifurcation requires careful reconsideration.

### Affectual Tone and the Predicament of the Original/Copy

In the movie *Ocean's 12* Rusty Ryan (Brad Pitt) in a conversation with Danny Ocean (George Clooney) comments, "there is new holographic technology where you can now produce anything in 3D." Danny Ocean replies, "127 million dollars' worth." Here an elaborate plot emerges in which a digital hologram of the famous Fabergé coronation egg created using 2,000 digital slithers and a mixture of tungsten and real light is used as a decoy in a burglary. Is this mere fictional illusion or has holographic technology developed to an extent where digital objects can challenge the authenticity of our responses to real objects?

Perhaps this question is one of the most contested areas of the copy-original debate and recalls Baudillard's concern of a loss of distance between the real and replicant where the former, through visual technologies, merges in meaning and hence poses a threat to the original and authentic.

Recurrent claims that new media technologies can externalize and objectify reasoning and that they can be used to augment or control it, are based on the assumption of the isomorphism of mental representations with external visual effects (Manovich 2001a). That is because the way we "view" and pay attention to both the digital and "real" visually in western societies is of a deeply historical character (Crary 2001). It is not a new argument, realism in painting, the reproduction of masterpieces, and the current quest for the perfect simulation all seek to deceive and merge mental processes, memory, and contextualization (Manovich 2001a, 71). Media theorist Sean Cubitt terms this desire as "The digital longs for the organic with the same passion in which the text longs for the reader" (1998, 35).

Therefore, what is the psychological standing or affectual tone of the historical object and its digital counterpart? Are digital simulations of historical objects able to convince the viewer and invoke equivalent affectual experiences or is materiality a defining factor?

With many digital analogons there is a reduction in size, proportion, and perspective. The more faithful the reproduction, the more credible the simulation becomes and hence the greater its persuasive power. But to achieve maximum affect, the technologies of production and display must remain invisible to the viewer. With the 3D digital representation of the statue of Zeus, the dream of its creators at the Powerhouse Museum was to produce a visible and intelligible simulation of the real object as close as possible in terms of scale, size, detail, color, texture, and perspective using spatial metaphors.[10] Although an impressive simulation, does this 3D surrogate really capture and play out the myths of authenticity and origin embedded in its real counterpart? Its role, like the real object, is intended to stimulate a wide range of feelings, thoughts, and memories of what the viewer has been taught about Ancient Greece and aesthetic conventions associated with the "reading" of Classical Greek Art. Here the deception however, is easily recognizable as the medium, a laser image viewed through shutter glasses, is at odds with what is depicted (Grau 2003, 15). Although lacking an authentic transcendence the digital Zeus replaces it with an illusionary spectacle as a new space of enchantment.

Indeed the ability for visual technologies to isomorphize the psychic and the visual as used by Rusty Ryan and the *Ocean's 12* team is not a fiction and made possible with development of new digital holographic techniques.[11] During a trip to H. Stern's jewelry factory in Rio de Janiero, Brazil, I encountered a collection of three watches displayed on a plinth. The Sapphire Collection, a digital hologram, was the ultimate in trickery, deception, and illusion thus challenging my conception of reality and the authenticity of my responses to them. It was not until the illusion was exposed by passing my hand through the image did its technologies of production and indeed its materiality become visible and the deception shatter. The digital watches initially offered an aesthetic pleasure equivalent to the real object and where my aesthetic judgments about the piece were based on the perceptions built in part by the hologram itself. Although once exposed as a digital replicant, it did not change its appearance but rather lost its status as a piece of jewelry and no longer had a direct link with its imagined maker. In this sense, a convincing visual simulation of a historical object, for example, has the potential to disrupt the connection to material evidence as a trace, and to history, as Benjamin predicted.

Art historian and museum director, Charles Saumarez Smith explains the nature of the encounter between the viewer and the museum object by suggesting that it "is more involved in the process of obtaining some form of felt relationship to the past, intuitive, sensual and three-dimensional." Arguably, digital objects have a power of affect somewhat similar to the "real" object in terms of the processes of alteration. Both the digital and "real" object offer a stimulus for a physical and emotional response, of coming to know, and for imagination to play a role (Stewart 1999, 17–38; Hooper-Greenhill 2000, 5). My physical reaction to the hologram, once its materiality was revealed, involved a felt emotional response that led to a greater degree of understanding, a change from unknowing to knowing and to one of awe, an aesthetic sense of astonishment and amazement.

Media theorist Ross Gibson (2005) relates the affect of media directly to the viewing of art and the role of museums as theatres of alteration. Gibson also refers to the process of alteration as two processes. The *huh*, is where our preconceptions are suspended, we feel off-course for a moment, and are enticed to fill in the cognitive vacuum that the wonder puts in place, and the *wow* factor where component parts come back together in a new way and you come out the other side feeling different. In the case of the watches, a narrative of origin, authenticity, and its related signification chains are suspended and replaced by heightened emotions relating to the illusion.

One of the most moving examples of this process of alteration in a museum setting was the exhibition *September 11 Bearing Witness to History* at the National Museum of American History. The "concreteness" of the objects and its metonymic relationship to the event offered tacit or felt knowledge at a deep reactive level producing powerful gut reactions, mobilizing feelings and emotions (Hooper-Greenhill 2000, 16). As cultural symbols, the affectual power of objects, such as the fragments of aircraft that crashed into the World Trade Center, for many elicited such strong emotions and led to a discussion about their inclusion in the exhibition by the curatorial team.

> Some objects were considered too emotional such as Ted Olsen's telephone, the man who received a call from his wife as her airplane hit the Pentagon and the airplane parts because people rode to their death in them, but without them we would have lost their stories.[12]

Moreover, responses to the exhibition elicited powerful performances of alteration. As one member of the curatorial team commented, "One man came out of the witness section, leaned his head against the wall and cried. It is not something we are accustomed to seeing."[13] And for myself, living on the other side of the world, an act beyond the realms of possibility was made tangible and concrete—the exhibition elicited proof, it confirmed that this horrific event did in fact happen, by the presence of the aircraft fragments juxtaposed with twisted pillars from the World Trade Center. My initial reaction was tacit and sensory. My alteration experience began with a sensory experience and the reconfiguration of what I perceived, an acknowledgment of being in the presence of objects that had a direct link to the event. The physical presence and material properties of the objects mobilized dramatic, immediate corporeal responses, shaped according to the deeply unsettled feeling of insecurity I experienced when watching TV footage of a second aircraft flying into the second tower at 4:30 AM on the morning of September 12, 2001. And most significantly, it elicited an empathetic response, for those who rode to their death in the aircraft.

The digital representation of the aircraft fragments and seat belt lacked the immediate transcendental, spiritual, or confrontational power of being in the presence of the original. Although it furnished recollections of 9/11, it operated more like a secondary source.

Clearly the material form is central to the meaning of analog and digital objects and hence their affectual tone. In discussing photographs as a media, curator Elizabeth Edwards looks less at their images and more at the messages conveyed by the medium itself: its appearance, smell, and texture as well as its setting in invoking affectual tone (Edwards 1999, 6). Similarly, Eilean Hooper-Greenhill (2000, 112) maintains that the interpretive framework brought to bear by the subject is largely dependent on the known physical character of the artifact, but also engages the cognitive and affective. The material form is central to their meaning, and hence their affectual tone (Edwards 1999, 223). Historian Daniel Miller elaborates on this by stating that the more mundane sensual and material qualities of the *object*, enable us to unpick the more subtle connections with cultural lives and values (1998, 8–9).

In this sense it can be argued that the digital surrogate is unable to carry the deep imaginary power and wealth of the physical object as a trace or as "The real object has a special psychological standing that the digital doesn't have, as primary evidence. Its concrete reality and density of stone, wood, or plastic as opposed to the materiality of digital codes and programming languages gives weight to the intensity of its interpretation, suggesting "this is how it is" and "what it means" (Hooper-Greenhill 2000, 114). The signs of history and time engrained on the physical object such as the patina on the bronze Zeus, contribute to its spiritual experience and aura (Benjamin 1968, 3). The digital reproduction however lacks the visceral thrill of being in the presence of the original, while its seduction operates as a visual sensation and as a phantom of the real, making reference to its place in time and history through the analogonic representation of its patina as a signifier of its life history. In this way the sociocultural values and the reification of original and authentic remain a defining factor in upholding object value, the way the past is constructed through material culture, and museum culture as a particular phenomenon. Here authoring and the exposure of the materiality of an "object" becomes a defining factor.

The experience of an object in the social and ritual space of the museum apart from the everyday also differentiates its psychological affect from its digital surrogate. As Elaine Heumann Gurian suggests, "the virtual experiences in the privacy of one's own home may be enlightening but, I think, are not part of the civilizing experience that museums provide. It is the materiality of the building, the importance of the architecture, and the prominence that cities give to museum location that together make for the august place that museums hold" (1999, 165).

Theorist Susan Stewart (1999) observes that although "the prospect of touching an object is real. . . . The desire to do so is suppressed in the ritualized space of the museum." Rather than relying on the triggering of a sympathetic response in a sense other than that which is directly stimulated—for example sight to evoke sound—as with physical collections, digital examples can evoke through a range of senses from sight to touch and sound. Museum scholar Simon Knell argues that in this sense technologies can open up possibilities for sophisticated interpretation it can come close to challenging the object in terms of information (Knell 2003, 132–146). This indeed is possible and is illustrated through the digital interactive of Picasso's masterpiece "La Vie" (1903) by The Cleveland Museum of Art and Cognitive Applications.[14] Here the interactive reveals a second painting of a female nude that lies below the surface, perhaps the lost work, *The Last Moment*, a key painting in Picasso's first exhibition in 1900.

## Rethinking Materiality: Severing the Original-Material/Copy Immaterial Binary

Not surprisingly, working definitions for digital historical objects outside their role as an analogon are in short supply. That is because the character of the virtual/material bifurcation is rarely disputed. A material object according to Manovich is a self-contained structure limited in space and time, is fundamental and clearly delineated from others semantically and physically (2001a, 163). By contrast, Pierre Levy interprets the virtual as something that is not quite there, something intangible with an absence of existence, as the complement of the real or tangible (1998, 23). The act of ignoring the materiality of the digital medium and responsibilities for its use by Levy, that is, not looking for meanings outside the digital object's role as an analogon, and by delimiting material objects as a self-contained structure by Manovich, continues to support established categorizations.

Graham's definition of the virtual, albeit in the context of Internet communities, provides a more useful theoretical starting point from which to approach digital historical collections. According to Graham, the "virtual" is not a semblance of something else, but an alternative type of entity with properties similar and dissimilar to those with which it is contrasted. They are entities in their own right and as such we must attribute to them a distinctive form of reality (1999, 159).

Media theorist Sean Cubitt supports this contention, linking this debate to previous reproductive technologies by arguing that every digital object, like the older technologies of print or film, is a unique object with its own physical and aesthetic qualities (1999, 31). In short, the digital historical object as surrogate is just one of its many roles, and it embodies its own material and aesthetic properties similar and dissimilar to physical collections. So what kind of reality (both metaphysical and metaphorical) do digital "historical" objects inhabit?

Digital materialism has been denied its own "materiality" because of the material—presence/absence—virtual divide. Theorist Katherine Hayles suggests that this dialectic, constructed through information theory during WWII, purposely privileged a hypothesis of information pattern as absence as opposed to presence, one that could conceptually be transmitted over telephone lines because it was metaphorically divorced from a material medium (1999, 70). Presence dominates in a similar way to material. Its analogy is founded in biological thinking where the body—the material medium—is information encoded in genes (1999, 69). Interestingly, Marshall McLuhan's now famous quote, "The medium is the message" (1964), makes no distinction between information and materiality.

Digital historical objects, like physical ones, are culture-encoded in a digital form and are made up of familiar cultural forms of meaning, significance, elements of perception, language, and reception (Manovich 2001a, 71). The materiality of the "real" and the digital historical object are different in essence and composition, the former analog made of continuous data from one point to another as well as discrete data, whereas the latter are digital codes, each representing their own unique processes of production.

From the physical collection, we might take as an example Michelangelo's *David*, in the Galleria dell'Accademia, Florence. Made from marble, it is subject to corresponding serial processes of production such as grinding, chipping, and polishing. The digital historical representation, a laser scan of the external shape and surface characteristics of David is described formally as a numerical representation of 0s and 1s.[15] Its sculptural form is defined by a mathematical function with its media elements—shapes, colors, or behaviors—represented as discrete samples (pixels, polygons, voxels, characters, and scripts) (Manovich 2001a, 30). Its medium and processes of production contain a hierarchy of levels: the digital camera, coding, the interface, the operating system, software applications, and its appearance created by operating systems and assembly language for digital compositing and machine language similar to film (2001, 11). The means of production—the computer—also becomes the means of distribution. But most importantly, its programmability sets digital objects apart from other simulations such as photographs and film.

Nevertheless, both modalities, David as a marble statue and David as numerical coding, are material objects by definition (Hooper-Greenhill 2000, 104). That is, they are both the result of human creativity, exist in real time, can be touched, can be looked at from many angles, and are the target for feelings and actions. Moreover, both the materiality of marble and the digital David are unstable and subject to change, the latter due to mutations of data. Whereas the marble *David* as a physical object is fixed, the digital David is no longer semantically and aesthetically discrete, can exist in potentially infinite versions, and be distributed in space and time due to its numerical coding and modular structure.

The analog and digital contain similar values of interactivity involving both imaginative and conceptual engagement, although in the digital world interactivity refers to technologically orientated concepts due to its inherent programmable materiality. Benjamin argues that the severing of the aura of the original artwork through reproduction can also be a liberating force (Benjamin 1968). The liberated object, for example the digital David, can be brought closer spatially and perceptually. Simulations before the introduction of computers focused on visual appearance almost exclusively but now appeal to other senses—touch, sensation, space—to simulate the behavior of inanimate objects and hence play to notions of the experiential (Manovich 2001a, 182). Aspects of the original David unavailable to tactility or the naked eye—such as chisel marks smaller than a millimeter—and overhead views instill new values of perception, space, and movement through zoomable technology and object movies.

Historical collections as possessions and as heritage items limit access. The bronze Zeus, for example, is retained in the ritual space of the museum, whereas the patterns and codes of the digital Zeus do not have a materially conserved quality, are not subject to wear, and have a ubiquitous, nonexclusive quality. For example, Zeus has been put into situations not available to the "real" in a way that photographs were previously (Benjamin 1968, 2). These are now accessed locally in the exhibition *1000 Years of the Olympic Games*, and dispersed through the Web site (Hayles 1999, 78). Simultaneous communication and the erasure of the concept of time and place through speed also presents a potential to exchange and act on objects through "telepresense" (McLuhan 1973). Clearly Baudrillard did not anticipate the imminent appearance of bidirectional and decentralized media with new opportunities for democratizing access, and for the reconstruction of the relationship between user and subject (Poster 1996, 19).

### Rethinking Benjamin's Aura

Given that aura is determined through links to materiality or original fabric and by tracing provenance, can Walter Benjamin's definition of aura still be present in digital historical objects or the media itself (Graham 2007)? Benjamin argues that this is not possible for reproductions, an idea widely accepted by theorists (Manovich 2001a, 174).

> Even the most perfect reproduction of a work of art is lacking in one element: its presence in time and space, its unique existence at the place where it happens to be. This includes the changes which it may have suffered in physical condition as well as the various changes in ownership. The traces of the first can be revealed only by chemical or physical analyses which it is impossible to perform on a reproduction; changes of ownership are subject to a tradition which must be traced from the situation of the original. (Benjamin, 1968, 2)

Looking more closely at digital historical objects, similar concepts—albeit a different history, materiality, and origin—are revealed. Like the analog, the materiality of the digital acts as a testimony to its own history and origin, and hence authenticity. And its provenance, chain of origin, and custody can be traced, albeit with some difficulty because of its mutable and distributive character, a process made easier with the use of authoring signatures (Lynch 2000). For example, digital Zeus has a history, a patina of human endeavor. It was commissioned as part of an exhibition of Greek treasures for the 2000 Sydney Games, formed by 3D imaging from the original in Athens, and the digital "original" produced at the Powerhouse Museum by known authors. Validating its claims of origin and history, like the analog, involves an examination of its characteristics and content in order to assess whether these are consistent with the claims made about it and its record of provenance. As with mass-produced and distributed published materials, validation involves determining the object's integrity by comparing it with other versions. Benjamin's assumption that aura is absent in reproductions appears not to hold up for digital media; rather it represents a lack of understanding of its materiality and thus denies the surrogate and indeed the digital historical object as a creative work in its own right, with a history and provenance.

Throughout the history of illusionism, the ability of designers to simulate the way that things look, and more recently, to digitally simulate the behavior of inanimate objects that invoke memory and emotions in the viewer raises the important question, should museums produce simulacra—copies without originals—or undertake restorative digital alteration? In the Australian Broadcasting Corporation program *Long Way to the Top* (a history of rock music in Australia), the label of a Slim Dusty release was superimposed on a Johnny O'Keefe record and passed off as a seemingly perfect rendition of an existing original. Arguably, the acknowledgement of its origins and authorship legitimizes such activities. Without visible authorship such a creative project has the potential to undermine the perceived trustworthiness of museums as cultural repositories, tricking the user into believing in the existence of a real counterpart. Significantly, the creators, designers, and interpreters of digital objects represent the emergence of a new set of power relationships, albeit in a different medium. The imprints of this authority stamped on the object are evident in the use of the camera and programming languages in shaping and manipulating it, its message and meaning.

## The Old, the New, and the Reconfigured: Mapping Meaning, Values, and Uses

The digital historical object can exist in many realms and perform many roles that go beyond reproduction, interpretation, education, documentation, and archive. Indeed its analogonic role is potentially diverse. As Deidre Brown in chapter 4 argues, in the context of indigenous Maori, the surrogate has the potential to connect objects with communities, facilitate repatriation, conservation, remote study, and curatorial research reconstruction (2007).

In creating new definitions for the digital historical object as an object in its own right separate from any referent, and as an entirely new creative project the materiality argument can no longer be given pre-eminence. Rather, user behavior and experience become key defining principles, which acknowledge the particular characteristics of digital medium (Witcomb 2007).

As with historical collections, photographs, and digital art, digital historical objects can be understood as independent creative works, acquire the status of objects in their own right, and be accessioned into museums. The UNESCO *Charter on the Preservation of Digital Heritage* articulates this turn by creating a new legacy—the digital heritage.[16]

Nonetheless, the status of copies from nondigital originals still remains ambiguous. Clearly definitions cannot be gleaned exclusively with reference to an institutional collection of static objects (Cook 2007). Museological acceptance is reliant on the need to include some commonly accepted attributes given to analog objects such as materiality origin, provenance, authorship, and aesthetics that justify their status as historical objects in a digital format, rather than as analogons or as information design. Beyond this, a range of expanded meanings, material characteristics, and behaviors emerge as representing a particular configuration of space, time, and surface, sequence of users' activities—a particular formal material and user experience like those attributed to digital art (Manovich 2001a, 66–67).

Drawing on classifications accorded to digital art by Manovich, Dietz, and Cook, material characteristics could be defined as Variability, Interactivity and Computability, Collaborative, Distributable,[17] and medium-independent behaviors described as Social, Networked, Encoded, Duplicated, and Reproduced. Representational experiential definitions could include those of simulation, control, action, communication, visual illusion, and information (Manovich 2001b, 16). All these underwrite the meaning of digital historical objects as cultural product, technology, philosophy, and experience. Such definitions pose a challenge for a documentation schema based on materialist epistemologies and canons of historical significance and value, as witnessed with digital art and the art canon.

## Conclusion

In referring to the uses of multimedia in museums, Andrea Witcomb suggests that museums are moving away from a referent in a reality constructed by objects (Witcomb 2007). While this may indeed be true for multimedia in the exhibition context, looking at the uses of digital historical objects in museums and their place as archive, I would suggest not. While digital historical objects potentially have multifarious roles to play in museums the analogon remains preeminent, operating within traditional relations of real-copy and classificatory frameworks, while reifying cultural and museological concepts of original and authentic. In this sense the original object preserves all its authority over the digital. Digital historical objects are tied up with the fantasy of seizing the real, suspending the real, exposing the real, knowing the real, unmasking the real.

Similarly, the digital historical object is subject to cultural politics—as with the analog—partially as it's analog's surrogate, but also as a creative work, where its values are shaped by the cultural attitudes, beliefs, and disciplinary values of its creators and curatorial practices. Like the analog, they are cultural constructs and have the power to shape cultural identities, engage emotions, perceptions, and values, and to influence the way we think.

Moreover, the prospect of a total identification between the analog and virtual is impossible. In Baudrillard's words "if so that would be a perfect crime—a crime that never happens" (Baudrillard 2000, 71). According to Baudrillard there is always a hiatus, a distortion, a rift that precludes this. That hiatus is constituted by their different modalities, notably their materiality and behavior even though they share a number of characteristics.

There is however a need to move away from formalist notions of technology and materiality, the original and authentic, and the desire to make digital objects fit into the rubric of replicant. Significantly, the digital historical object has been undervalued and subject to suspicion because its labor of production has been concealed and therefore bears less evidence of authorship, provenance, originality, and other commonly accepted attributed to analog objects. For these reasons the digital object's materiality is not well understood.

By understanding the modality and materiality of digital historical objects, new roles and a set of defining characteristics emerge beyond their role as servant to the "real" as representation, presence, affect, experience, and value in a museum context. Both modalities, the analog and the digital, are material objects by definition, each acting as testimony to its own history and origin, and hence authenticity and aura.

The acceptance of digital historical objects as independent creative works in their own right separate from any referent, and worthy of a place in the museum collection will be based on the ethical choices of curators and museum management and the wider cultural sector. One potent example of this transition is the history of the photograph over the last 100 years. No longer seen solely as a simulacrum, it has now attained a status of its own as a creative work as well as value as a historical document, along with its archival and interpretive roles. Photographs, according to Roland Barthes, do radiate an aura, a distance, reference to a past and memory.[18]

*Notes*

1. For a discussion on classifications of digital art see Cook and Graham 2004 and Dietz 1999. On multimedia and museums, see Witcomb 2007.

2. See Hooper-Greenhill 2000, Kavanaugh 1996, Pearce 1996, and Stewart 1984.

3. See Gottdiener 1994, 33, and Pearce 1992.

4. Love tokens were made from coins and given to loved ones in Britain prior to a convict's transportation to Australia in the late eighteenth and early nineteenth centuries. Powerhouse Museum collection, at ⟨http://www.powerhousemuseum.com/opac/87-1494.asp⟩ (April 19, 2005).

5. Barthes describes how artists inscribe themselves on space, covering it with familiar gestures, memories, customs, and intentions. See Sontag 1983, 63, and Leyton 1992.

6. For a discussion of this in terms of the development of photography, see Walsh 2007.

7. Powerhouse Museum: "Behind the Scenes," at ⟨http://projects.powerhousemuseum.com/virtmus⟩ (April 19, 2005).

8. The Paul J. Getty Museum at ⟨http://www.getty.edu/art/collections/⟩ (March 15, 2005).

9. Cycladic female figure, ⟨http://www.getty.edu/art/collections/objects/o15192.html⟩ (April 19, 2005).

10. 1000 years of the Olympic Games: Treasures of Ancient Greece at ⟨http://projects.powerhousemuseum.com/ancient_greek_olympics/⟩ (19 April 2005).

11. Spatial Imaging Ltd. ⟨http://www.holograms.co.uk/⟩.

12. Contested Sites Staff Focus Group Transcript USA#b, Exhibition team, 9/11 exhibition. National Museum of American History, Washington, DC. September 16, 2002, University of Sydney.

13. National Museum of American History, 2002, 9.

14. Information at ⟨http://www.clevelandart.org/exhibcef/picassoas/html/9476994.html⟩ (April 14, 2005) and ⟨http://www.cogapp.com/home/clevelandPicasso.html⟩ (April 14, 2005).

15. "The Digital Michelangelo Project" ⟨http://graphics.stanford.edu/projects/mich/⟩ (April 27, 2005).

16. *UNESCO Charter on the Preservation of Digital Heritace* at ⟨http://portal.unesco.org/ci/en/ev.php-URL_ID=8967&URL_DO=DO_TOPIC&URL_SECTION=201.html⟩ (November 1, 2005).

17. See Dietz 1999, and Cook and Graham 2004, 84–91.

18. See Roland Barthes's reference in Susan Hazan, "The Virtual Aura: Is There Space for Enchantment in a Technological World?" Museums and the Web 2001 at ⟨http://www.archimuse.com/mw2001/papers/hazan/hazan.html⟩ (April 19, 2005).

## *References*

Appadurai Arjun. 1986. *The Social Life of Things: Commodities in Cultural Perspective*. Cambridge, UK: Cambridge University Press.

Barthes, Roland. 1977. *Image-Music-Text*. Trans. S. Heath. London: Fontana.

Barthes, Roland. 1981. *Camera Lucida, Reflections on Photography*. Trans. Richard Howard. New York: Hill and Wang.

Baudrillard, Jean. 1982. *Simulations*. Trans. P. Foss, P. Patton, and P. Beitch. New York: Semiotext(e).

Baudrillard, Jean. 1994. *Simulacra and Simulation*. Trans. Sheila Faria Glaser. Ann Arbor: University of Michigan Press.

Baudrillard, Jean. 1996. *The System of Objects*. Trans. James Benedict. London: Verso.

Baudrillard, Jean. 2000. *The Vital Illusion*. New York: Columbia University Press.

Benjamin, Walter. 1970. *The Work of Art in the Age of Mechanical Reproduction*. Trans. H. Zohn. London: Jonathon Cape.

Best, Steven. 1994. "The Commodification of Reality and the Reality of Commodification: Baudrillard, Debord, and Postmodern Theory." In *Baudrillard: A Critical Reader*, ed. D. Kellner. Oxford: Blackwell.

Bourdieu, Pierre, and Alan Darbel. 1991. *The Love of Art: European Art Museums and Their Public*. Oxford: Polity Press.

Brown, Deidre. 2007. "Te Ahua Hiko: Digital Cultural Heritage and Indigenous Objects, People, and Environments." In *Theorizing Digital Cultural Heritage: A Critical Discourse*, ed. F. Cameron and S. Kenderdine. Cambridge, Mass.: MIT Press.

Cameron, Fiona. 2000. "Shaping Maori Histories and Identities: Collecting and Exhibiting Maori Material Culture at the Auckland and Canterbury Museums, 1850s to 1920s." Ph.D. diss., School of Global Studies, Dept of Social Anthropology, Massey University, New Zealand.

Cook, Sarah. 2007. "Online Activity and Offline Community: Cultural Institutions and New Media." In *Theorizing Digital Cultural Heritage: A Critical Discourse*, ed. F. Cameron and S. Kenderdine. Cambridge, Mass.: MIT Press.

Cook, Sarah, and Beryl Graham. 2004. "Curating New Media Art: Models and Challenges." In *New Media Art: Practice and Context in the UK, 1994–2004*, ed. L. Kimbell. London: Arts Council of England.

Crary, Jonathan. 2001. *Suspension of Perception: Attention, Spectacle, and Modern Culture*. Cambridge, Mass.: MIT Press.

Crimp, Douglas. 1993. *On Museum Ruins*. Cambridge, Mass.: MIT Press.

Cubitt, Sean. 1998. *Digital Aesthetics*. London: Sage.

Davison, Graeme. 2000. *The Use and Abuse of Australian History*, Sydney: Allen and Unwin.

Dietz, Steve. 1999. "Why Have There Been No Great Net Artists?" In the exhibition Through the Looking Glass: Critical Texts. Available at 〈http://www.voyd.com/ttlg/textual/dietz.htm〉 (accessed 19 April 2005).

Edwards, Elizabeth. 1999. "Photographs as Objects of Memory." In *Material Memories*, ed. M. Kwint, C. Breward, and J. Aynsley. Oxford: Berg.

Fyfe, Gordon. 2004. "Reproduction, Cultural Capital and Museums: Aspects of the Culture of Copies." *Museum and Society* 2, no. 1 (March): 47–67.

Gibson, Ross. 2005. "Theatres of Alteration." Paper presented at the Sites of Communication 2 Symposium, March 18–19, Art Gallery of New South Wales, Sydney, Australia.

Gottdiener, Mark. 1994. "The System of Objects and the Commodification of Everyday Life: The Early Baudrillard." In *Baudrillard: A Critical Reader*, ed. D. Kellner. Oxford: Blackwell.

Graham, Beryl. 2007. "Redefining Digital Art." In *Theorizing Digital Cultural Heritage: A Critical Discourse*, ed. F. Cameron and S. Kenderdine. Cambridge, Mass.: MIT Press.

Graham, Gordon. 1999. *The Internet: A Philosophical Inquiry*. London: Routledge.

Grau, Oliver. 2003. *Virtual Art: From Illusion to Immersion*. Cambridge, Mass.: MIT Press.

Gurian, Elaine Heumann. 1999. "The Many Meanings of Objects in Museums." *Daedalus* 128, no. 3 (Summer): 163–183.

Hall, Stuart. 2005. "Whose Heritage? Un-Settling the 'Heritage': Reimagining the Post-Nation." In *The Politics of Heritage: The Legacies of "Race,"* ed. J. Littler and R. Naidoo, 23–35. London: Routledge.

Hayles, N. Katherine. 1999. "The Condition of Virtuality." In *The Digital Dialectic: New Essays in New Media*, ed. P. Lunenfeld. Cambridge, Mass.: MIT Press.

Hazan, Susan. 2001. "The Virtual Aura: Is There Space for Enchantment in a Technological World?" Paper presented at Museums and the Web conference, March 14–17, Seattle, Washington. Available at ⟨http://www.archimuse.com/mw2001/papers/hazan/hazan.html⟩ (accessed 19 April 2005).

Hodge, R., and W. D'Souza. 1999. "The Museum as a Communicator: A Semiotic Analysis of the Western Australian Museum Aboriginal Gallery, Perth." In *The Educational Role of Museums*, 2d ed., ed. E. Hooper-Greenhill. London: Routledge.

Hooper-Greenhill, Eilean. 2000. *Museums and the Interpretation of Visual Culture*. London: Routledge.

Kavanagh, Gaynor. 1996. *Making Histories in Museums*. London: Leicester University Press.

Kenderdine, Sarah. 2001. "1000 Years of the Olympic Games: Treasures from Ancient Greece, Digital Reconstruction at the Home of the Gods." Paper presented at Museums and the Web conference, March 14–17, Seattle, Washington. Available at ⟨http://www.archimuse.com/mw2001/papers/kenderine/kenderdine.html⟩ (accessed 19 April 2005).

Knell, Simon. 2003. "The Shape of Things to Come: Museums in the Technological Landscape." *Museum and Society* 1, no. 3 (November): 132–146.

Kwint, Marius. 1999. "Introduction: The Physical Past." In *Material Memories*, ed. M. Kwint, C. Breward, and J. Aynsley. Oxford: Berg.

Levy, Pierre. 1998. *Becoming Virtual: Reality in the Digital Age*. Trans. Robert Bononno. New York: Plenum Trade.

Leyton, Michael. 1992. *Symmetry, Causality, Mind*. Cambridge, Mass.: MIT Press.

Lubar S., and D. W. Kingery, eds. 1995. *Learning from Things: Method and Theory of Material Culture Studies*. Washington, D.C.: Smithsonian Institution Press.

Lunenfeld, Peter, ed. 1999. *The Digital Dialectic: New Essays in New Media*. Cambridge, Mass.: MIT Press.

Lynch, Clifford. 2000. "Authenticity and Integrity in the Digital Environment: An Exploratory Analysis of the Central Role of Trust." Available at ⟨http://www.clir.org/pubs/reports/pub92/lynch.html⟩ (accessed 19 April 2005).

MacDonald, George. 1992. "Change and Challenge: Museums in the Information Society." In *Museums and Communities: The Politics of Public Culture*, ed. I. Karp, C. Mullen Kreamer, and S. D. Lavine. Washington, D.C.: Smithsonian Institution Press.

Marwick, Alun. 2001. *The New Nature of History, Knowledge, Evidence and Language*. London: Palgrave.

Manovich, Lev. 2001a. *The Language of New Media*. Cambridge, Mass.: MIT Press.

Manovich, Lev. 2001b. "Post-media Aesthetics." In *Dislocations*, ed. Zentrum für Kunst und Medientechnologie. Karlsruhe: ZKM. Available at ⟨http://www.manovich.net⟩ (accessed 19 April 2005).

McLuhan, Marshall. 1964. *Understanding Media: The Extensions of Man*. New York: McGraw-Hill.

Miller, Daniel. 1998. "Why Some Things Matter." In *Material Cultures: Why Some Things Matter*, ed. D. Miller. Chicago: University of Chicago Press.

Munslow, Alun. 2001. "Book Reviews, Institute of Historical Research, *Reviews in History link, Discourse Section*. Available at ⟨http://www.history.ac.uk/reviews/discourse/index.html⟩ (accessed 19 April 2005).

Ned Kelly: Australian Ironoutlaw, info available at ⟨http://www.ironoutlaw.com/⟩ (accessed 5 April 2005).

Paul, Christiane. 2003. *Digital Art*. London: Thames and Hudson.

Pearce, Susan, M. 1992. *Museums, Objects and Collections: A Cultural Study*. London: Routledge.

Pearce, Susan, M., ed. 1996. *Interpreting Objects and Collections: A Cultural Study*. Leicester: Leicester University Press.

Poster, Mark. 1996. *The Second Media Age*. Cambridge, Mass.: Polity Press.

Powerhouse Museum, "Love Tokens." Available at ⟨http://www.powerhousemuseum.com/opac/87-1494.asp⟩ (accessed 19 April 2005).

Powerhouse Museum. "Behind the Scenes." Available at ⟨http://projects.powerhousemuseum.com/virtmus/⟩ (accessed 19 April 2005).

Powerhouse Museum. "1000 Years of the Olympic Games: Treasures of Ancient Greece." Available at ⟨http://projects.powerhousemuseum.com/ancient_greek_olympics/⟩ (accessed 19 April 2005).

Robinson, Helena, and Fiona Cameron. 2003. "Knowledge Objects: Multidisciplinary Approaches in Museum Collections Documentation." Department of History, University of Sydney.

Schlereth, Thomas, ed. 1994. *Material Culture Studies in America*. Walnut Creek, Calif.: Altamira Press.

Sontag, Susan. 1982. *Barthes: Selected Writings*. Oxford: Oxford University Press.

Spatial Imaging Ltd. Available at ⟨http://www.holograms.co.uk/⟩ (accessed 27 April 2005).

Stanford University. "The Digital Michelangelo Project." Available at ⟨http://graphics.stanford.edu/projects/mich/⟩ (accessed 27 April 2005).

Stewart, Susan. 1984. *On Longing: Narratives of the Miniature, the Gigantic, the Souvenir, the Collection*. Baltimore: Johns Hopkins University Press.

Stewart, Susan. 1999. "Prologue: From the Museum of Touch." In *Material Memories*, ed. M. Kwint, C. Breward, and J. Aynsley, 17–36. Oxford: Berg.

The J. Paul Getty Museum. Available at ⟨http://www.getty.edu/art/collections/⟩ (accessed 15 March 2005).

Walsh, Peter. 2007. "Rise and Fall of the Post-photographic Museum: Technology and the Transformation of Art." In *Theorizing Digital Cultural Heritage: A Critical Discourse*, ed. F. Cameron and S. Kenderdine. Cambridge, Mass.: MIT Press.

Witcomb, Andrea. 2003. *Reimagining the Museum: Beyond the Mausoleum*. New York and London: Routledge.

Witcomb, Andrea. 2007. "The Materiality of Virtual Technologies: A New Approach to Thinking about the Impact of Multimedia in Museums." In *Theorizing Digital Cultural Heritage: A Critical Discourse*, ed. F. Cameron and S. Kenderdine. Cambridge, Mass.: MIT Press.

# 4 Te Ahua Hiko: Digital Cultural Heritage and Indigenous Objects, People, and Environments

Deidre Brown

*I heard it at Awataha Marae in te reo—waka rorohiko—"computer waka," about a database containing whakapapa. Some tapu information, not for publication. A dilemma for the library culture of access for all, no matter who, how, why. A big Western principle stressing egalitarianism. My respects. However, Maori knowledge brings many together to share their passed down wisdom in person to verify their inheritance; without this unity our collective knowledge dissipates into cults of personality.*
—Robert Sullivan, *Star Waka*

In his poem "Waka Rorohiko" ("Computer Vessel"), Robert Sullivan anticipates many of the issues that face the cultural sector and indigenous peoples as the digital divide narrows and electronic high-technologies spread out from the First World. The emergent techno–middle class of the developing world has capitalized on favorable exchange rates and fluency in its former colonizers' languages to provide everything from Web site design to tax consultancy for Western clients. However, the challenge for the Fourth World is to shape new electronic waka (vessels) to meet their cultural needs within a First World economic system. This chapter examines the possible heritage applications of three-dimensional augmented and virtual reality (AR and VR) to New Zealand Maori treasures, bodies, and landscapes, examining the potential benefits and problems this technology presents for institutions and indigenous people. It draws experience from bicultural pilot projects involving the collaboration of museum professionals, curators, Maori participants, software and hardware industries, and academics.

## Technology and Cultural Development

There is a perception in Western societies that new technologies impact on the "authentic" and "traditional" aspects of customary indigenous culture, and products made with new tools by indigenous hands are somehow "hybridized." This perspective does not take into account the centuries of technological appropriation from the non-West to West that has occurred as a result of imperial expansion and globalization. Indeed, without the developing and colonized worlds, Western culture would not exist in its present form. Notions of "authenticity" and "tradition" have been under review in Anthropology and Art History for the last twenty years, and are founded on the same beliefs of cultural purity that underlie the term "hybrid." Instead, it could be argued that indigenous peoples, like their Western counterparts, have always been interested in the possibilities of new technology and have never deliberately appropriated it without considering its impact on customary culture. This is certainly the case for Maori. One only needs to look to tool technology and the movement from softstone, to hardstone, to metal, to mechanical, and most recently to digital devices, over the last eight hundred years. The nineteenth and twentieth centuries, in particular, were a time of rapid technological appropriation for Maori, so that today the lives of most Maori have the appearance of that of their *Pakeha* (non-Maori) neighbors, although the cultural values of each group remain distinct.

The continuing development of three-dimensional software, scanners, studio-based recording facilities, printers, and head-mounted displays offers humanity a wide range of possibilities, as well as significant issues to consider. A recent graduate research project by Barbara Garrie on the application of AR and VR in museum settings found that we are now at the point where technology can create anything, and more than we can imagine, leaving cultural heritage workers straddling a conceptual digital divide (2003). The history of technology would suggest that it is better for indigenous people to appropriate AR and VR for their own purposes before the technologies are applied to them, and their culture, by another group.

## Cultural Significance and the Replicant

The most important question, when reflecting on the cultural heritage potential of AR and VR, is whether the inherent and essential qualities that give an object, person, or environment their meaning and significance are transferred to their digital copy. If this is the case, then these technologies offer tremendous opportunities for indigenous peoples to recover and record their cultural heritage.

In his study of the meaning of *taonga* (cultural treasures) Paul Tapsell, Tukuaki Maori (Director Maori) of Auckland War Memorial Museum, identified many interrelated qualities: *mana* (authority, power, prestige); *tapu* (protected, sacred, prohibited);

*korero* (oratory, narratives); *karakia* (recitation, incantation); *whakapapa* (genealogy, systematic framework); *wairua* (everlasting spirit); *mauri* (life force, life essence); *ihi* (spiritual power); *wehi* (to incite fear and awe); and *wana* (authority, integrity) (1997, 326–331). None of these qualities are visual, the property that AR and VR singularly relies on, although ideas of craftsmanship and beauty are inherent in all of them.

It is my proposition that some, if not all, of these cultural values are transferred by digital replication, to a lesser or greater degree, depending on circumstance. This seemingly bold claim is founded on the observation that Maori and national cultural heritage institutions already treat moving and still images of Maori objects, people, and environments as if they embody the same original qualities. Depictions of ancestors and culture-heroes still dominate Maori wood carving, and at the time of photography's introduction to New Zealand in the mid-nineteenth century, these carvings were regarded as embodying all of the qualities that Tapsell defines except, perhaps, *mauri*. The carved replicant is regarded as having its own life force separate from that of the ancestor. When confronted with portrait photography, Maori initially responded by hiding from the camera, fearful that their *mauri* would be lost, but later began to see the new medium as an effective method of embodying the *wairua* of a person. Today rows of photographs of deceased relatives line the walls of *wharenui* (meeting houses) as a way of maintaining their presence within the architectural bosom of the community (King 1991, 2). Augmented and virtual reality, using images captured from the real world, is the next step in the *whakapapa* (sequence) of depiction, which has already moved from carving to photography and, more recently, to video. As AR and VR become commonplace they too will be regarded as *taonga* with their own *mauri*.

While the question of whether the essential qualities of a *taonga* are transferred to its replicant will ultimately be decided by customary processes of consensus within the Maori community, there remain other issues about AR and VR that need to be reviewed from a cultural perspective. Is there, for example, a preexisting *kawa* (protocol) that would ensure the safety of an object, person, or environment while their form and essential qualities are being replicated? *Kaumatua* (elders) may find that similar *kawa*, related to the process of opening a new *wharenui* and the crossing of the *pae* (the transitional space between inside/outside, life/death, heritage/future) may be suitable; as could be the ritualized *karanga* (call) performed when a child enters the world; or *karakia* when an ancestral carving is made. There exists *kawa* for the opening and closing of Maori art exhibitions, from formal *powhiwi* and *poroporoaki* (ceremonial welcome and conclusion) to *whakawatea* (blessings) and *mihi* (greetings). *Kawa* are already in place for the storage and accessibility of Maori library collections, as are protocols for the treatment of indigenous literature in libraries around the world, and the commentary about this is extremely helpful with regard to AR and VR indigenous replicants. As Karen Worcman, Founder and Director of the Museum of the Person, wrote in a 2002 article for *D-Lib Magazine*,

> [one] great challenge to consider while planning collaborative digitization projects is determining how to successfully create and preserve digital history in such a way that it will be incorporated into and used by the community from which it comes. It is important, on one hand, to discover channels through which the members of these communities (even the youngest) can master digital technology. But it is equally essential to be aware of the form in which this new media can be incorporated in a consolidated process of oral transmission of values. It may be that the most important factor of the digitization project is not the creation of the "digital collection" as such, but the group's engagement in the process that motivates new generations to value their history. (2002)

Robert Sullivan, a senior Maori librarian and acclaimed poet whose 1999 book *Star Waka* explores the relationship between technology and the customary world, has noted that digital archiving also presents opportunities for Maori. He writes, "Many communities want training and employment opportunities. Building a global digital library requires first infrastructure, and then content. Various technical protocols and standards must be met to ensure that the resource is accessible—and accessible in the manner intended. Dealing with these technical issues alone provides an enormous employment opportunity" (2002).

After facilitating two AR and VR brainstorming sessions with Canterbury Museum (Christchurch) staff in 2002, followed by personal interviews with museum staff working across curatorial, management, educational, and technical areas, the three exhibit types of objects, people, and environments emerged, with a fourth non-exhibit type consisting of a virtual gallery guide (an AR version of an audio cue) and internet-based virtual museum.[1] It is envisioned that these technologies would be audio and visual experiences, using head-mounted interfaces, and therefore located in the spatial world, rather than keyboard and screen arrangements. The nature of the exhibits proposed remains confidential as it is intimately linked with the mission and collections of the museum; however two experimental projects sought to illustrate the potential for the cultural heritage sector and Maori communities.

Figure 4.1   Virtual *Patu*, visualized using "Magic Book"–augmented reality technology developed by the Human Interface Technology Laboratory. © HIT Lab (NZ) 2002.

## *Objects*

In 2002 a pilot project to produce a virtual reality whalebone *wahaika* (cleaver) (figure 4.1), digitized from an unprovenanced original held in Canterbury Museum, was undertaken by myself, Eric Woods from the HIT (Human Interface Technology) Laboratory New Zealand, Roger Fyfe from Canterbury Museum, and Mark Nixon from ARANZ (Applied Research Associated New Zealand). The ARANZ fast-scan handheld scanner was used to contour-map the *wahaika*, (figure 4.2), while photographic images from a digital camera were used to texture-map the coloration of the *taonga* onto the contour. "Virtual *Patu*," a generic name for cleaver, as the *wahaika* came to be known, is made manifest in the HIT Laboratory's "Magic Book" format. A number of techniques are currently under development for scanning three-dimensional objects from handheld to fixed devices, as are scanners that marry the currently separate tasks of contour and texture mapping. It is envisioned that scanning devices will eventually

Figure 4.2    Mark Nixon from ARANZ, Christchurch, contour-mapping the Canterbury Museum wahaika in the first stage of creating the *Virtual Patu*. © HIT Lab (NZ) 2002, photography Eric Woods.

become either so simple to operate or so commonplace that collection management staff will be able to digitally replicate objects without the assistance of a technician.

AR and VR are most effective as cultural heritage devices when they replicate objects, events, and scenes that are difficult or impossible to realize or access, as simulation for the sake of repetition offers no value and incites little interest. Digitized replicants of unique objects are likely to appeal to a general audience; however VR's greatest contribution will be in the replication of objects for remote study, curatorial work, conservation, and repatriation. A substantial number of *taonga Maori*, over 90,000 objects, are held by New Zealand museums, and of this number almost seventy percent are in Auckland War Memorial Museum and the Museum of New Zealand Te Papa Tongarewa (Wellington).[2] Major collections of *taonga* also exist in the British Museum and the Peabody Essex Museum (Salem, Massachusetts) with other important small collections and individual items scattered around the globe. Funding for New Zealand-based researchers to travel to these institutions is limited, when compared to that for their North American and European colleagues, which restricts opportunities for specialized research. If three-dimensional scanning were to become as commonplace as digital photography within museums, the data of replicant objects could be sent through the post, or even cyberspace, for comparative study. Curators and curatorial students could also insert replicant *taonga* into digital museum spaces, desired or extant, to experiment with different display possibilities. The 1999 Philip Guston exhibition at the Kunstmuseum in Bonn is an early example of a virtual prototype realized (2001, 171–182). When the properties of different materials can be digitized, medical virtual reality software—in which experimental surgeries and treatments are performed on virtual bodies—could be translated to the cultural heritage arena allowing experimental conservation practices to be applied to replicant *taonga* in preparation for work on the original. These would augment, and even possibly replicate, second-generation conservation techniques that move away from manual intervention towards using bacterial agents, radiological imaging, and laser beams. Light-based three-dimensional scans, which can then be materialized through three-dimensional printers, are a much less invasive form of modeling objects than liquid-based molds. Conservation applications could also be extended so that irreversibly incomplete objects could be rebuilt virtually, as an educative and display tool. But this process raises the controversial issue of the creative, rather than non-restorative, digital alteration of *taonga*, with accompanying changes in cultural value and commercial values.

The most controversial application of AR and VR will be in repatriation, a politically sensitive issue, which is nonetheless central to the process of reconciliation between indigenous communities and cultural heritage institutions. Due to nineteenth- and early-twentieth-century collection policies most *taonga* are not provenanced, and are unlikely to be returned to their people of origin, but those that are located are part of a growing dialogue between Maori and museums about ownership and custodianship. For some Maori communities the ideal solution would be repatriation to a tribal heritage institution, and in this scenario AR and VR would permit digital replicants to remain part of the civic or private collection. If a tribe lacks the resources to house and conserve their *taonga* in an acceptable manner (this notion of "acceptability" itself a contestable cultural issue), then a digital archive could serve an interim purpose.

All archives grapple with access issues, and these are magnified if the archive is Web-based. Metropolitan organizations, such as Auckland War Memorial Museum, Auckland Art Gallery, Christchurch Art Gallery, and the Museum of New Zealand Te Papa Tongarewa, display *taonga* on their Web sites. But these offerings are purposefully limited, when compared to the number of *taonga* in their collections, in response to Maori concerns about the unrestricted accessibility of Internet-based information. In the physical realm, local museums have tried to minimize the possibility of *whakanoa* (making ordinary) that occurs when bringing *tapu taonga* into contact with *noa* (free from *tapu*) elements, particularly food and money. With Internet-based collections offering access from any domestic or commercial situation there is a high probability that *tapu taonga*, for instance those associated with death, may be viewed in a transgressive context. With regard to the internet and digital cultural heritage, foreign Web hosting of digitized indigenous knowledge places *taonga* under the political jurisdictions of other cultures which have no mandate to respect values such as *tapu* and *mana* (Smith 1997).

## *People*

In late 2002, *Te Ahua Hiko* (The Digital Form), a proposal to digitize customary Maori performance, was forwarded to the Ministry of Research Science and Technologies inaugural "Smash Palace" art and science funding round.[3] Being the first three-dimensional digital project involving indigenous people, the project *kaumatua* selected *Te Timatanga o te Ao Marama*, the Maori creation narrative, as the performance concept since AR and VR, like Maori creation, begins with *te kore* (the void) to build *te ao* (the world of light). AR and VR using real people has been limited to short demonstrations of people in motion, although the technology is ready for application to performance arts and narrative, and lends itself to playback in informative public display settings like the *marae* (open air forum) and the museum. With these new tools there is the possibility of recording tribal narratives from *kaumatua* in a way that is more tangible than audio or video, of three-dimensionally visualising Maori arts that have few practitioners, and of demonstrating narrative-based performance for wider audiences. Standard methods of digitization, such as photography and video, always impact on the qualities of customary Maori performance by flattening its appearance, whereas AR and VR maintains the integrity of the *ahua* (three-dimensional form).

Since the publication of Donna Haraway's "A Cyborg Manifesto" in the mid-1980s there has been much debate over whether the virtual body is emancipated from gender and culture (2000, 291–324). The more recent discussions, which have been informed by the study of Internet communities (something that Haraway could not have envisioned at the time of her publication), have generally concluded that cyberspace reflects the divisions and prejudices of Western society. Not so clear is whether parallel cyberspaces exist where indigenous peoples, who are able to overcome the digital divide, can create their own social systems which run independently of colonial, diasporic, and global restrictions. The indigenous digital replicant can exist in many realms—and as an agent of its original and an entity with its own *mauri*—requires protection from exploitation.

Many of these rights will be determined by the original human participant and their *whanau* (extended family) or community. The amount of memory required to three-dimensionally digitize a person would, for the foreseeable future, prevent such images from being situated in highly accessible cyberspaces like the World Wide Web, allowing participants to have a high degree of control over the replication and replay of their image. Restrictions could be placed, for example, on the physical and social context in which the replayed replicant is located, as site and its relationship to *noa* elements has an effect on *tapu*. Customary Maori portraiture, in wood and paint, is hierarchical, with large images denoting *mana* (Neich 1996, 91), so image size will be a consideration, as is viewing angle, with front and sideviews more desirable in some performance and visual arts, since backviews are associated with *whakapohane*, a customary insult. Death is the most *tapu* of states, and many restrictions are placed on institutionally archived images of deceased Maori—a consequence likely to be transferred to digital replicants. Clauses in recording contracts may allow for the return of replicants, in the form of their uncompressed and compressed data, to their *whanau*, these recordings themselves becoming *tapu* through their status as *taonga*. In general, for consensual participants, regulation of their replicants will be closely modeled on the types of procedures that are already used to control the publication of archived photographic, video, and audio recordings, in which participants, and their *whanau*, have the final approval on applications for the reproduction of images.

If reproduction rights remain in the hands of the participant they may be able to demand remuneration for replay, as is the standard for other performance-based activities in theater, film, television, and radio. Often this takes the form of a lump-sum payment or a fee for each use of the performance. Broadcast-standard remuneration rights sometimes stipulate that the producers and other contracted recording staff can also claim a fee, as authorship of recorded performance is a complex issue. Participants would need to carefully consider the relationship between live and recorded performances as part of their remuneration, in the same way that thespians acknowledge the difference between theater and film. A number of Maori are employed as full-time performers at museums and heritage sites, and often the intensity of a performance is a direct response to the reaction of the audience, hence the "live" aspect of the performance is a critical consideration, both financially and artistically. Remuneration should then be set at a level in accordance with film-based acting, and replay possibly restricted to a limited number of locations. For AR and VR to be most effective digitization must offer something beyond simple replication. To this end *Te Ahua Hiko*'s project *kaumatua* established an important point of difference, by suggesting an additional digitally animated background layer that would move the emphasis away from simulation to a three-dimensional visual artwork involving performance.

## Environments

By layering or replacing the real environment, AR and VR offer a challenging alternative to the built world. The challenge is not only architectural/environmental, but also financial. Virtual environments are more cost-effective than real environments, and for indigenous peoples offer the ability to realize spaces of significance without the financial, social, and cultural burdens and conflicts that restrict such schemes to two-dimensional "paper architecture." The reconstruction of heritage sites for remote location and on-site display has already been realized, and extensive research has been undertaken into improving the production techniques, image and spatial quality, and accessibility of this type of rendering. One possible application is that existing *marae* spaces could be digitized and packaged as interactive programs to instruct tribal members of the architecture, heritage, and *tikanga* (customs) of their cultural landscapes. Such an application would particularly benefit the descendants of Maori who left their rural homes after the Second World War in search of better educational and employment opportunities in cities, as well as the new generation of Maori born overseas (Pool 1991, 153). Embedded in this concept are issues relating to the significance of a replicant environment that would have to be the subject of extensive consultation. In one of the founding narratives that inform *tikanga*, the culture-hero Maui embarks on a long journey to look for the mother who abandoned him at birth and to reconnect with her *marae* and his maternal *whanau*. Epic journeys of discovery and rediscovery are a recurring theme in Maori ancestral stories, as are the feelings of *ihi*, *wehi*, and *wana* when their characters are finally confronted with the reality of their destination. It might be suggested that an introduction to one's tribal *marae* via CD-ROM oversimplifies and devalues the importance of such life-changing undertakings. But it could also be argued that few, if any, people would confuse the concept of *ahi ka* (literally keeping the home fires burning) with a mouse click on a disk icon, or believe that the reality of a homecoming can be replaced with AR and VR. Certainly, the *marae* videos and photographs that already circulate among diasporic tribal communities have made people more inclined to visit their ancestral homes, and AR and VR—framed in an appropriate manner—may assist in personal identity recovery.

The increasing affordability and accessibility of interactive AR and VR environments also has the potential to undermine the business of theater-inspired museum design, which translates the lighting and optical effects of the dramatic stage into educational space. Indeed, if the diorama is interpreted as sitting on the threshold between photography and three-dimensional still life, then AR and VR could be conceived as the high-tech development of the former, while contemporary museum display design is the logical conclusion of the latter. The built environment, which supports the cultural heritage industry, is by comparison physically inflexible. It has long construction times, requires closure during periods of remodeling, is not transportable, and is often no more "real" or authentic than a virtual environment in its attempts to replicate space or experience.

There is an extensive literature on the complexities of exhibiting indigenous environments in cultural heritage spaces that is as pertinent to the virtual as it is to the real world. In the study of Maori environments, the commentary has focused on the decorated meeting house (variously known as the *wharenui*, *whare runanga*, and *whare whakairo*) and storehouse (*pataka*) as objects of imperialistic gaze and desire, which are culturally and physically consumed by the museum.[4] In the late nineteenth and early twentieth centuries no museum collection of *taonga* was considered complete unless it included examples of these two buildings and a *waka taua* (vessels), all of which lose their original meanings and functions when relocated in a non-Maori environment. While AR and VR could recreate the appearance of the original situations of these *taonga*, the exact purpose for doing this would have to be carefully considered. On the one hand, digital technology has the ability to place *whare* (houses) and *waka* (boats) in a sympathetic and educative simulated context if this is seen as a desirable outcome by institutions and tribal groups; on the other, it could decontextualize these *taonga* by concealing their actual institutional situation.

## Cultural Values and Legal and Moral Protections

Maori are in the process of making legal inroads into the digital domain in order to ensure that inequity will not be a barrier to access. The current and far-reaching Wai 262 Flora and Fauna[5] claim to the Waitangi Tribunal[6] has implications for the future of museum-held *taonga Maori* and genetic cloning, which is of relevance to the digital replicant. This action, brought by the Ngati Kuri, Ngati Wai, Te Rarawa, Ngati Porou, Ngati Kahungunu, and Ngati Koata tribes, asserts that the Crown, through legislation, policy, and international agreements, breached Maori rights to *taonga*, specifically flora, fauna, and *matauranga Maori* (cultural knowledge and property), that are protected under Article Two of the 1840 Treaty of Waitangi (2005b). In this Article the Crown guarantees Maori "unqualified exercise of their chieftainship over their lands, villages and all their treasures."[7] Wai 262 localizes more general debates on indigenous culture and intellectual property rights heard in the World Trade Organization and World Intellectual Property Organization (2005a). While the original claim was lodged in 1991 before digital technologies like the internet, cameras, and two- and three-dimensional scanning devices became commonplace, the forthcoming findings will hopefully impact on some, if not all, of the issues outlined in this chapter.

A culturally responsive intellectual property law may also provide some protection, but it cannot address the ownership and custodianship of digital cultural heritage, in the form of data, as the laws were originally developed to protect physical entities. The very understanding of a singular intellectual property is under threat from digital technology itself, due to its lack of materiality, dependence on copying as the only route of access, and ability to achieve seemingly perfect and infinite simulacra (NRC 2000). Solutions to these problems currently lie outside the legal domain, and depend on community morality. Beginning in the early 1990s, the New Zealand government and its agencies initiated bicultural policies—that now permeate every aspect of civic and civil life—in order to engender respect for Maori values and opinions. This includes the training, recruitment, and retention of Maori staff working in cultural heritage institutions, and bicultural learning at all educational levels. Therefore, unless there is a radical revision of intellectual property law, it seems that the wider community will have to take responsibility for the sensitive management of indigenous digital cultural heritage, beyond the protections that the law can provide.

## Conclusion

In 2003, I completed and published a four-year research project documenting all the museum-held *whakairo rakau* (wood carvings) from my tribal district, Tai Tokerau (Northland), which I could locate. Extensive travel to museums around the country and the world was involved and, despite the kindness and hospitality of the curators, these were not the epic journeys of self-discovery undertaken by Maui that I had desired, nor was I able to undertake invaluable comparative studies of scale and proportion (not possible with two-dimensional images). If the collections were digitized, using a head-mounted display, a computer and some tracking cards I could have compared, for example, a *wakaika* from the Auckland Museum with one from Te Papa in Wellington and another from the Peabody Essex Museum in Salem, Massachusetts, in my own culturally located environment, in my own time, and at much less expense. My *kaumatua* could have been asked for their interpretations. These copies could have been returned to the museums under copyright licensing agreements or even returned to Tai Tokerau. Once back home they could be materialized with stone or bone powder using a three-dimensional printer, or virtually curated into environment-specific exhibitions. Performative elements may have been added to bring their use as weapons and ritual *taonga* to life and reconnect them with expert hands, in anticipation of the originals perhaps one day returning to their *turangawaewae* (place of belonging). In the space of only a few years, the *rorohiko waka* has been realized and is awaiting launch, as a new generation of indigenous people embark on an epic and challenging journey through cyberspace to rediscover their cultural heritage.

*Notes*

1. The brainstorming sessions were facilitated by Deidre Brown and Kevin Fisher, both staff members of the University of Canterbury at that time; Doug Rogan of Canterbury Museum; and Eric Woods of the HIT Laboratory NZ. Barbara Garrie conducted the interviews.

2. See O'Regan 1997, 38, 42. These figures take into account the nonresponses to O'Regan's surveys.

3. *Te Ahua Hiko* was comprised of a production team including myself, Eric Woods from the HIT Laboratory New Zealand, Adrian Cheok from the National University of Singapore, and the Ngai Tahu tribal performers Leo Hepi and Mere Edwards, under the leadership of the project *kaumatua* Te Ari Brennan. Motion capture was to occur in Singapore using Real 3D technology, with further digitization employing ARToolkit in Christchurch. The project was short-listed but not selected for funding, and the search for resources continues.

4. See, for examples, Brown 1996, Ihimaera 2002, and Mane-Wheoki 1992.

5. Waitangi Tribunal, "The Crown and Flora and Fauna," ⟨http://www.waitangi-tribunal.govt.nz/research/wai262/⟩ (accessed 5 April 2005).

6. See ⟨http://www.treatyofwaitangi.govt.nz⟩ (accessed 5 April 2005).

7. This is the generally agreed translation of the signed Maori language version, which reads "te tino rangatiratanga o ratou wenua o ratou kainga me o ratou taonga katoa."

## References

Battiste, Marie, and James (Sa'ke'j) Youngblood Henderson. 1990. *Protecting Indigenous Knowledge and Heritage: A Global Challenge*. Saskatoon: Purich.

Brown, Deidre. 1996. "Te Hau-ki-Turanga," *Journal of the Polynesian Society* 105, no. 1: 7–26.

Brown, Deidre. 2001. "Navigating Te Kore." In *Techno Maori: Maori Art in the Digital Age*, ed. D. Brown and J. Mane-Wheoki, CD-ROM. Wellington: City Gallery Wellington and Pataka Porirua.

Brown, Michael. 1998. "Can Culture be Copyrighted?" *Current Anthropology* 39, no. 2: 192–222. Available at ⟨http://www.journals.uchicago.edu/CA/journal/contents/v39n2.html⟩ (accessed 5 April 2005).

Copyright Council of New Zealand. 2005. Available at ⟨http://www.copyright.org.nz/⟩ (accessed 5 April 2005).

Eckel, Gerhard, and Steffi Beckhaus. 2001. "ExViz: A Virtual Exhibition Design Environment." In *Virtual and Augmented Architecture*, ed. B. Fisher, K. Dawson-Howe, and C. O'Sullivan, 171–182. London: Springer.

Garrie, Barbara. 2003. "Pilot Study to Determine the Feasibility of Introducing Augmented/Virtual Reality Applications to the Canterbury Museum." Research project, University of Canterbury and Canterbury Museum.

Haraway, Donna. 2000. "A Cyborg Manifesto: Science, Technology and Socialist-feminism in the Late Twentieth Century." In *The Cybercultures Reader*, ed. D. Bell and B. Kennedy, 291–324. New York: Routledge.

Ihimaera, Witi. 2002. "The Meeting House on the Other Side of the World." In *Te Ata: Maori Art from the East Coast, New Zealand*, ed. N. Ellis and W. Ihimaera, 89–98. Auckland: Reed.

Intellectual Property Office of New Zealand. 2005a. "Copyright Protection in New Zealand." Available at ⟨www.iponz.govt.nz/pls/web/dbssiten.main⟩ (accessed 5 April 2005).

Intellectual Property Office of New Zealand. 2005b. "Treaty of Waitangi Claim Wai 262," Intellectual Property Information Sheets, ⟨www.med.govt.nz/buslt/int_prop/info-sheets/wai-262.html⟩ (accessed 5 April 2005).

King, Michael. 1991. *Maori: A Social and Photographic History*, 4th ed. Auckland: Reed.

Mane-Wheoki, Jonathan. 1992. "Imag(in)ing our Heritage: Museums and People in Aotearoa." *New Zealand Museums Journal* 25, no. 1: 2–8.

National Research Council. 2000. *The Digital Dilemma: Intellectual Property in the Information Age*. Washington, D.C.: National Academy Press. Available at ⟨http://www.nap.edu/html/digital_dilemma/ch1.html⟩ (accessed 5 April 2005).

Neich, Roger. 1996. "Wood-carving." In *Maori Art and Culture*, ed. D. C. Starzecka, 69–113. Auckland: David Bateman & the British Museum.

O'Regan, Gerard. 1997. *Bicultural Developments in Museums of Aotearoa: What is the Current Status?* Wellington: Museum of New Zealand Te Papa Tongarewa/National Services and Museums Association of Aotearoa New Zealand.

Parker, Brett. 2001. *Maori Access to Internet Technology*. Wellington: Ministry of Maori Development Te Puni Kokiri.

Pool, Ian. 1991. *Te Iwi Maori: A New Zealand Population Past, Present and Projected*. Auckland: Auckland University Press.

Smith, Alastair. 1997. "Fishing with New Nets: Maori Internet Information Resources and Implications of the Internet for Indigenous Peoples." In *INET'97*, Kuala Lumpur: INET'97. Available at 〈http://www.isoc.org/isoc/whatis/conferences/inet/97/proceedings/E1/E1_1.HTM〉 (accessed 5 April 2005).

Sullivan, Robert. 1999. *Star Waka*. Auckland: Auckland University Press.

Sullivan, Robert. 2002. "Indigenous Cultural and Intellectual Property Rights: A Digital Library Context." *D-Lib Magazine* 8, no. 5. Available at 〈http://www.dlib.org/dlib/may02/sullivan/05sullivan.html〉 (accessed 5 April 2005).

Tapsell, Paul. 1997. "The Flight of Parerautututu." *Journal of the Polynesian Society* 106, no. 4: 326–331.

Waitangi Tribunal. 2005a. "Report of the Waitangi Tribunal on Claims Concerning the Allocation of Radio Frequencies." Available at 〈http://www.waitangi-tribunal.govt.nz/reports/generic/wai26_150/〉 (accessed 5 April 2005).

Waitangi Tribunal, 2005b. "Wai 262 Reports." Available at 〈http://www.waitangi-tribunal.govt.nz/research/wai262/〉 (accessed 5 April 2005).

Williams, David. 2001. *Matauranga Maori and Taonga*. Wellington: Waitangi Tribunal.

Worcman, Karen. 2002. "Digital Division is Cultural Exclusion. But is Digital Inclusion Cultural Inclusion?" *D-Lib Magazine* 8, no. 3. Available at 〈http://www.dlib.org/dlib/march02/worcman/03worcman.html〉 (accessed 5 April 2005).

# 5 Redefining Digital Art: Disrupting Borders

Beryl Graham

---

"Redefining" digital art is somewhat difficult in a field where firm definitions are in short supply. This chapter, therefore, seeks to round up some current working categories, and to look at how the definitions are affecting how the art is shown. Starting with some argumentative binary divisions, some more complex categories from the real world of art are explored.

Any new medium tends to be necessarily disruptive of safe categories, and the particularly "fluid" characteristics of new media are perhaps even more inherently worrying to the departmental territories of arts institutions. Led by the art itself, this chapter also looks at how art challenges concepts of materiality, history, and institutions in ways that digital interpretation may not.

To start with a definition of "digital art," my own working definition is art made with, and for, digital media including the Internet, digital imaging, or computer-controlled installations. The more usual contemporary term within the arts is "new media art," although this name, of course, may itself date relatively quickly, and is prone to confusion with the commercial new media (interactive quiz shows and the like)[1]. The history of the field has included the namings of "computer art," "cybernetic art," "virtual art," "hypermedia art," "unstable media," "emergent media," and art "that you can plug in."[2] Each has a variable meaning in itself, very much depends on the historical period being referred to, and the background of the author, be that art, science or cultural studies.[3] Lev Manovich has even argued that we are now "post-media" and that the categories which best describe the characteristics of post-digital art concern user behavior and data organization, rather than medium *per se* (2005). These definitions of digital art will, therefore, be looked at through the wider lens of categories from the world of art.

## Interpretation or Art?

> The Susan Collins artwork *Audio Zone* is spread around the exhibition space. The audience must wear infrared headphones, which at certain points receive seductive voices urging you to "touch" and "stroke" the triggered video projections of nipples, lips, and keyboard buttons. The desk staff who issue the headphones quickly noticed a very common misconception in the audience, and now carefully explain to each person that this is NOT "a guide to the exhibition." Beryl Graham (2000)[4]

On seeing a piece of new media technology in a gallery or museum, a member of the public is justifiably likely to assume that it is some kind of interpretive aid rather than an artwork in itself.[5] The use of digital media as a tool for interpretation, education, promotion, or archiving is relatively common, whereas new media art is still rarely shown in mainstream art venues. As reflected in this book, digital interpretation is relatively well researched, has international standards, and benefits from regular expert conferences such as ICHIM and Museums and the Web. The same cannot be said for digital art. Even in the field of media theory, the hegemony of debate lies with the more accessible forms of popular culture—chat rooms, mobile phones, and virtual reality—rather than with the reality of current, critical art practice.

A primary taxonomical binary for digital art is, therefore, Interpretation/Art: are digital media being used to interpret, reproduce, or archive the art, or is the art actually made using the particular characteristics of digital media? This may sound like a very obvious difference, but there are still surprising confusions for the audience, and even for curators. The Walker Art Center in Minneapolis, for example, had for many years a particularly well-developed series of digital art exhibitions and events, including one of the first exhibitions of Net art, *Beyond Interface: Net Art and Art on the Net* in 1998, and *Art Entertainment Network* in 2000. The Walker also had a particularly integrated approach to interpretation on the Web, including *ArtsConnectEd*, where visitors to the site could select their own "collection" from the database of digitized images of artworks in the museum's collection. In 2003 it emerged that the Walker's plans for a new building no longer included a curator of digital art, and the argument for this was partially based on the factor that the use of digital media as a tool for interpretation or archiving was to be retained.[6] This elision of digital interpretation and digital art is by no means confined to the Walker, and has had dramatic repercussions for budgets and departmental relationships at several arts venues. A certain amount of "productive confusion" between interpretation and art, however, does not always work to the disadvantage of art, as long as the art is allowed to lead.

Graham Harwood's *Uncomfortable Proximity*,[7] for Tate Britain in London, was the institution's first Net art commission, and the artist very deliberately used the work to question the role of digital media in promotion and collection. The artist visited the collections, took digital photographs of famous paintings, and then montaged the images to make contemporary references to class and race. He then made a Web site that copied the Tate publicity site, and inserted his own content, including his own montaged "collections" and a historical section that took the educational intent of the original site but chose very different aspects of the history of the Tate, such as its past proximity to prison ships on the Thames. This artwork caused substantial institutional disruption around the marketing department, because the Tate's Web site, in common with many art museums, was seen primarily as a marketing tool, then perhaps as interpretation, but never before as a venue for digital art (Cook 2001).[8] The Tate has continued to commission Net art, and has also continued to question ideas of "venue" with its *Tate in Space* commission.

The Science Museum in London is a venue with much experience in making and maintaining interpretational interactive computer-based installations and Web sites. Its educational pieces include a fingerprint-operated device which tracks users through the exhibits, and generates a personal Web page of data. It is also, ironically, the only museum in London (including art museums) with a permanently installed collection of digital artworks (including physical works by Gary Hill, Christian Moller, and David Rokeby).[9] The museum has a very diverse, high-throughput, and demanding audience, yet still manages to show the artworks successfully, alongside interpretational installations. The artworks are "signaled" by having differently colored labels, but otherwise are treated to the same commendably clear style of label-writing which asks questions, as well as providing information. The museum employed a specialist new media art curator from the early stages of planning the new Wellcome Wing exhibits, and although the artworks are clearly chosen for their relevance and accessibility, they do function as artworks, while benefiting from the technical installation and durability knowledge of the museum staff. The questions that the artworks might raise are seen as a valuable part of their function, rather than a "problem," which has happened in some art museums.

### Binaries and Borderlands

The "boundary-subject" that theorist Gloria Anzaldúa calls the *Mestiza*, one who lives in the borderlands and is only partially recognized by each abutting society. (Stone 1991, 112)

The binary of interpretation or art is only one of the borderlands that digital art inhabits. This position in the borderlands triggers, in certain contexts, the tiresome question, "is it art?" Sometimes, it is simply the newness of the media technology that stimulates this question, as it was in the 1800s when photography was emerging (it is fortunate, perhaps, that photography did not wait for an answer from the critics before proceeding, as a firm decision has yet to be formally announced). Sometimes, however, it is the lineage of activist, participative, or conceptual art which causes the question to be raised (for example, around Heath Bunting's work discussed in chapter 6 in this book).

One way in which digital art can differ from the existing debates around photography, and mechanical reproduction of images, is that certain technologies, such as computers networked to the Internet, are both a means of production and a means of distribution. That is, artists may make their artwork using exactly the same reasonably accessible equipment that they use to make their art available to their audience. Indeed, "New Media as Computer Technology Used as a Distribution Platform" is one of the eight answers that Lev Manovich has to the question "What is new media?" (2003, 13–25). Thus, the border of Production/Distribution is another site where digital art sits in a provocative way. Much has been made of the potential for Net art to distribute itself outside of the traditional cultural gatekeepers of curators, and while Net artists certainly can and do that, they are also maintaining a dialogue with museums, which are still very much in charge of the publicity, promotion, and collecting that bring digital art into the canons of Art History.

Yet another binary is that of Art/Science, or Art/Engineering. The development of digital art owes much to early projects such as EAT (Experiments in Art and Technology) in the USA in the 1960s,[10] which roamed the boundaries of art, design, architecture, and science. Initiatives including Sian Ede's book on art and science, and the Play Garden project[11] have more recently critically examined the relationship between art and science. Natalie Jeremijenko, for example, has used a range of media and technology to articulately deal with current scientific issues, including *One Tree*, which used cloned trees as installations in public and gallery spaces (1999). In 2004, Critical Art Ensemble's art work with genetic modification gained attention from the US law enforcement bodies,[12] putting the Art/Science debate into the most heavily contested borderlands that exist. However, if strict definitions of digital art are examined, then some artworks which examine technology are not actually made with digital media, but rather with living cells, or scientific equipment. Thus, some authors have explicitly excluded such projects from consideration as digital art, while Stephen Wilson in his book *Information Art* has used a series of chapter headings including: Biology, Physics, Kinetics, Telecommunications, and Digital Information Systems.

Strongly related to the Art/Science binary is Lev Manovich's 1996 bifurcation of digital art into "Turing-land" and "Duchamp-land (1996)." The former, he argues, is about technology, and takes that technology very earnestly indeed. The latter is more about content, and is often ironic about its own media. While the former tends to inhabit research labs and specialist digital art events such as Ars Electronica or ISEA, the latter, often in the form of Net art, has made it into art venues more quickly (albeit via the less-guarded backdoor of the museum Web site, or interpretational program). Examples of this include Mark Napier's Net art at the Whitney, or Harwood's at the Tate. Manovich considered that "What we should not expect from Turing-land is art which will be accepted in Duchamp-land. Duchamp-land wants art, not research into new aesthetic possibilities of new media" (Manovich 1996). Others have disagreed with these boundaries,[13] but there is a growing pattern of division of expertise within digital art into the rough bifurcation of "Net and Not Net" (Graham and Cook 2002, 44–45). This division is certainly linked to the binaries of Production/Distribution and Turing-land/Duchamp-land mentioned above, and to the pragmatic differences in the technical knowledge and equipment needed to show a Net artwork compared to, say, an interactive video installation. The practical rigors of both material and immaterial forms bring us back to a recurring theme in this book: it may be possible to map Immaterial/Material very approximately onto Net/Not Net, but the effects of this for practical art curating may be very different than for archiving. It may be conceptually difficult to persuade a curator of the value of the immaterial, but it may also be aesthetically difficult to persuade a curator of the value of material objects in gallery territory:

> It's only under huge pressure that a visual arts curator would agree to hang a video projector, and only if it is agreed the projector will project an image on the wall and take us back to painting. Only under threat of torture will a visual arts curator put a computer in the galleries. (Philippe Verge 2000)[14]

While Verge was somewhat satirizing his own profession, these issues of physical space are very serious ones for museums, and are further explored on the CRUMB Web site and discussion list.[15]

## Some Working Categories

> The necessary complement to this is the development of a critical vocabulary and a willingness to examine practices self-critically within the field of new media arts/culture itself. Our discussion . . . showed that we have not even begun to develop such a vocabulary, nor a proper framework in which to develop it in the first place. (Eric Kluitenberg 2004)[16]

While bifurcations may be good for starting the debate about the position and history of digital art, more complex categories are needed in the longer term. In order for the critical and journalistic coverage of digital art to move beyond the "can computers be art?" phase, it is necessary for some kind of shared critical vocabulary to emerge. Such is the fluid, new, and hybrid nature of the range of digital arts, however, that many are justifiably wary of the current task, and especially of the hierarchical structure that traditional taxonomies infer. As Christiane Paul says, "While definitions and categories may be helpful in identifying certain distinguishing characteristics of a medium, they can also be dangerous in setting up predefined limits for approaching and understanding an art form, particularly when it is constantly evolving, as is the case with digital art" (Paul 2003, 8).

Existing definitions and descriptions such as the Dublin Core[17] form an international starting point, and yet, with the categories of date, size, and materials, it is fundamentally materially oriented—concerning art objects. Taxonomies and mapping infer a strongly hierarchical structure, which tends to produce a whiff of the ludicrous when applied to the distributed, immaterial, anarchic nature of Net art, in particular. This factor is consciously toyed with in several artworks that contrast the obsessional data-logging of technology with a playful, yet rigorous approach. The artist Saul Albert, for example, collaborated on setting up a "Faculty of Taxonomy" within the *University of Openess* [sic], which uses wiki Web sites to gather tools and collections concerning taxonomy, both academic and imaginative. Alternatively, David Rokeby's *Giver of Names* involves a computer connected to a camera, which, when shown an object, assigns a name that is sometimes illuminating, but often simply illustrates the fascinating absurdity of machine logic.

The categories used by the media arts field are, therefore, currently a diverse collection of different strategies. A range of these will be identified here, ending with those that have proved most workable in the context of curating. The categories referred to in table 5.1 are just five from those tabulated online as a starting point for the CRUMB discussion list theme of September, 2004.[18]

**Table 5.1**
Five Sets of Categories of Digital Art

| Multimediale Award Competition (pre-2004) | Prix Ars Electronica (2004) | Frieling and Daniels CD-ROM: *Media Art Interaction* (2000) | Christiane Paul: *Digital Art* book (2003) | Steve Dietz: article (1999) |
|---|---|---|---|---|
| Image | Digital communities | Users can search using three fields: | *Digital Technologies as a Tool:* Digital imaging; photography and print; sculpture | Interactivity |
| Interaction | Computer animation/Visual effects | 1. *Medium/Context* Keywords include: | *Digital Technologies as a Medium:* Installation; film, video and animation; Internet art and nomadic networks; software art; virtual reality . . . ; sound and music | Connectivity |
| Software | Digital music | Public art | | Computability |
| | Interactive art | Multimedia | *Themes in Digital Art:* Artificial Life; Artificial Intelligence . . . ; telepresence . . . ; body and identity; databases . . . ; beyond the book; gaming; tactical media, activism and hacktivism; technologies of the future | |
| | Net vision | Stage | | |
| | | Film | | |
| | | Installation | | |
| | | Environment | | |
| | | Internet | | |
| | | 2. *Themes/Content* Keywords include: | | |
| | | East/West | | |
| | | Feminism | | |
| | | Closed Circuit | | |
| | | GDR | | |
| | | 3. *Dates* | | |

To start with some competitions of digital art, the *Transmediale Award Competition* awarded, until 2004, the three categories of Image, Interaction, and Software. However, they then made the deliberate decision that their categories should be abolished, and that artists should define their own areas of work. As can be seen from both *Transmediale* and *Prix Ars Electronica*, their categories include specific media (Computer Animation or Image) and characteristics that might range across various media (Interaction may be a characteristic of sound, image, new, or old media). The interesting category of "Digital Communities" acknowledges the wider interpretive role of digital media.

When it comes to databases of digital art, the keywords used for searching reflect other ways of categorizing: *The New Media Encyclopaedia* simply lists the artists by name, whereas Rhizome's *Artbase*[19] uses a long list of keywords ranging from "artificial life" to "death" (i.e., the keywords concern both media and content). The Frieling/Daniels CD-ROM database of artworks separates the keyword searches into three options. The date field is an obvious art-historical tool, and the other keywords concern either media/context or content/theme. This division may seem a simple one, but it helps to clarify whether the artwork simply has technology as a subject matter, as some art-science projects do, or whether it uses digital media, and hence can be classified as digital art.

Curators and writers including Christiane Paul and Steve Dietz have developed categories that reflect the needs of curating, where it is very important to be aware of both the artistic process as well as the end product. In the world of art organizations, it is also very important to be able to differentiate media, (which may affect which museum department or funding body one deals with), from themes that may form the intellectual basis for a theme show, and also from the particular *characteristics* of digital art that may work across several media, but which present particular challenges for curator and audience. Christiane Paul has usefully separated "Themes" from "Media," and divided the digital technologies into those that are "Media," and those which function as "Tools": ". . . paint is a medium and the brush is a tool. . . ."[20] In 1999, Steve Dietz concentrated on the particular characteristics of digital art that present most new challenges to curators: Interactivity, Connectivity, and Computability. In a book chapter in 2004, Sarah Cook and I explored the challenges under these three headings and, taking examples from current exhibition practice, noted issues of control, immateriality, and authorship, respectively (2004, 84–91). Sarah Cook has gone on to identify three further Duchamp-land art-historical equivalents of Dietz's categories, and has named them "Collaborative, Distributable, and Variable."[21] She has also reordered

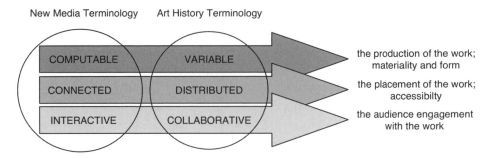

Figure 5.1    Sarah Cook's taxonomic diagram. © 2004.

and placed the categories in relation to key issues for curators, which usefully reintroduces factors of production and distribution (as shown in figure 5.1).

Whatever the exact names, those working in arts are beginning to make their own useful categories which identify "differences" in characteristics and behaviors, rather then being tied to the technological categories that label digital art as Turing-land. Can digital art, for example, usefully relate to the history of "time-based art"? Curator Benjamin Weil has pointed out that in addition to the gallery being able to offer the material comforts of a well-lit place with a free Internet connection, it also offers a conceptual "time slot"[22] for the audience. Weil has shown Net art both in physical venues and strictly online, but has refused to adopt a dogmatic approach to either position. The reluctance of most curators to name "movements" or hierarchical taxonomies for digital art, at this stage, also reflects a need for the structures to be led by the art itself.

### What Is Digital Art Doing That Digital Interpretation Is Not?

> For a long time we have assigned to machines our dirty laundry whilst maintaining the image of their enameled white veneers. (Graham Harwood 1999)

Contemporary art often works under the inherited responsibility to *épater le bourgeois*, whereas interpretation's burden is to patiently explain and communicate. However, sometimes what causes the shock is simply digital art's tendency to hybridize and stray across media boundaries, (another digital art naming from the 1990s was, of course, "multimedia"), contemporary artists tend to use any means or media necessary to make their work, making it particularly difficult to categorize. Jeremy Deller and

Alan Kane's *Steam Powered Internet Computer*, for example, was made in collaboration, and demands the help of local model steam engine enthusiasts wherever it is shown. Nina Katchadourian's *Talking Popcorn*, which uses a computer program to translate the noises made by popcorn, via morse code, into text, illustrates again how crossing old and new media makes points about each.

Concerning materiality, early digital artists such as Linda Dement, and theorists including Allucquère Rosanne Stone and Donna Haraway, resolutely refused to let technology forget the materiality of the physical: the body, the "dirty laundry" of money and power. Graham Harwood's work, even when using immaterial media, resolutely brings the body into close contact with the audience through the content of his work, which has included a work made with patients at a high-security mental hospital (1996). As well as the physical aura of the body and the hand which is the content of some artwork, Josephine Berry has argued that Walter Benjamin's definition of "aura" can still be present in the media itself: within the "unpredictable mutations and instability of digital information" (2001) lies the aura of Net art such as Lisa Jevbratt's *1:1*,[23] which quixotically tries to map the fluid and crumbling landscape of the Internet.

In the binary of Turing-land/Duchamp-land, although many artists may be fascinated by the serious issues of technology, they are also often highly ironic: Natalie Bookchin's *Metapet* game, for example, takes the engagement of online games and Tamagotchis, and thoroughly satirizes both technological industries and power games. The aim of the game is for the player (the boss) to keep their "pet" (a worker in a genetic modification company) productive and obedient, by using perks and threats. If the boss is unsuccessful, then the pet may leave to set up a competing startup company, or worse. Artists can even play with the sacred categories and "interactive" conventions of the digitized art archive, itself. Cohen, Frank, and Ippolito made *The Unreliable Archivist* as a real database-driven Web site where the user can mess with the archive in various disruptive or tasteless ways. As a museum employee involved with archiving digital art and "Variable Media," Jon Ippolito has helped create an artwork both witty and thoughtful. Duchamp may well have approved, in the historical tradition of artists who cross the borders between artist and collector, including Marcel Broodthaers and Susan Hiller.

Artists can also be highly critical of each of the named characteristics of digital media. "Interactivity," for example, was much hyped in the early 1990s, but several artists have questioned "how interactive?" Diller and Scofidio's *Indigestion*, for example, allows the user to choose the gender, class, and masculinity of the characters in a dinner-table narrative, but traps the characters in an inescapable repeating plot. The food and the dialogue change in clever ways, but the user has no real control over the story.

Far from being frustrated with the problems of materiality, artists have moved comfortably between them. Nick Crowe's body of work includes early Net artworks, and physical installations, including *The New Medium* (2000), which took Web sites concerning death and online memorials, and etched them onto sheets of glass. By physically appearing in unexpected public places, digital art can challenge ideas of "history," "heritage," and "culture." For example, KIT's artwork *Joyriding in the Land that Time Forgot* provocatively placed tents in a traditional sculpture garden set in the gardens of an eighteenth-century stately home. The tents were digitally printed with landscapes from computer games including *Jurassic Park*, and asked questions concerning the artificiality of the "natural" landscape surrounding the digital art.

These artworks illuminate how digital media can challenge several borders and boundaries, and it could even be argued that the nature of the media means this is a necessary part of "truth to materials." The characteristics of unpredictability, disruption, or mutation that so challenge archivists, are an essential part of some artworks. While the practice of art often resists categorization (at least in its early stages), when art comes into contact with institutions categories become important in influencing what can and cannot be shown, or in the case of competitions, what is rewarded. The research of CRUMB has also looked particularly at the ways it might affect the manner in which museums work.

### Changing the Working Categories of the Museum?

> The whole show had this sense of blur to it, that was really nice in terms of curatorial blur, in terms of chronological blur, in terms of media blur, and in terms of the way pieces were either newly commissioned or loaned. I think that, to me, was the strength of the show; that it developed.[24]

The most obvious question for art museums is: "Whose department should this be in?" At SFMOMA in 2000, director David Ross took the radical step of getting curators from the departments of Media Arts, Painting and Sculpture, Architecture and Design, and Education and Public Programs to work together on *010101: Art in Technological Times*. Research on this show (Graham 2002) illustrated that as well as the obvious cross-fertilization between knowledge areas, the work demanded a critical crossroads where the Web site was used as both an interpretive tool and as a venue for the Net artworks. Like the Walker Art Center, SFMOMA is a museum with a good reputation for digital interpretational work, including extensive Web sites on Bill Viola, and the sculptor Eva Hesse. This crossroads caused a certain amount of anxiety over roles; for example, the question of how the Net art commissions could be found amidst the interpretation in their only "venue," (the Net art was shown only on the Web site, not in the galleries). Were conventional concerns over "usability" affecting the intent of the artwork? Were the interpretation screens in the galleries being confused with the art? The binaries of Art/Interpretation and Production/Distribution were, therefore, recurring themes in the show.

The *010101* exhibition also illustrated the potential and challenges of communication between all departments of the museum. The need for excellent relationships with installation, technical, and archiving staff is an obvious factor for digital media, and one that SFMOMA successfully developed collaboratively over a number of years. There is also, however, a crucial point at the time of marketing any exhibition that may be immaterial. SFMOMA took much care to aim different press releases at different fields of the media, whether technology-, Net art- or Fine Art-oriented, and managed to get extensive coverage for the Net artworks, in particular. Other curators and artists, including Matthew Gansallo and Pope and Guthrie,[25] have vouched for the importance of quality press coverage for any process-based, interactive, Net-based, or immaterial work.

The process and timescale of curating has also been influenced by certain artworks. Participative artworks such as Heath Bunting's often "evolve" over the duration of an "exhibition," and mean that a curator can be much more involved during an exhibition than is usual for static object-based works. At the very start of the process it is also interesting to note that curators can choose to share their research—for example Barbara London of MoMA New York made a *Stir Fry* Web site during a studio-visit trip to China (Cook 2001).

This approach, which necessarily crosses the boundaries between departments in museums, can be an anxious one for institutions, but the collaboration between archivists, technicians, and curators can help to achieve more long-term stability than the shifting grounds of curating alone. The history of curating digital media at art institutions has been a rather unstable one, often driven by visiting curators, or the strong enthusiasm of individual curators such as Jean Gagnon at the National Gallery of Canada. What tends to remain after institutional change is the heritage of critical coverage, and the archives of the institutions themselves.

## Collecting and Archiving—The End and Beginning of the Line

> *Lev Manovich:*   I understand your position regarding museums, art institutions, preserving, archiving, databasing—but it's so different from the Futurists who said "shoot the painters, burn the museum." Here we are—the avant-garde—and we want to keep all the stuff. . . . Maybe we should be looking towards the future.
>
> *Sara Diamond:*   It's different when a canon is being created, as opposed to a movement.[26]

While art movements need to be free-ranging, evolving, and border-crossing, art documentation needs to be structured and defined. If contemporary art is not documented, then it is very difficult to build a critical history of any media, unless the critics were at the very place an event happened, or present at the moment before the Web site ceased to function. For collections in particular, conservators have been working on the problem for some time: Pip Laurenson of the Tate, for example, has made very detailed case studies as part of her research into video installation, including Gary Hill's *Between Cinema and a Hard Place* (2002, 259–266).

What is unusual about digital art is the involvement of artists/curators in this work rather than strictly conservation specialists, which again reflects the pattern of more collaboration and communication across museum departments. Benjamin Weil, for example, came to San Francisco Museum of Modern Art with the experience of the *äda'web* collection of Net art. The conservators also actively developed knowledge alongside the installation staff, with seminars such as *TechArcheology* in 2000. Consequently, SFMOMA's *espace* online gallery of Net art and design remains one of the longest-running institutional collections of such work, despite the departure of the curators who were originally involved.

Rhizome's *Artbase* is a specialist database of digital art, and uses tactics of preservation as useful categories. The artworks are either Linked (linked to an URL) or Cloned (an archival copy of the artwork is stored on the Rhizome server). The naming of their tactics for preservation in the wider field are: ". . . **migration** (updating code), **emulation** (running outdated software on new platforms), or **reinterpretation** (recreating the work in new technological environments)" (Depocas et al. 2003).

Jon Ippolito, a curator, and one of the artists involved in the *Unreliable Archivist* project mentioned here, was also one of the founders of the *Variable Media* project. This network has brought together some key examples of the problems of showing work from collections, including Felix Gonzalez-Torres's *Public Opinion* (where the audience can take and eat from a pile of candies) and Mark Napier's *net.flag*, a partici-pative Net art piece. The naming of "Variable Media" is an important one, for it cov-ers a range of challenges that cross physical media. In drawing up a questionnaire that aimed to preserve the intent of the artist for future showing of the work, the project made a list of "Medium-Independent Behaviors" (2003, 48) including: Networked, Encoded, Duplicated, Reproduced, Interactive, Performed, Installed, and Contained. These "behaviors" are important namings because, like the "characteristics" of Steve Dietz, they work across media and yet summarize the major new differences for arts workers. "Interactivity," for example, is something which is seldom adequately documented in the traditional gallery installation shot, and is an intent which must be discussed with the artist. As Jon Ippolito says: "Among the important questions for interactive behavior is whether traces of previous visitors should be erased or retained in future exhibitions of the work" (2003, 50). As well as being independent of medium, Caitlin Jones, a Daniel Langlois Fellow in Variable Media Preservation, makes the point that the skills needed may also be "occupationally independent" (2003, 61), which echoes the cross-departmental experience of SFMOMA and others. Caitlin Jones has also managed to question the boundaries by successfully curating *Seeing Double: Emulation in Theory and Practice*, an exhibition at the Guggenheim, which was both a digital art show and an exploration of emulation as a tactic for preservation.

In "redefining digital art," this chapter has illustrated some of the ways in which digital media are redefining themselves by being characteristically in a state of flux. While this might appear to be essentially at odds with the needs of collecting and archiving, there is at least the start of a structure. The particular namings of "Medium-Independent Behaviors" and "occupationally independent" skills can be use-fully applied to the widest range of contemporary artwork, and form useful tools for a wide scope of institutions. The creative cross-disciplinary approaches adopted by some organizations mean that collecting and archiving can be the beginning, rather than at the end for the critical life of an artwork. The cross-media nature of some of the most useable categories for digital art also reflects the ways in which some disrup-tion of borders can lead to workable and creative categories that can be led by the art itself.

For a field which typically "flows around and avoids institutions" (Schleiner 2003), digital art has, to some extent, already molded some aspects of the way that even conservative art institutions work. The blurring of departmental boundaries, and the evolving timescale of the process, are just two of the ways in which this happens. In consistently challenging notions of technology, heritage, or materiality, the content of the artwork itself will, no doubt, keep the definitions and the institutions in a healthy state of flux for some time to come.

*Notes*

1. The terms *new media art* and *digital art* are used in very variable ways, although the former may generally infer a history of "media" as in video, whereas the latter may relate more to the history of computers. For example, two books in the Thames and Hudson series are Rush 1999, which uses the chapter headings "Media and Performance," "Video Art," "Video Installation Art," and "Digital Art," and Paul 2003 (chapter headings summarized in table 5.1).

2. *Unstable media* is a phrase used by the V2_media art organization, ⟨http://www.v2.nl⟩. Benjamin Weil jokingly described himself as "curator of everything that you can plug in" in Graham 2002. See also next note.

3. Those who have traced the definitions of digital art in historical terms include: Sarah Cook (2004), Charlie Gere (2002), Martin Lister et al. (2003), Randall Packer and Ken Jordan (2001), and Oliver Grau (2002).

4. Beryl Graham, unpublished notes from observational case study of the exhibition *V-topia*, July–September 1994 at The Tramway, Glasgow.

5. This theme is expanded on in Graham and Cook 2001.

6. See ⟨http://www.nytimes.com/2003/05/13/arts/13ARTS.html⟩ and ⟨http://www .mteww.com/walker_letter/halbreich_letter.html⟩ (both accessed 7 April 2005).

7. ⟨http://www.tate.org.uk/netart/mongrel/⟩ (accessed 7 April 2005).

8. Details of the process can be found in Cook 2001a.

9. See ⟨http://www.sciencemuseum.org.uk/on-line/wellcome-wing/digitopolis/art.asp⟩ (accessed 7 April 2005).

10. Some good insights into EAT can be found in Lovejoy 2004.

11. See Ede 2000 and *Play Garden* (London: Arts Council of England), available at ⟨http:// www.newaudiences.org.uk/playgarden/⟩ (accessed 7 April 2005).

12. See ⟨http://www.critical-art.net/⟩ and ⟨http://www.washingtonpost.com/wp-dyn/ articles/A8278-2004Jun1.html⟩ (both accessed 7 April 2005).

13. Including Greene, Sutton, and Fuller (2001) and Gillman (2001).

14. Sins of Change: Media Arts in Transition Again Conference at the Walker Art Center, Minneapolis, April 2000. Transcriptions of panel discussions by Sarah Cook.

15. See ⟨http://crumbweb.org/⟩ (accessed 7 April 2005).

16. See Kluitenberg 2004.

17. ⟨http://www.dublincore.org⟩ (accessed 7 April 2005).

18. Beryl Graham, *A Table of Categories of Digital Art* (2004), available at ⟨http://www.newmedia.sunderland.ac.uk/crumb/phase3/append/taxontab.htm⟩ (accessed 7 April 2005). Substantial quotes from the discussion on taxonomies (subsequent to submission of this chapter) are used in Graham 2005.

19. ⟨http://rhizome.org/⟩ (accessed 7 April 2005).

20. Christiane Paul, "Re: FEED Article: "The Demise of Digital Art," *NEW-MEDIA-CURATING Discussion List* (29 March 2001). Available via e-mail at ⟨NEW-MEDIA-CURATING@JISCMAIL.AC.UK⟩.

21. See Cook 2004, 40 *ff*.

22. See Mirapaul 2001.

23. ⟨http://jevbratt.com/⟩ (accessed 19 April 2005).

24. Kathleen Forde, quoted in Graham 2002.

25. Cook 2001 and Pope and Guthrie 2001.

26. Sins of Change conference.

## *References*

Berry, Josephine. 2001. "The Thematics of Site Specific Art on the Net." PhD thesis, University of Manchester. Available at ⟨http://www.metamute.com/look/mfiles/josie_thesis.htm⟩ (accessed 23 April 2005).

Cook, Sarah. 2001a. "An interview with Matthew Gansallo." *CRUMB* (2001). Available at ⟨http://www.newmedia.sunderland.ac.uk/crumb/phase3/nmc_intvw_gansallo.html⟩ (accessed 7 April 2005).

Cook, Sarah. 2001b. "Multi-Multi-Media: An interview with Barbara London." *CRUMB* (2001). Available at ⟨http://www.newmedia.sunderland.ac.uk/crumb/phase3/ilondon.html⟩ (accessed 7 April 2005).

Cook, Sarah. 2004. "The Search for a Third Way of Curating New Media Art: Balancing Content and Context in and out of the Institution." Ph.D. thesis: University of Sunderland.

Cook, Sarah, and Beryl Graham. 2004. "Curating New Media Art: Models and challenges." In *New Media Art: Practice and Context in the UK 1994–2004*, ed. L. Kimbell, 84–91. London: Arts Council of England.

*CRUMB*. Available at ⟨http://www.crumbweb.org⟩ (accessed 7 April 2005).

Depocas, Alain, Jon Ippolito, and Caitlin Jones, eds. 2003. *Permanence through Change: The Variable Media Approach*. New York: Guggenheim Museum. Available at ⟨http://www.variablemedia.net/e/preserving/html/var_pub_index.html⟩ (accessed 7 April 2005).

Dietz, Steve. 1999. "Why Have There Been No Great Net Artists?" *Through the Looking Glass: Critical Texts*. Available at ⟨http://www.voyd.com/ttlg/textual/dietz.htm⟩ (accessed 7 April 2005).

Ede, Siân, ed. 2000. *Strange and Charmed: Science and the Contemporary Visual Arts*. London: Calouste Gulbenkian Foundation.

Frieling, Rudolf, and Dieter Daniels, eds. 2002. *Media Art Interaction: The 1980s and 1990s in Germany*. Vienna: Springer.

Gere, Charlie. 2002. *Digital Culture*. London: Reaktion Books.

Gillman, Clive. 2001. "Re: Net Art and Large Museums: Assignment Number One." *New-Media-Curating Discussion List* (8 March). Available via e-mail at ⟨New-Media-Curating@JISCMAIL.AC.UK⟩.

Graham, Beryl. 2002a. "An interview with Benjamin Weil." *CRUMB* (2002). Available at ⟨http://www.newmedia.sunderland.ac.uk/crumb/phase3/iweil.htm⟩ (accessed 7 April 2005).

Graham, Beryl. 2002b. *Curating New Media Art: SFMOMA and 010101* (2002). Available at ⟨http://www.newmedia.sunderland.ac.uk/crumb/phase3/append/sfmoma.htm⟩ (accessed 7 April 2005).

Graham, Beryl. 2005. "Taxonomies of New Media Art: Real World Namings." In *Museums and the Web 2005*, ed. D. Bearman and J. Trant. Pittsburgh: Archives & Museum Informatics. Available at ⟨http://www.archimuse.com/mw2005/⟩ (accessed 7 April 2005).

Graham, Beryl, and Sarah Cook. 2001. "A Curatorial Resource for Upstart Media Bliss." In *Museums and the Web 2001: Selected Papers from an International Conference*, ed. D. Bearman and J. Trant, 197–208. Pittsburgh: Archives & Museum Informatics. Also available at ⟨http://www.archimuse.com/mw2001/papers/graham/graham.html⟩ (accessed 7 April 2005).

Graham, Beryl, and Sarah Cook. 2002. "Curating New Media: Net and Not Net." *Art Monthly* (November): 44–45.

Grau, Oliver. 2002. *Virtual Art: From Illusion to Immersion*. Cambridge, Mass.: MIT Press.

Greene, Rachel, Gloria Sutton, and Matt Fuller. 2002. "Voiceover." *Afterimage* (March/April).

Harwood, Graham. 1996. *Rehearsal of Memory*. CD-ROM. London: Bookworks.

Harwood, Graham. 1999. *Creativity and Consumption: New Media Arts in Advanced Technology Culture*. Available at ⟨http://www.luton.ac.uk/creativity/abstracts/harwood.html⟩ (accessed 7 April 2005).

Jeremijenko, Natalie. 1999. OneTrees Documentation. Available from ⟨http://www.onetrees.org⟩.

Jevbratt, Lisa. *1:1*. Available at ⟨http//:jevbratt.com⟩ (accessed 19 April 2005).

Kluitenberg, Eric. 2004. "Re: Art and science..." *New-Media-Curating Discussion List* (1 March). Available via e-mail at ⟨New-Media-Curating@JISCMAIL.AC.UK⟩.

Laurenson, Pip. 2002. "Developing Strategies for the Conservation of Installations Incorporating Time-Based Media with Reference to Gary Hill's 'Between Cinema and a Hard Place.'" *Journal of the American Institute for Conservation* 40, no. 3: 259–266.

Lister, Martin et al. 2003. *New Media: A Critical Introduction*. London: Routledge.

Lovejoy, Margot. 2004. *Digital Currents: Art in the Electronic Age*. 3d ed. London: Routledge.

Manovich, Lev. 1996. *The Death of Computer Art*. Available at ⟨http://www.thenetnet.com/schmeb/schmeb12.html⟩ (accessed 7 April 2005).

Manovich, Lev. 2001. "Post-media Aesthetics." In *(Dis)locations*. DVD-ROM, ed. ZKM. Ostfildern: Hatje Cantz Verlag. Also available at ⟨http://www.manovich.net⟩ (accessed 7 April 2005).

Manovich, Lev. 2003. "New Media from Borges to HTML." In *The New Media Reader*, ed. N. Wardrip-Fruin and N. Montfort, 13–25. Cambridge, Mass.: MIT Press.

Mirapaul, Matthew. 2001. "O.K., It's Art. But Do You View It at Home or in Public?" *New York Times*, 19 March.

*New-Media-Curating Discussion List*. Available via e-mail at ⟨NEW-MEDIA-CURATING@JISCMAIL.AC.UK⟩.

Packer, R., and K. Jordan, eds. 2001. *Multimedia: From Wagner to Virtual Reality*. New York: Norton.

Paul, Christiane. 2003. *Digital Art*. London: Thames and Hudson.

Pope, Nina, and Karen Guthrie. 2001. "On Being Curated: A Critical Response." In Presentation at *Curating New Media Seminar*, May 10–12. Gateshead: Available at ⟨http://www.newmedia.sunderland.ac.uk/balticseminar/popgut.htm⟩ (accessed 7 April 2005).

Rush, Michael. 1999. *New Media in Late 20th-Century Art.* London: Thames and Hudson.

Schleiner, Anne-Marie. 2003. "Curation Fluidities and Oppositions among Curators, Filter Feeders, and Future Artists." *Intelligent Agent* 3, no. 1. Available at ⟨http://www .intelligentagent.com/archive/Vol3_No1_curation_schleiner.html⟩ (accessed 7 April 2005).

Siân, Ede, ed. 2000. *Strange and Charmed: Science and the Contemporary Visual Arts.* London: Calouste Gulbenkian Foundation.

Stone, Allucquère Rosanne. 1991. "Will the Real Body Please Stand Up? Boundary stories about cyberspace." In *Cyberspace: First Steps*, ed. M. Benedikt, 81–118. Cambridge, Mass.: MIT Press.

Wilson, Stephen. 2001. *Information Arts.* Cambridge, Mass.: MIT Press.

# Online Activity and Offline Community: Cultural Institutions and New Media Art

Sarah Cook

---

*Last year I was at a lake turnout, in Banff, Alberta, going towards Radium on Highway 93. I stopped at a beautiful scenic view where you overlook the lakes, it's just amazing. Lo and behold as I walked to the edge, which is maybe a five or six feet drop, not even, I came across a huge grizzly bear, probably easily 1000 pounds, like the size of a car. My heart was pounding, talk about heart-wrenching! I went up to the bear and I said 'hey, how's it going?' The grizzly looked up, looked at me, and looked back down and started backing off down the hill.*[1]

This story—which could well be a parable for the encounter between new media art and the museum—was recorded by someone who used their mobile phone to call a toll-free phone number connected to a computer, which then stored the recording in an online database. His call, together with that of other callers, generated a sonic field guide to the locale. The project, "Mobile Scout"—by the artists Marina Zurkow (a storyteller), Scott Paterson (an architect), and Julian Bleecker (a technologist)—was commissioned by the Walter Phillips Gallery at The Banff Center for the exhibition *Database Imaginary*.[2] As Banff is a National Park one would expect call stories about bears, mountains, lakes, and northern lights to dominate the database. Callers, who spoke to an automated park ranger and squirrel, were asked to first choose their mission and then two attributes of the location (or habitat) where they were, before leaving their messages (figure 6.1). These attributes then enabled visitors to the project's Web site to sort and select the "pages" of the field guide they wanted to hear. However, as information about the project was distributed online as well as in print brochures, and as the toll-free number worked across North America, there were also many callers from New York, where the artists are based, commenting on their urban habitat as well. As the artists write: "'Mobile Scout' defines place as being made of social habitats, not geography."[3]

Figure 6.1    Screenshot of Mobile Scout: A Sonic Field Guide, 2004, by Marina Zurkow, Julian Bleecker, and Scott Paterson ⟨www.mobilescout.org⟩.

In this chapter I argue that working on the Internet is, for many new media artists, not just a novel use of a new technological medium, but a strategic practice in their engagement with the social, with culture. Some forms of Net-based art can be seen as manifestations of actual, and located, social networks—whether conceptually or in participatory or activist terms. As the above example shows, artists use the Web—a kind of nonplace—as a way to engage with and even sometimes create offline communities as much as, if not more so, than online ones. The Internet might only play one role within the experience of the artwork as a whole, and the place of the encounter might be more about a social use of technology than about the technology itself. Technologies like the Internet have a more rapid rate of change than the social communities that use them. For these reasons, among others, engaging with Net-based art has proven a challenge for cultural heritage institutions. In this chapter I suggest that, first, this is due to the way art history is written, around forms and genres of work, and second that the politics and economics of online space play a role. In considering how to curate technology-driven projects I argue that a consideration of Net-based art should not be about the Net—its obvious form—as much as about the community-generating practice of engaging with networks.

### The Landscape of New Media Art

> Technologies often tend to develop faster than the rhetoric evaluating them, and we are still in the process of developing descriptions for art using digital technologies as a medium—in social, economic, and aesthetic respects. (Paul 2003, 67)

Digital art is, in my opinion, a problematic term, evoking 2D graphics made using computers—namely *artifacts* that are aesthetic—more than *events* that are social, political, or economic. The field is, in fact, rich and diverse, comprised of engaging virtual (digital) and real objects, as well as actions, interactions, and interventions. However, thematic division by medium (i.e. digital or analog) has emerged as a recurrent methodology in these early stages of writing an art history of computer-driven art. These attempts to include digital art in the canon of Art History have drawn heavily on media theory, and many have mapped its development against time-based, screen-based art forms such as film and video. Whereas those art forms have, to a limited degree, acquired the status of "objects" and been accessioned into museum collections (bizarrely, for reproducible media, in limited editions), the wider landscape of new media art, where the distribution of the work relies on the digital as much as its production does, has proven more complicated to map.

Much new media art demands collaboration, whether between artist and engineer, or computer programmer and writer, as the example of *Mobile Scout* proves. New media art, as a category, contains within it a number of genres, ranging from art-science collaborations to interactive interpretational installations. New media art is dynamic and interactive, and often encounters its public audience beyond the borders of the museum (for instance, on your phone by the side of the highway). As such, its history cannot easily be gleaned with reference to an institutional collection of static objects.

As Beryl Graham writes in chapter 5 (in this volume), it is helpful in defining the category of new media art to consider separately what might be the form of the work (i.e., what media technology it uses—that it is screen-based, for instance), the theme of the work (i.e. biotechnology), or the characteristics of the work (i.e. interactivity). This means also distinguishing between the ways in which technology is used as either a tool for the creation of the work, as the medium of the work itself, or as a defining practice made evident in the work.[4] I would suggest going even a step further and approaching works by examining alternative non-medium-specific *practices* employed by artists. As curator Steve Dietz has also pointed out, the use of the Internet in new media art is better characterized as a *practice* than a medium (2001, 50). To my mind, *practices* exemplified in the work differ from *characteristics* of the work because they suggest the social and cultural framework informing the making of the work. Thus, other practices might include:

• the use of networks (such as the Internet)
• conceptual art
• installation
• performance
• gaming.

As one technological form quickly outstrips its predecessor, an investigation of practices, which have longer traditions, is one way around the museological impetus to evaluate new media art chronologically, on the basis of the development of the medium. Another way to move beyond the institutional debates concerning aesthetics (of the objects in museum collections, of what is worth saving and what is not), to the important debates around the social and economic implications of new media art, would be to approach it through an investigation of the strategies employed by artists in the production and dissemination of the work. Ursula Frohne writes,

> So far, most critics and theorists have reflected on new media art in classic iconographic categories. That is a problem, since moving images cannot be described with those traditional tools. Moreover, the close relationship of media art to certain everyday cultural phenomena makes it necessary to deal with these kinds of artworks from a cultural perspective. (2000, 124)

Little scholarship has been undertaken to examine new media art historically in terms of cultural practices or strategies, with the exception of Charlie Gere's book, *Digital Culture*, Craig Saper's book, *Network Art*, and the curatorial efforts of Simon Pope and Matt Fuller in the exhibition, *Art for Networks*, and Beryl Graham in her exhibition, *Serious Games*.[5] And yet, considering the cultural context of the work might get one closer to an understanding of the experience of the artwork than judging it by way of a museologically informed chronological mapping of its form and medium.

Thus, in this chapter I will discuss predominantly one *practice* within the expansive field of new media art, namely, Net-based art, or art that exists and is encountered online and was created by and for online and offline communities.

The first question to ask is whether the Internet, in Net-based art, is more than just the work's medium. The answer is to be found in the nature of the Internet itself. The fact that the Internet can subsume or mimic other traditional media is key. As art historian Julian Stallabrass writes, "the Internet is not a medium, like painting, print or video, but rather a transmission system for data that potentially simulates all reproductive media" (2003, 12). Thus, as a "system," the Internet is formally both where and how the work is made and displayed. In the next section I will consider how working on the Internet is a culturally inflected and strategic practice for new media artists.

### Internet-based Art: What Is It?

Web-based works are often, somewhat inaccurately, considered "virtual" (as opposed to "real" or "actual" works of art), because of their seeming immateriality. Internet-based art, as much as other, more tangible forms of new media art, nevertheless can be seen to exhibit the same characteristics as other non-Net-based forms of media arts, such as connectivity, computability, and interactivity.[6] With network-driven projects, a computer is used not just as a means of production of the work but also as the means of its distribution. So to say that the computer itself (the hardware) is the materiality of the work would be an oversimplification. The work is made manifest (becomes material) through one's interaction with it. Theorist Lev Manovich writes, somewhat controversially:

> Net art projects are materialisations of social networks. These projects make the networks visible and create them at the same time. . . . So-called Net art projects are simply manifestations of social, linguistic, and psychological networks being created or at least made visible by these very projects, of people entering the space of modernity. . . . Exchanging person-to-person communication for virtual communication (telephone, fax, Internet); exchanging close groups for distributed virtual communities.[7]

The question of the materiality of Net-based art (the materialization of the network, hence the "look" of the medium) is thus quite nuanced, and deliberately made aesthetic by some Net-based artists. Stallabrass (2003, 24) argues that early Net-based art practice did have the "appearance of a medium"—i.e., the browser window—

even if the Web itself (and computer code more generally) was understood, in an object sense, as immaterial.

So if the Internet might look like a medium but is indicative of a practice, what kind of practice is it and how is it manifest?

## Conceptual Practices

Given that Net-based art has at its core a manipulation of data, particularly digital data, it has clear links to conceptual art (Stallabrass 2003, 33). Many Net-based artists have gone a step further and sought to allow the machine to take over the process of manipulation, to "generate" the work, in a playful, Dadaist fashion. The notion that conceptual art practices dematerialize the art object is, in some sense, reversed here—Net-based art might never have been material in the first place; "protomaterialization" perhaps?

## Participative Practices

The number of Net-based artworks that encourage the visitor's interaction in the completion of the work—through clicking, linking, and entering data—suggests not just the conceptual art origins of Net practice, but also its socially engaged or participatory origins. This is one area where Net-based art overlaps with contemporary visual art more generally, finding new ways to engage with the world (Bourriaud 2002). In the field of new media, for many the software interfaces of information technology (Internet browsers, e-mail, online databases) are overly familiar, and artists have simply intervened to point out alternative and surprising uses for them, and for our perspectives on them.

## Activist Practices

Most of the problems encountered when trying to identify whether Net-based art is, in fact, art or not can be traced to its activist approach. Net-based projects often seek to access, frame, and share data for some other end purpose than merely aesthetic contemplation. Yet this is no different from other forms of mainstream visual arts, such as Fluxus art or performance art—joining art and life through an activation of or intervention into the social fabric.

The conceptual, participatory, and activist tendencies of Net-based art have been argued as demonstrative of some of new media's most "avant-garde" characteristics. As Stallabrass writes:

> Many of the actual conditions of avant-gardism are present in online art: its anti-art character, its continual probing of the borders of art, and of art's separation from the rest of life, its challenge to the art institutions, genuine group activity, manifestos and collective programmes. (2003, 35)

Some of the early Net-based artworks were concerned with creating interventions that were manifest in a physical presence. In 1994, Heath Bunting, one of the founders of the online bulletin board system (BBS), "Cybercafe," posted the phone numbers of all the public telephones in Kings Cross railway station in London and invited members of the public from all over the world (via BBSs such as Cybercafe) to phone in on a particular date and time.[8] The result was mild chaos—a mess of people answering phones, dancing to the sound of them ringing, and, by some reports, eventually the police threatening to shut down the station and delay trains if the crowd did not disperse.

Indeed, many artists think of the Web as a real space, a site in which to place their work, suggesting that Net-art practice is, above all, site-specific (Berry 2001). Which brings us to the question of place and the notion that while the Internet might not be material (logically, then, supposing that Net art is not either), the effects of Net art in the real world are very material.

### Politics and Economics 1: Historical Characteristics of the Online World that Necessitate Artistic Collaboration, or the Internet is a Culture

> It would be more accurate to suggest that digital technology is a product of digital culture, rather than vice versa. (Gere 2002, 13)

Culture is formed by a community and the argument that the practice of creating Net-based art is culturally strategic, and that Net-based artworks themselves are materializations of "the social" (of communication, of participatory exchanges) presupposes that there must be online communities generating the culture to begin with. But can there be such a thing as an online community? As theorist Eric Kluitenberg writes:

In the early 1990s high hopes were placed on networking technology to offer new tools for shaping communities translocally, as well as strengthening localised communities. In the US, during 12 years of Republican rule, the public sphere was effectively slaughtered. . . . The explosion of drug abuse and small-scale crime . . . and an anxiety campaign about the dangers of public space were the final blows to public discourse and community. In this barren desert of social isolation any tool that could recapture something of this lost *socios* was embraced eagerly. (2000, n.p.)

In this light, Heath Bunting's ongoing projects with ⟨www.irational.org⟩, the successor to Cybercafe—from setting up shared art servers and online pirate radio stations to keeping up-to-date databases of information for the itinerant Net-worker about which medialabs have free places to sleep—are perhaps some of the strongest cases for demonstrating how Net-based practice is culturally mediated and manifest in the real world. Former museum director David Ross (1999) has spoken at length about how the emergence of networked new media art in the early 1990s was like the emergence of video art practices in the early 1970s, but at the same time reached farther back to a collaborative impetus in art making:

I've been trying to understand this thing as it develops. From a text-based BBS to the graphical browsers that have developed from Mosaic through Netscape and Internet Explorer. They all have the same idea: to enable people to communicate. The idea of art taking place in this new way is astounding to me. It seems to be re-opening the potential that I thought had disappeared. (Ross 1999)

It seems clear that the online activity would not have emerged were it not for the realities of life in the offline community. Another example of the politics and economics of Net-based art and the impetus towards collaborative practice can be found in the exhibition "Art and Money Online" (2001) at Tate Britain. This group exhibition sought, through the display of three new works, to demonstrate the "online clash of culture and commerce"—noting that Net-based art has grown up in tandem with the commercialization of the Web. Included was the work of the Redundant Technology Initiative, an artist group that accepts donations of old computers and either reconfigures them into sculptural installations (stripped down, like old style banks of televisions), or refits them (with free open source software) for their open access computer lab in Sheffield.[9] Hosting their project at the Tate galleries meant that all of the Tate's old computers awaiting disposal were, in fact, diverted to RTI, to comprise their most technologically advanced donation to date. Considering, globally, how few people are online, or have technological access to get online (those that are online are among the elite minority), participatory, activist, anticommercial, anticorporate projects such as RTI's become all the more important. As they write:

> The many-to-many nature of the Internet challenges the assumption that media is dominated by professionals and can only give voice to powerful interests. [Our work] proves there's nothing inherent in . . . technology that makes this so. (Kimbell 2004, 198)

The shifting politics and economics of the Internet play a substantial role in the creative process of Net-based artists. They recognize that the realm of computer-mediated communication is a culture, and that shared networks have codes and signifiers all their own. UK-based collaborators Jon Thomson and Alison Craighead take on the roles of archaeologists of the World Wide Web in their work, saving for posterity outdated tropes of Web page design, from midi (music) files to "this site under construction" animated.gifs (graphic image files). They have recycled these into objects, such as temporary tattoos or badges, that you can buy through their online ⟨www.dot-store.com⟩ shop. The inherent multiplicity of agendas prevalent in the online world (that it has gone from "hopeful utopia to online shopping centre," as Thomson and Craighead (2004, 174) have noted) has brought about a necessity for collaborative art creation, as a way to combat the increased commercialization and information overload that the Internet has brought into each of our lives. In fact, in the case of artists working online, it is difficult to determine if they are creating new content or creating (excavating, illuminating) new contexts for artistic activity.

### The Pairing of Online Activity with Offline Community: The Need for Collaboration

> Net art is ubiquitous, it has been created to be seen by anybody, anywhere in the world at any time, and I would like to see it in multiple contexts, the museum being one of them. It would be equally strange if it would be left out of the museum when you can access it in a shopping mall or in a cyber cafe.[10]

Internet-based art is defined by the fact that it is distributed—often simultaneously. Not only can you connect to the work from more than one location or timeframe, you can collaboratively create and change the experience (not just perception) of the work by interacting with it, alone or in a group, from within your own social, political, or geographic context. The "artness" or "objectness" of the new media artwork can vary from one context and one timeframe to the next. Given the possibilities of distribution, and connection from afar, it is continually intriguing that the more interesting projects of Net-based art are still so rooted in the real world—in located (rather than translocal) physical communities more than in virtual ones.

The early examples above suggest one reason for this—that Net-based practice arises out of an activist participatory tendency that eschews the increased commercialization of the space of the Internet, and the market's tight control on access to technology. There have been moments when it has felt like a war against the Web becoming a global TV, making us, its "viewers" (rather than users), its greatest homogenous cultural product. Maintaining a connection to the real world and actual communities has been one tactic in that battle.

The San Francisco-based team of Margaret Crane and Jon Winet have collaborated in online space for over ten years. Trained in photography and journalism, their projects are hybrid creations employing new media research, social science methods, documentary techniques, and hypertext fiction (2004, 142). Their recent project, "Conventional Wisdom," is a Web-project tracking the 2004 US Presidential election campaigns.[11] Running from an edited database rather than a completely open "blog," the site is refreshingly original in that it is engineered to be fast and content-rich even on a slow computer with a 56kbps dial-up modem. An alternative to the mainstream media coverage, "Conventional Wisdom" depends on the contributions of a team of collaborators uploading site-specific text, photos, and video clips "at the speed of news" that are then made accessible worldwide (figure 6.2). As the artists write:

> In our calculus, collaboration produces arithmetic in which one plus one equals any number except two. Professional boundaries are blurred. Process trumps product. Community works best as a verb. Things get interesting. (quoted in Kimbell 2004, 142)

### Politics and Economics 2: Questioning the Success of the Cultural Institution in Bridging the Gaps Between Online and Offline Artistic Collaborative Communities

In part due to this emphasis on collaboration between the artist and the contributors to the work (who may or may not be the same people as the audience to the work), curators dealing with Net-based art have had to be as much focused on the work's production as on its distribution and exhibition. This is a significant shift for arts institutions (see chapter 5). Grant Kester writes:

Figure 6.2    Screenshot of Web site for 2004—America and The Globe, 2004, by Margaret Crane | Jon Winet ⟨www.america-the-globe.net⟩.

> Net-based culture holds out an even more challenging possibility; to force us to rethink the conventional identity of the artist as someone who develops projects or works that are then administered to a receptive viewer. (2001)

"Administered" is a word with very different meanings in the information technology world and in the museum world. You do not notice good IT administration in an office environment, whereas in a museum environment curators often cannot get the art out on display without first navigating through levels of museum administration; the former is designed to be seamless and invisible to the user, the latter to be transparent to the public. Either way, when museums engage with online art it is certain that administering the work to a viewer is not as straightforward as it is with more static, object-based works of art. Even in time-based arts such as video or performance, there is still often a straight relationship of viewer to perception of a completed work. By contrast, with net-based art, access can be from anywhere, and the work is not complete unless you access it and interact with it.

The Tate Gallery's commission of Heath Bunting's online work "BorderXing" is a good example.[12] The piece distributes information about moving across international borders via routes other than those that go through the usual guarded checkpoints. Miming the content of the work itself, the site can only be accessed from recognized computers at authorized locations. Negotiating between the work of art and its audience has added an extra layer of administration to the curatorial task of "exhibition making." Thus a need has emerged for alternative distribution mechanisms to be adopted by our museums and cultural heritage institutions.

Given that artists deliberately blur the line between the technology being the tool, medium, or practice of the work, education and accessibility are key areas in which museums can support Net-based art practice. The museum can garner an audience to Net art that might not find it otherwise. Yet a museum audience is by definition a passive one, and Net-based art is interactive—the viewers are more than contemplative "visitors," they are participants, users. As such, educational programming can entail not only supporting the exhibition of the work, but also the online and offline communities the artists are collaborating with in the creation and sustenance of their works. As Matt Fuller writes:

> For artists working via the nets to now involve museums as one of the media systems through which their work circulates, what is crucial is, alongside the avoidance of being simply nailed down by the spotlight, to attempt to establish, not a comfy mode of living for the museum on the networks, but a series of prototypes for and chances at something other and more mongrel than both. (2000)

Thus cultural institutions (whether museums, galleries, or educational establishments) could attempt to move away from formalist notions of technology—trying to make Net art fit into the familiar rubric of art as an aesthetic experience—to showing it as a means of sustaining engagement between offline communities. This might be more truthful to the complicated boundary areas where the art exists and to its activist intentions which suggest that you should use the online art to engage at a deeper level with the locality you actually live in, or the international community you feel you belong to.

Curators thus inevitably have to follow the artists (the avant-garde), especially those whose activities within the network include curating and distributing their own projects alongside those of others. This demands openness and flexibility, and a loosening of the authorial and authoritative voice that museums have traditionally based their activities on. As curator Illiyana Nedkova writes:

> How can curators ensure that the artist-led community projects, now proliferating into the new media sector, are being sensitive and morally being responsible to the particular community or partner's needs? How can we possibly maintain the precarious balance between being very proactive and responsive through process-led projects whereby the product comes second, yet, is nonetheless of equal importance? (2002, 103)

Graham Harwood's intervention into the Tate Gallery's Web site, "Uncomfortable Proximity,"[13] exposed the difficulties cultural heritage institutions have in questioning their own authority and their accessibility to audiences. By seeing the Web site project as a "product," it has been argued that the Tate neglected the ongoing process of engagement with the community whose consciousness about the museum and its art collection was raised through the work. According to traditional notions of the art object, Net-based projects and their resulting shared intellectual property—in Harwood's case the rethinking of British history, slavery, and systems of patronage which legitimized the art in the collection of the Tate—do not necessarily qualify as art; the product and the process cannot be commodified or distributed via the ways in which art is usually circulated. The success of these works is as much conditioned by its users, as by its observant audience.

### An Alternative: Curating in Emergence

Given the distributed nature of online art and the shifting goalposts of technological know-how, museum curators are hard pressed to keep up with the diverse range of art activity on the Net. One potential model of curatorial practice is to curate "in emergence," as suggested by Clive Gillman, an artist who worked in a consultancy capacity with FACT (the Foundation for Art and Creative Technology) in Liverpool. FACT was formerly a non-gallery-based commissioning agency and is now housed in a new building in a city embroiled in a government-initiated "cultural regeneration" agenda.

Gillman described FACT's curatorial role as primarily to support "emergent art" and defines emergent art as "creativity emerging through new forms of media," (thus neither "heritage" art nor "contemporary" art), and therefore where cultural currency is most valuable (1999, 126–128). What is interesting about Gillman's distinction is that he sees emergent forms of art—such as Net-based art—arise not so much through the specificities of the media (i.e., the Internet) as through the process of emergence itself (i.e., how the Internet is used—for instance, blogging or streaming). Therefore the window of opportunity to work with or, more specifically, curate emergent forms keeps moving. It is not so much about keeping up with the technology as keeping up with the practices of technology's use. He charges that "we need to make sure this window remains open and accessible, therefore we need new paradigms of support, new infrastructures" (2000). Gillman proposes this model as one way to answer the perennial question:

> How do we face up to the issues of cultural currency—and not simply adopt the manner of existent contemporary art forms (i.e., putting work into the gallery when perhaps that's not the place for it)? (2001)

In the case of FACT, the social and cultural context is, in part, the regeneration agenda for the city of Liverpool put forth by the government (Liverpool has been named European Capital of Culture for the year 2008). Thus one limitation of curating in emergence is the shifting goalpost of what is considered emergent and what is considered established or regenerated ("heritage" or "contemporary" in Gillman's terms) in this context. As the city evolves, the role of a curator here, as Gillman has characterized it, involves placing oneself between the government bodies and the arts institution, trying to avoid or promote the agendas of each in order to best support the artist, focusing on the period of emergence and commission, researching new forms of art practice, and, most of all, keeping up with the artists!

For FACT this has meant a strong "collaborations program" running for over ten years, joining educational initiatives with new art production. The collaborations program is well known internationally for establishing a Net-based television station in Liverpool's oldest tower block and teaching the residents how to run it (the "Superchannel" Tenant-spin project by artists Superflex).[14] More recently, the program paired an artist, Nick Crowe, with city policemen to create Web-radio, giving voice to an interesting and often depersonalized facet of local society.[15] In both cases, the Web is used to bring together and support communication and the creation of culture in an offline community more than an online one.

A curator working in a model of emergence has to be aware of the tacit assumptions that are often made about the power of technology and new media in relation to social politics—as learned from the observation by Kluitenberg cited above. As Gillman has pointed out (2001), the art field is "dominated by canon and peer consensus, therefore it is exclusive and culturally specific; do you put new media in this?" In other words, do you segregate new media further from the mainstream of visual arts by arguing its newness, its uniqueness, and then risk it being tied into political agendas of providing access to technology; do you risk new media being seen as nothing more than a tool to be used in educational and interpretive contexts? How can new media be incorporated into the culturally specific canon of art practice? Who is deciding the criteria of "emergence" and of the "cultural currency" of a particular art form over another? These questions are all the more pressing when arts funding is tied to political agendas.

> Why are we interested in new media? Is it because it opens up a space that extends beyond the frame (the art context) to the rest of the world in the way that the fields of multiculturalism and public art do? (Gillman 2001)

Perhaps. But in the meantime, projects like "Police Radio" and "Tenant-spin" suggest that there is a middle ground that recognizes that (offline) communities will, with some assistance from our art institutions, find new culturally specific uses for (online) technologies that fit their social, economic, and political contexts.

## Conclusion

> When working with Mongrel, what was of primary importance was the development of new works with and by communities. To go the next step and make links across communities requires a huge amount of follow up both for the artists and for organisers. The artists of necessity often go back to their own locations. Building virtual networked communities I think takes a lot of follow up and communication, much of which really still needs to be undertaken in real time in physical follow up sessions.[16]

This quote, from curator Amanda McDonald Crowley, suggests both the challenges of working with communities as well as the challenges of curating in emergence—both technological emergence and community emergence. In this essay I have shown that working on the Internet is, for many new media artists, not always about the new technology at hand, but forms a part of a strategic practice in their engagement with culture—understood as an integral part of social community. For the artist team behind the project "Mobile Scout," that community was as much in New York as where the work was "located," in an art gallery in Banff. Net-based art can be seen as a manifestation of social networks—whether conceptually or in participatory and activist terms. On the Web, artists have learned to accommodate and resist both political and economic agendas in the creation and distribution of their site-specific artworks, engaging with offline communities as much as, if not more so, than with online ones. As a result, the collaborations entailed in the production of Net-based work, and the communities involved in the sustenance of Net-based work, add another layer to the administrative tasks of the institution. The example of works of Net-based art that specifically engage communities can serve as a model for a new method of curating—"curating in emergence," a method that recognizes that Net-based art is not about the Net so much as it is about the practice of engaging with the Net. A closer examination of Net-based art could be one way to discover the transformations required in institutional mission, curatorial practice, and in a museum's relationships with its audience when dealing with emergent cultural forms.

## Notes

1. Recording number 7600 in the *Mobile Scout* database recorded at 20:43 on 22 November 2005. Available at ⟨www.mobilescout.org⟩ (accessed 20 April 2005).

2. *Database Imaginary*, curated by Sarah Cook, Steve Dietz and Anthony Kiendl, November 2004 at the Walter Phillips Gallery, The Banff Centre, Banff, Alberta. See ⟨http://databaseimaginary.banff.org⟩.

3. From the Web site for *Mobile Scout*.

4. See Beryl Graham's chapter in this volume.

5. See Saper 2001, Pope 2002, and Brown and Graham 1998. Also available at ⟨http://www.newmedia.sunderland.ac.uk/serious/rcessay.htm⟩ (10 April 2005).

6. These three characteristics were first detailed in relation to net art by Steve Dietz (2000).

7. See Manovich 2001, 54.

8. Heath Bunting, "King's Cross Phone In," 1994. See the description in Greene 2004, 34–35.

9. The Web site of RTI can be found at ⟨www.lowtech.org⟩ (accessed 20 April 2005).

10. Christiane Paul in Sarah Cook, An interview with Christiane Paul. *CRUMB* (2001), available at ⟨http://www.newmedia.sunderland.ac.uk/crumb⟩ (accessed 20 April 2005).

11. Conventional 2004 is at ⟨www.america-theglobe.net⟩.

12. Heath Bunting's project "BorderXing" (2002) can be found at ⟨http://www.tate.org.uk/netart⟩ (accessed 20 April 2005).

13. As discussed in Beryl Graham's chapter in this volume. Harwood@Mongrel's work "Uncomfortable Proximity" (2000) can be seen at ⟨www.tate.org.uk/netart⟩ (accessed 20 April 2005).

14. Superflex's project "Superchannel" can be seen at ⟨http://www.fact.co.uk/main/collaboration/projects/super_channel⟩.

15. Nick Crowe's work "Police Radio" (2003) can be streamed and heard at ⟨www.policeradio.org.uk⟩.

16. Amanda McDonald Crowley, in an interview with Steve Dietz. Cited in *Be-Coming Community* by Steve Dietz (ARCO, Part of Coming Communities organized by Peter Weibel, Saturday, 12 February 2005), available at ⟨http://www.yproductions.com/writing/archives/000686.html⟩ (accessed 20 April 2005).

## *References*

Berry, Josephine. 2001. "The Thematics of Site-specific Art on the Net." PhD thesis, University of Manchester.

Bourriaud, Nicolas. 2001. *Relational Aesthetics.* Paris: Les Presses du Reel.

Brown, C., and B. Graham, eds. 1998. *Serious Games.* London: Barbican Art Gallery/Tyne and Wear Museums.

Dietz, Steve. 2000. "Signal or Noise? The Network Museum." *Webwalker* #20 Art Entertainment Network. Available at ⟨http://www.walkerart.org/gallery9/webwalker/ww_032300_main.html⟩ (accessed 10 April 2005).

Dietz, Steve. 2001. "Curating New Media." In: *Words of Wisdom: A Curator's Vade Mecum on Contemporary Art*, ed. C. Kuoni, 49–52. New York: ICI/Independent Curators International.

Frohne, Ursula. 2000. "Illusions of Experience in Screen-Based Art." In *Lier en Boog Series of Philosophy of Art and Art Theory*, vol 15, ed. A. W. Balkema and H. Slager, Amsterdam: Lier en Boog.

Fuller, Matthew. 2000. *Art meet Net, Net meet Art.* London: Tate. Available at ⟨http://www.tate.org.uk/netart/mat1.htm⟩ (accessed 10 April 2005).

Gere, Charlie. 2002. *Digital Culture.* London: Reaktion.

Gillman, Clive. 1999. "The Flashing Prompt: New Media Centers and Regeneration." In *New Media Culture in Europe: Art, Research, Innovation, Participation, Public Domain, Learning, Education, Policy*, ed. F. Boyd, C. Brickwood, A. Broeckmann, L. Haskel, E. Kluitenberg, and M. Stikker, 126–128. Amsterdam: Uitgeverij De Balie and the Virtual Platform.

Gillman, Clive. 2000. "Momentum." Unpublished notes from presentation at Momentum DA2 workshops, Watershed, Bristol.

Gillman, Clive. 2001. "Re: Installing It. June Theme of the Month." In *New-Media-Curating Discussion List* (6 June). Available at ⟨http://www.jiscmail.ac.uk/lists/new-media-curating .html⟩ (accessed 20 April 2005).

Greene, Rachel. 2004. *Internet Art*. London: Thames and Hudson.

Kester, Grant. 2001. "Curating." In *New-Media-Curating Discussion List* (17 April). Available at ⟨http://www.jiscmail.ac.uk/lists/new-media-curating.html⟩ (accessed 20 April 2005).

Kimbell, Lucy, ed. 2004. *New Media Art: Practice and Context in the UK 1994–2004*. London: Arts Council England.

Kluitenberg, Eric. 2000. "Mediate Yourself!" In *Supermanual: An Incomplete Guide to the Superchannel*. Liverpool: FACT.

Manovich, Lev. 2001. "Subject Re: . . ." In *Interaction: Artistic Practice in the Network*, ed. A. Scholder and J. Crandall. New York: Distributed Art Publishers.

Nedkova, Illiyana. 2002. "Presentation." In *Curating New Media. Third Baltic International Seminar, May 2001*, ed. S. Cook, B. Graham, and S. Martin. Gateshead: BALTIC.

Paul, Christiane. 2003. *Digital Art (World of Art)*. London: Thames and Hudson.

Pope, Simon, ed. 2002. *Art for Networks*. Cardiff: Chapter Arts.

Ross, David. 1999. "Net.art in the Age of Digital Reproduction" (also sometimes titled "Art and the Age of the Digital"). Transcript of a lecture at San Jose State University (Cadre), 2 March. Available at ⟨http://switch.sjsu.edu/web/v5n1/ross/index.html⟩ (accessed 10 April 2005).

Saper, Craig J. 2001. *Network Art*. Minneapolis: University of Minnesota Press.

Stallabrass, Julian. 2003. *Internet Art: The Online Clash of Culture and Commerce*. London: Tate.

# II  *Knowledge Systems and Management: Shifting Paradigms and Models*

# 7 A Crisis of Authority: New Lamps for Old

Susan Hazan

Museums are producing electronic scenarios that are enacted both inside and beyond the museum walls; in study rooms, placed adjacent to the collections, in computer kiosks, located in the gallery, and a plethora of activities that are disseminated over the Internet. The digital applications and environments resemble traditional museum practices, in that they facilitate a range of hands-on and minds-on scenarios, but through their flexibility, and relentless cloneability, they are often able to disseminate narratives more resourcefully, both within the museum, as well as beyond the museum walls. Where digital narratives are employed in the gallery, there may be a saving of human resources. When disseminated online, they may save on printed publications, and traditional distribution.

Questioning the role of new media in the museum, this chapter will look at new media, not just for their efficiency, but whether they modify the relationship between the museum and the visitor in any meaningful way. The idea of a single, generic institution that may be termed "the museum" is problematic. In reality the term covers discovery centers and ecomuseums, art galleries, and encyclopedic museums where each kind of institution conjures up a different image, and represents a different kind of experience for visitors. What they all do have in common, however, according to the AAM (American Association of Museums) *Code of Ethics for Museums*, is their "unique contribution to the public by collecting, preserving, and interpreting the things of this world."[1] The series of dialogues, described in the AAM publication, *Civic Engagement: A Challenge to Museums*, that took place between museums and communities across the United States in 2002, critically challenged the museum in its relationship with its public. Drawing on these dialogues, and the prolific literature on visitor research that has emerged from the museum profession over the last decade, this chapter investigates the museum, and its relationship to its visitors in response to the criticisms cited in the publication. The museum in the dialogues was described as "floating above the community," and the idea that the museum's positive self-image is not fully endorsed by the community was raised in these discussions. Questions about authorship and ownership were also broached, indicating a call to museums to present a variety of perspectives, rather than a singular, institutional voice.

This chapter describes new media intervention—not as "new lamps for old," but as new iterations of traditional strategies of display and interpretation. Looking into the museum as an Aladdin's cave of wondrous objects, this chapter considers museum collections as a wealth of potential learning scenarios, but unlike Aladdin's magician who stood outside the flying palace crying "new lamps for old!," this chapter does not prioritize the *novelty* of the electronic experience over the embodied museum experience, nor does it advocate a bandaid solution for the charges outlined by the museum community dialogues. New media applications, like their analog predecessors, have been developed to accomplish a number of institutional goals that extend and interpret the material collections. Together with traditional practices, they facilitate innovative hands-on, minds-on scenarios, extend user-driven experiences for both the local and remote visitors, open up new opportunities to contextualize the museum experience, and generate novel scenarios for life-long learning. New kinds of interactive experiences, not possible before the Internet, are also evolving for remote visitors such as online interrogation across a museum's collections, or a synchronous or asynchronous discussion with the institution through an online forum or via a videoconference.

I will argue that new media applications integrated into museum practice do not seek to either displace or distract from the museum mission, or to collect, display, and interpret the material collections for the visitor. Rather, they serve to enhance and extend the museum mandate in novel ways, and even open up new possibilities for those who may have conceptualized themselves outside of the museum, to be able to find a way in.

## Museum Dialogues

The museum as defined by ICOM,[2] the International Council of Museums, describes first and foremost an institution in the service of society:

> A museum is a non-profit-making, permanent institution in the service of society and of its development, and open to the public which acquires, conserves, researches, communicates and exhibits, for purposes of study, education and enjoyment, material evidence of people and their environment.[3]

Drawing on the ICOM definition this chapter seeks to explore the ways in which museums address their institutional goals to provide a service to society. It also questions whether, on the cusp of the third millennium, *society* in fact expects, or even desires, to receive this service in the way it is being offered. The AAM publication represented a call-to-action to reflect on how the museum community contributes to civic engagement and to evaluate the assets museums bring to the shared enterprise of building and strengthening community bonds. The publication was essentially acknowledging a time of crisis for museums, and the AAM seized this as an opportunity to set out guidelines to readdress community agendas.

Throughout the series of six dialogues between museums and communities that took place in 2001 across the US in Providence, Tampa, Los Angeles, Detroit, Wichita, and Bellingham, participants were asked to examine the role of the museum in their communities, and the ways in which museums contributed to civic engagement. More than half of the participants in the meetings were corporate leaders, educators, social service representatives, philanthropists, politicians, and other community opinion leaders and, together with representatives from the museum, questioned what it means to build and sustain communities (Archibold 2002, 2). The dialogues, according to consultant Ellen Hirzy, illustrated the different ways in which community members viewed the museum, and the museum mission that underlined a capacity to learn and master civil engagement. The criticisms that the discussions revealed were set out in the AAM publication:

- Museums are limited by the public's perceptions that they control knowledge, expertise, and learning, that floats above or passes through the community, and that they are not as "public" as libraries. These perceptions are mixed with enough reality to make them hard to dispel.
- Resistance from within museums is often disguised by references to mission. Instead of reviewing mission to ensure its relevance and vitality, some museums use the perceived limitations of mission as an excuse to avoid the kind of ongoing self-examination and change that connects a museum to a community.
- The same assets that people respect are also liabilities. For example, museum's reputation for accuracy and authenticity inspires trust, but it also endangers doubt about their ability to reflect a variety of perspectives, especially when they are telling the stories of particular cultures.
- Museums' positive self-image as educational institutions is not fully endorsed by the community. Embedded in the image of educator is the attitude that "we know and want to share" with the museum controlling knowledge, expertise, and learning. However, the public sees an institution that devalues their knowledge and what they, too, can teach. (Hirzy 2002, 16)

The booklet, published soon after September 11, 2001, referred to program-based relationships and audience development, and essentially revealed how the museum's attitude to the (U.S.) community could be seen as "token" or "patronizing," especially in connection to activities that community representatives felt were not sufficiently reciprocal or cocreated. Partner organizations also reported how they felt manipulated, exploited, and sceptical of the museum's motives, and called for a rethinking of relationships that would connect the assets and agendas of the organizations with the museum (2002, 16). These were pretty uncomfortable messages to museums and, rising to the challenge to face this crisis, the publication included guidelines for appropriate action, setting out examples of good practice that included, for example, rethinking the front door of the physical museum to make it more welcoming, or forging links with other institutions such as public libraries. Other than integrating public TV broadcasts into museum programs, the AAM publication refrained from suggesting new media solutions in response to the criticisms. I would argue, however, both through my own practice and in an engagement with the literature, that new media practice and the dissemination of the digital museum may be a catalyst for thinking through these problems, and suggest an alternative approach to deal with these challenges in innovative ways.

Clearly the digital museum does not attempt to replace the material object with an electronic surrogate, but instead opens up new possibilities to harness and to enact reciprocal, user-driven scenarios, as well as new opportunities for the remote visitor to be able to interact with the museum. While the implementation of technology in the museum has been theorized through the rhetoric of the displacement of the treasure house,[4] or as a privileging of information over the object (Fahy 2001), this chapter looks to a digital museum that neither replaces nor prioritizes traditional modes of collection and display, and suggests that new media practices and methodologies offer new possibilities for novel hands-on, minds-on scenarios.

## Floating above the Community

The complaint, voiced in the dialogues, that the museum is "floating" above the community recognizes that the museum is not as hospitable as we would like or expect it to be. These kinds of visitor responses are hard to distinguish, let alone acknowledge, but to respond to these criticisms we need to consider both the physical setting of the museum, as well as intellectual access to collections. The museum has often been described as an ideological institution, where the hegemony of the museum is articulated behind the scenes, resulting in a problematic reading of the museum, both for the visiting public as well as for those who might sense this patronizing attitude, but who never even make it through the doors. The political discourses of power and knowledge that run through critical literature on the museum have often been inflected by a Foucauldian reading. Museum scholar, Andrea Witcomb, argues that this critical reflection has been mostly concerned with the museum as "inculcating bourgeois civic values that serve the needs of the emerging nation-state and the dominant interest within it" (Witcomb 2003, 14). All these subtle, or not so subtle, cues and clues suggest a patronizing attitude towards visitors who might then be expected to absorb historically, or politically driven narratives, evident not just through the display of carefully curated objects, but in the very fabric of the institution itself.

While internationally known as "The British Museum," the institution is more about universal culture than anything intrinsically British, and in fact bills itself on all marketing material as "illuminating world culture." The micronarratives played out in the galleries, while impressive in their scope and lavish extravagance, collapse past and present, and are full of slippages in their historical and geographical narratives, leaving no time or space to illuminate land occupation, border disputes, and changing ownerships. The Enlightenment Gallery, "Enlightenment," according to the Web site commentary, "is a rich new exhibition using thousands of objects from the Museum's collection to show how people understood their world in the Age of Enlightenment. Their view was different from ours, but our knowledge has been built on the foundations they laid."[5]

In spite of this retroactive commentary on the founders of the museum, here the monarchy still reins over the nations, and a walk around the museum represents a walk around the world—with Britain at the epicenter. The British Museum, founded by Act of Parliament in 1753, and now governed under the British Museum Act, formally opened the Queen Elizabeth II Great Court in December of 2000. The circular dedication carved in enormous letters in white marble to Her Majesty Queen Elizabeth II was inaugurated on the eve of the new millennium, and runs around the exterior wall of the Reading Room, serving to focus the galleries that revolve around the central hub as a formidable bastion of learning, with the Queen enthroned in her Great Court.

A walk around the collections presents the opaque metanarrative of colonial history, but visitors might be puzzled as to what their role in this scenario might be when they are invited into the king's (or queen's) treasure house. The British Museum, like its sister encyclopedic institutions across the world, offers a range of visitor services, gallery talks, lectures, workshops, and study days, which serve both to make the collections accessible, and to offer an active role for the visitor, who is conceptualized as someone who has come to the museum in order to learn something. In universal museums rooted in postcolonial histories these kinds of anachronous attitudes still run deep and the inculcation of bourgeois civic values are not that easy to dispel (Lyons and Papadopoulos 2002, 5). Describing how these ideologies are inscribed in museum practice, Sharon Macdonald reminds us how "'politics,' in other words, lies not just in policy statements and intentions (though these are important) but also in apparently nonpolitical and even 'minor' details, such as the architecture of buildings, the classification and juxtaposition of artifacts in an exhibition, the use of glass cases or interactives, and the presence or lack of a voiceover on a film" (1998, 3). These intentions, in fact, resonate in the physical constitution of the building, as the British Museum's Great Court exemplifies, as well as in the positioning of the collections, and it is a combination of these spatial and ideological attitudes that determines how visitors are welcomed, or not, into the museum.

As keepers of the material artifact, it falls to the responsibility of the museum management to secure valuable collections from harm. With security issues in mind, new methods and rules of behavioral management are enforced through a strict regime of conduct that prevents disorderly or rowdy behavior. While security management is clearly of paramount concern for the institution, especially since 9/11, demands made by the museum staff on visitors, encrypted by curators, and articulated by museum educators, serve to produce power relations that often alienate the visitor from the museum (figure 7.1).

Museum authorities often require visitors to remain within roped-off areas, and enforce behavioral patterns such as no running, eating, or drinking near the collections, or touching the artworks, as the notice in figure 7.1 indicates. These kinds of restrictions set by the institution may seem puzzling, for example, when there is a prohibition on pens, but not on pencils. One cannot help wondering how a loud conversation might disturb a Rembrandt's oily gaze, and how anything but a reverent silence or muted whispers would in any way distract from the auratic magnetism of an artwork. Behavior is moderated by the watchful guards on duty in the gallery, who also control items that may be allowed into the museum area, such as bags over sixteen inches, umbrellas, etc., as if the visitor is regarded with suspicion as a potential thief or vandal. George Hein, museum educator and proponent of the constructivist museum, recognizes how the physical space of the museum is critical to the successful visit, noting how children on field trips need to "mark out their surroundings" and "take owner-

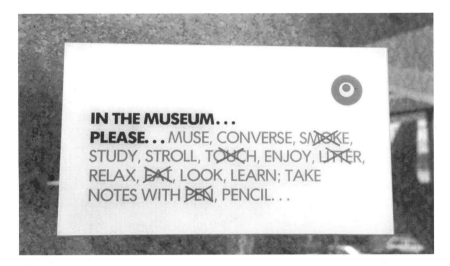

Figure 7.1    Notice to the public in the Hirshorn Gallery, Smithsonian Institution, Washington, D.C. © Hazan, 1999.

ship" before they can engage in the educational programs.[6] Setting the stage for a safe, rewarding, and meaningful experience demands an insightful comprehension of spatial arrangements and social management, but just as crucial is the intellectual access to collections. All the subtle clues and cues that come together to shape the visit remind the visitor that the museum experience is not one to be confused, for instance, with a shopping expedition, or a stroll through the park.

Zahava Doering and her team from the Institutional Studies Office at the Smithsonian Institution have questioned what it means to be a visitor. Looking closely at the museum experience, she describes three interpretive categories to summarize the ways museums conceptualize visitors: as *Strangers*, *Guests*, or *Clients*, each, in turn reframing the relationship between the visitor and the institution. Doering notes how rather than illustrating a sequential progression from one dynamic to the next, the three attitudes, styles, and approaches often coexist across the [Smithsonian] Institution or even conflict within a single institution (1999, 75). The first paradigm suggests that when the museum receives the public as strangers the museum's primary responsibility is to the collection and not to the visitor, emphasizing "object accountability," and views the public (at best) as strangers and (at worst) as intruders. This inevitably produces a setting where the public is expected to acknowledge that by virtue of being admitted, he or she has been granted a special privilege. This is perhaps sensed by the visitor in the opulent setting of the British Museum, where the museum may be understood as taking on the active role of national benefactor as it subsumes the role of absent king (or queen). In this scenario the visitor may "be awed by the sheer magnitude of the treasure" and in so doing may be swept up by the national narrative (Duncan 1991, 95).

According to Doering, where the museum sees the public as guest, this assumes a certain responsibility towards the visitor. There is a sense that the host wants to "do good" which is usually expressed as "educational" activities, and assumes that the visitor-guest is receptive to this approach. Doering describes a third paradigm, that of the status of the visitor as client, where the museum is accountable to the visitor, who is no longer subordinate to the museum. This suggests that the museum no longer seeks to impose the visitor experience that *it* deems most appropriate, rather, the institution acknowledges that visitors, like clients, have needs and expectations, and anticipate that the museum will understand, and respond to these needs. It seems that visitor's expectations are now more complex and sophisticated, and members of the public no longer simply see themselves as passive learners (and perhaps never did). Attitudes to the visitor conceptualized as client are evident in the range of services that the museum now offers the public—competitive cafes and restaurants, book and gift shops, and a hospitable environment from car park to car park. Electronic study rooms, information kiosks in the gallery, and institutional Web sites in the same way perceive of the visitor as client, with his or her own needs and expectations. In the new media application, however, this service enhances and extends institutional goals of the museum mission to interpret and educate, enabling the visitor to be proactively engaged in his own self-directed learning, while providing new intellectual entry points to the collections.

## *A Variety of Perspectives*

Reiterating the idea that museums tend to send out mixed messages to their public, Elaine Heumann Gurian, museum consultant, comments—"we espouse the goal of enlarging our audiences to include underserved populations and novice learners . . . and yet we continue not to accommodate them: we demand that they accommodate us and then wonder why they do not visit our galleries" (1991, 76). Gurian asked this provocative question in 1991, and more than a decade later this issue has mostly gone unresolved. The project described below, *Moving Here*, represents one of the ways museums may reach out to new audiences. Once the archived material that resides in museums is disseminated over the Internet (and in this case thirty museums have pooled resources), these knowledge bases may be harnessed to reach out to under-served populations, serving to extend museum activities in new ways.

Sonia Livingstone, social psychologist, and Leah Lievrouw, Professor of Information Studies, remind us that ICTs (information communication technologies), extend far beyond the obvious arenas of entertainment and the workplace. Banking systems, utilities, education, law enforcement, military defense, healthcare, and politics, for ex-ample, are all dependent on extensive ICT systems for recording, monitoring, and transmitting information—activities that affect anyone who deals with these services or activities (2002, 9). Where the Internet is available, accessing the bank account from home prevents unnecessary waits in bank queues, and scouring catalogs to see if a book is back on the library shelf without leaving our chairs means a better manage-ment of our time. Taking action across the Internet may even save wasted travel and effort, such as purchasing a new chair without even having to leave the house. Like other systems, the field of cultural production is undergoing tectonic shifts, and the print media, radio, and television are already locating their production either directly within ICTs, producing Web sites that indicate, promote, or connect to their activ-ities, or through synchronous applications to connect directly to visitors. One of the ways that the museum is already taking up new schema is evident in the explosion of outreach programs now disseminated through museum Web sites around the world.

Much has been written about inclusion in museum literature, usually in terms of encouraging new visitors to come into the museum, and museums are now turning to electronic solutions as a platform from which to embrace a plurality of voices. *Moving Here* is an online database of digitized photographs, maps, objects, documents, and audio items from local and national archives, museums, and libraries across the UK, which record migration experiences of the last 200 years. It tells the story of Carib-bean, Irish, Jewish, and South Asian people leaving their homelands to move to En-gland over the last 200 years. The lead partner of *Moving Here* is the National Archives (TNA), an association launched in the spring of 2003 when the Public Record Office and the Historic Manuscripts merged to form one organization. The thirty partner institutions that came together to author and manage *Moving Here* recognize that the public has much to contribute to museums, and encourage remote visitors to upload their stories, photographs, and records, or to share the accounts of others who have already contributed their own microhistories to the collective narrative. Museums pro-mote themselves as places of life-long learning, but when it is felt that it is the mu-seum that is controlling knowledge, expertise, and learning this patronizing attitude goes against the grain of an agenda of self-directed learning, and individual agency, that has, literally, the Internet at its fingertips. The kinds of activities that the Web site *Moving Here* provides go some way toward addressing the criticisms that the AAM dialogues broached, specifically towards the kind of curator who has the patronizing attitude that "we know and want to share," discussed in the AAM dialogues. These tools may be harnessed not only to enable self-directed learning, but also to transfer agency from the institution to the individual, opening up the potential for active participation and coauthored narratives.

To return to Gurian's query about visitors not coming into the gallery, we could draw on the French sociologist, Pierre Bourdieu, who suggests that one of the more daunting modes of domination and the most successful ideologies are those that have no need of words. Bourdieu describes a cultural field, an analysis of the economy of symbolic capital, that conceals within itself cultural power relations that shape the sociology of culture. Bourdieu situates artworks within the social conditions of their production, circulation, and consumption, arguing that individuals may access and appreciate art according to the cultural capital at their disposal, and through learned abilities that have enabled them to master the social code. The institution of the museum, and especially art museums, embody powerfully coded experiences, where Bourdieu suggests one needs a studious comprehension [of iconography of the many schools and styles] to gain intellectual access to these kinds of collections. Unlike Bourdieu's connoisseurs, people who sense they don't have the cultural capital to take part in the experience may perceive the museum as a highly intimidating space, and consequently may not wish to come into the museum at all. With collections now presented online, such as the British Museum's *COMPASS*, the British Museum's database of images and text-based resources opens the collections to local visitors in the museum's Reading Room and to remote visitors on the museum's Web site. The user-friendly interface offers new entry points into the museum directly from home, from the office and from schools, twenty-four hours a day—seven days a week, inviting individuals who perhaps might have never even come into the museum in their own lifetime an opportunity to reset their social and cultural codes, and to extend their cultural capital through the online learning scenarios.

Bourdieu first turned his attention to art and literature and the field of cultural production in the 1960s in France, at a time when there were a limited number of terrestrial television stations, and decades before the Internet was to pervade the field of cultural production. The Internet now offers a full range of subject/object positions for the remote visitor and, as Livingstone and Lievrouw remind us, the term "audience," when applied to online visitors, can be understood to mean many different kinds of engagement—*playing* computer games, *surfing* the Web, *searching* databases, *responding* to e-mail, *visiting* a chatroom, *shopping* online, and so on. Etymologically, they argue, "the term 'audience' only satisfactorily covers the activities of listening and watching" (2002, 10–11). Museums, in fact, have taken to the electronic arena with enthusiasm, and already offer vast grazing grounds for both the connoisseur to be able to savor his or her own favorite cultural delight, albeit in miniature and as electronic surrogate, as well as an opportunity for the uninitiated, who may wish to explore these culturally enriched narratives and interfaces by searching, playing, or even shopping in the museum. New media architectures, such as the Tate's online resources, *The Value of Art* (aimed at community leaders) or their prize-winning *I-Map* (a resource for visually impaired people), both enable self-directed, life-long learning twenty-four hours-a-day, seven days-a-week, and are accessible to the visitor from the privacy of his or her own home or office. These kinds of scenarios do not seek to displace the *treasure house*, nor do they necessarily presume to replace the physical visit. Rather, they serve to make the unfamiliar familiar, and offer a method for the uninitiated to be able to reset his or her cultural compass in reciprocal and cocreated schema.

### Hands-on, Minds-on: The Active Participant

One of the key ways the "active visitor" is inscribed in the museum experience is through hands-on interactivity. Quoting Claire Bayard-White (1991), and referring to British Audio Visual Society research, Anne Fahy notes "whilst we only remember ten percent of what we read, we remember ninety percent of what we say and do" (2001, 89). Encouraging interactive participation in this way enhances the learning experience, and hands-on learning may be well served by the electronic applications that can be easily programmed to engage the user, or hard-wired to instigate and monitor a response. Whether analog or digital, the labels, texts, and various display techniques come together to engage not only the hand, but also the mind, in order to enable access to the intellectual scaffolding of the curatorial message.

When active participation is conceptualized as a minds-on experience, visitors are so engaged in the activity that they may even forget that they are in the gallery as they mentally project to other places and other times, in much the same way as we get lost in literary space, or when watching a movie. These kinds of immersive environments, of course, can be greatly enhanced with augmentative technologies, such as those encountered by visitors to the Foundation for the Hellenic World that exhibits material collections which are extended by the new media-driven activities. In this cultural heritage institution/museum in Athens visitors learn about *4000 Years of Hellenic Costume* incorporated in an electronic mosaic from the Byzantine Period, or fly towards a simulation of *The Temple of Zeus at Olympia* aided by a tracking cap and handheld wand. Participants in this 3-D world interact with the projection in a CAVE, an acronym for the CAVE Automatic Virtual Environment, and registered trademark of the University of the Illinois Board of Trustees. CAVE is a room-sized, advanced visualization tool that combines high-resolution, stereoscopic projection and a 3D illusion of complete sense of presence.

Even without the aid of stereoscopic glasses and tracking wands, curators have used other kinds of traditional immersive environments, such as dioramas and installations, to engage audiences in the galley and pique their curiosity. The electronically driven, hands-on, minds-on applications may extend these kinds of experiences in novel ways. I argue, however, that through the incorporation of new media activities, the Foundation for the Hellenic World, and other institutions who turn to new media applications are not in fact prioritizing "old lamps for new," but choose to display material objects in conjunction with electronically driven immersive environments, combining, and enhancing the thematic exhibitions. Working side-by-side in this way, the objects in the gallery, and electronic augmentation of the thematic content, work together to articulate the museum mission to display and interpret the material collections.

## Conclusion

The publication of the AAM dialogues on mastering civil engagement set out substantial criticisms of the museum, and the fact that they immerged from within the museum culture was courageous. The idea that educational activities could be perceived by some "as floating above" or "passing through the community," or that museums were perceived "not as public as libraries," were propositions not often raised in museum discourse, and the AAM publication's stance on these critical issues should be recommended. Victor Burgin described how the master discourse organizes the art field of what is generally thinkable, and in so doing reflects the ideology of the institution:

> The discourse allows the fiercest debates (as proof that it is open and spontaneous) but cannot recognize dissent—in the dispute over the number of angels who may gather on the head of a pin the existence of angels cannot be brought into doubt; in the art world also, to question the existence of certain ideological "angels" is to commit self-exile, to disappear over the discursive horizon. (1986, 159)

In writing this chapter I was gratified to read the AAM's inquiry into civic engagement setting out the possibility that the "museum's" positive self-image as educational institution was not fully endorsed by the "community" (Hirzy 2002, 16). Thinking from within the ideology of the institution, as the AAM dialogues describe, requires the recognition of the existence of certain ideological "angels" and a confidence of institutional mission to face these criticisms. I argue that solutions may be found both through traditional practices as well as from innovative new media interfaces that open up intellectual access to the museum in novel ways. This chapter suggests how new media intervention—not as "new lamps for old," but as new iterations of traditional strategies of display and interpretation—may serve to confront some of these criticisms, without displacing the material object as the central pivot of the museum mission, as the Foundation for the Hellenic World illustrates.

Museums are already actively producing electronic scenarios that are enacted both inside and beyond museum walls, thereby offering new opportunities for those who may conceptualize themselves as outside of the museum to be able to find a way in. Shifting the point of entrance of the personal narrative away from the physical museum to the home, the office, or the school may entice those who perhaps feel lacking in cultural capital to enter into dialogue with the museum, and thus become connected. Once connected, remote visitors may find a way to make the unfamiliar familiar, and when invested with adequate cultural capital they may realize that they have already been initiated into the museum. Through electronic connectivity, remote visitors may also discover a place for cocreated and reciprocal activities, as the *Moving Here* Web project suggests, and realize that the museum values their knowledge and is aware that they too have something of value to contribute.

These kinds of innovative scenarios open up innovative avenues of connectivity between the museum and its audiences. When new media applications are conceptualized as augmentations of the museum mandate (rather than as a distraction from traditional practice or as actively displacing the material object), they can be implemented with confidence to extend the museum mission.

*Notes*

1. American Association of Museums, ⟨http://www.aam-us.org/aboutmuseums/whatis.cfm⟩ (accessed 10 April 2005).

2. ICOM was founded in November 1946 under the auspices of UNESCO, ⟨http://icom .museum/⟩ (accessed 19 April 2005).

3. ICOM Statutes, available at ⟨http://www.icom.org/statutes.html⟩ (accessed 19 April 2005).

4. See Witcomb 2003, 108; Frost 2002, 79–94; and Falk and Dierking 1992 and 2000.

5. The British Museum, ⟨http://www.thebritishmuseum.ac.uk/enlightenment/⟩ (accessed 17 March 2005).

6. Balling and Falk 1980, quoted in Hein 1998, 160.

## References

American Association of Museums. Available at ⟨http://www.aamus.org/aboutmuseums/whatis.cfm⟩ (accessed 10 April 2005).

Archibold, Robert. 2002. "Introduction." In *Mastering Civic Engagement: A Challenge to Museums.* Washington, D.C.: American Association of Museums, 1–6.

Balling, J. D., and J. H. Falk. 1980. "A Perspective on Field Trips: Environmental Effects on Learning." *Curator* 23, no. 4: 229–240.

Bayard-White, Claire. 1991. *Multimedia Notes*. London: Chrysalis Interactive Services.

Bourdieu, Pierre. 1999. *The Field of Cultural Production: Essays on Art and Literature*. Cambridge, UK: Polity Press.

British Museum. Available at ⟨http://www.thebritishmuseum.ac.uk⟩ (accessed 10 April 2005).

Burgin, Victor. 1986. *The End of Art Theory*. London: Macmillan.

Clarke, Peter. 2001. *Museum Learning on Line*, Resource: The Council for Museums, Archives and Libraries. Available at ⟨http://www.mla.gov.uk/documents/muslearn_printout.pdf⟩ (accessed 10 April 2005).

Doering, Z. 1999. "Strangers, Guests or Clients? Visitor Experiences in Museums." *Curator* 42, no. 2: 74–87.

Duncan, Carol. "1991. Art Museums and the Ritual of Citizenship." In *Exhibiting Cultures, the Poetics and Politics of Museum Display*, ed. I. Karp and S. D. Lavine, 88–103. Washington, D.C.: Smithsonian Institution Press.

Fahy, Anne. 2001. "New Technologies for Museum Communication." In *Museum: Media: Message*, ed. E. Hooper-Greenhill. London: Routledge.

Falk, J., and L. D. Dierking. 1992. *The Museum Experience*. Washington, D.C.: Whalesback Books.

Falk, J., and L. D. Dierking. 2000. *Learning from the Museum: Visitor Experience and the Making of Meaning*. Washington, D.C.: AltaMira Press.

Fine Arts Museums of San Francisco. "ImageBase." Available at ⟨http://www.thinker.org⟩ (accessed 10 April 2005).

Foundation for the Hellenic World. Available at ⟨http://www.fhw.gr⟩ (accessed 10 April 2005).

Frost, Olivia. 2002. "When the Object Is Digital: Properties of Digital Surrogate Objects and Implications for Learning." In *Perspectives on Object-centered Learning in Museums*, ed. S. G. Paris, 79–94. Mahwah, N.J.: Erlbaum.

Gurian, Elaine Heumann. 1991. "Noodling around with Exhibition Opportunities." In *Exhibiting Cultures, the Poetics and Politics of Museum Display*, ed. I. Karp and S. D. Lavine, 176–190. Washington, D.C.: Smithsonian Institution Press.

Hein, George. 1995. "The Constructivist Museum." *Journal for Education in Museums* 16: 21–23.

Hein, George. 1998. *Learning in the Museum*. London, New York: Routledge.

Hirzy, Ellen. 2002. "Mastering Civic Engagement: A Report From the American Association of Museums." In *Mastering Civic Engagement: A Challenge to Museums*, Washington, D.C.: American Association of Museums, 9–21.

International Council of Museums. "ICOM Statutes." Available at ⟨http://www.icom.org/statutes.html⟩ (accessed 10 April 2005).

Livingstone, S., and L. Lievrouw. 2002. *Handbook of New Media: Social Shaping and Consequences of ICTs*. London: Sage.

Lyons, K. L., and J. K. Papadopoulos. 2002. *The Archaeology of Colonialism*. Los Angeles: Getty Research Institute.

Macdonald, Sharon. 1998. "Exhibitions of Power and Powers of Exhibitions: An Introduction to the Politics of Display." In *The Politics of Display: Museums, Science, Culture*, ed. S. Macdonald, 1–24. London: Routledge.

*Moving Here, 200 Years of Migration to England*. 2003. Available at ⟨http://www.movinghere.org.uk⟩ (accessed 10 April 2005).

Witcomb, Andrea. 2003. *Re-Imagining the Museum: Beyond the Mausoleum*. London: Routledge.

# 8 Digital Cultural Communication: Audience and Remediation

Angelina Russo and Jerry Watkins

---

This chapter discusses the potential for convergent new media technologies to connect cultural institutions to new audiences through community cocreation programs. This connection requires more than the provision of convergent technology infrastructure: the cultural institution must also consider the audience's familiarity with the new literacies and supply and demand within the target cultural market.

The framework within which a cultural institution can engage in audience connection through community cocreation using new media platforms and new literacy training is termed Digital Cultural Communication. In order to establish this framework successfully, the cultural institution must seek to expand its curatorial mission from the exhibition of collections to the remediation of cultural narratives and experiences.

## Convergent Technologies: Literacy, Cocreation, and Distribution

The issues surrounding the "digital divide" contribute to a familiar and well-established debate which quite justifiably challenges the proposal that convergent technologies are an effective force for popular digital liberation on the basis that much of the global population does not have access to a telephone.[1] In some ways, the digital divide debate parallels some of the important questions raised by the public health debate, and quite often proposes the same solution: universal access. For example, Dan Schuler[2] argued in 1994 for universal access to new technologies in order to support community cohesion, including easy and inexpensive connection to community network services using open standards. Since then, Internet protocols have certainly gone a long way to achieving open access for online communities—although the emerging higher-bandwidth-crossplatform services such as Internet on TV currently remain locked within closed proprietary middleware systems, rather than open MHP-based systems (Carroll 2002).

However, although a policy focus on universal access to online services is worthy, it equates in utility to a public health program that provides medical drugs without the education and information programs to support deployment and usage. Funding Information Communication Technologies (ICTs) for schools which can surround the infrastructure with training programs is one thing; but providing low-/no-cost access in cultural institutions without comprehensive training and support may be less cost-effective. Increased ICT access should be optimized by providing individuals and communities with sufficient training to become familiar with the new literacies required by the new media. Furthermore, the universal access solution misses a major factor which continues to elude participants in the cultural communication debate: providing access does not simultaneously create supply and demand for digital products and services. Specifically, simply investing substantial amounts of capital into community Internet access has little effect on the cultural "market"—it is comparable to building a department store and not inviting buyers or sellers. Cultural institutions should consider the desires of the audience before committing significant public funds to ICT investment.

### The New Literacy

Literacy has historically been a field beloved by educationalists, however, media and cultural studies have recently focused their attention on the "new literacy," a field of studies which describes the skills demanded of audiences as they negotiate the potential of expanding digital services. For example, Helen Nixon (2003) proposes that forces such as the global cultural economy and public policies regarding ICTs are now so deeply embedded in our daily lives—at home, work, and school—that in many places they are shaping a "new landscape of communication" and "new learning environments." Leu et al. (2005) suggest that "The new literacies of the Internet and other ICTs include the skills, strategies, and dispositions necessary to successfully use and adapt to the rapidly changing information and communication technologies and contexts that continuously emerge in our world and influence all areas of our personal and professional lives." The new literacy is a readily recognizable phenomenon: the enormous impact of technology on cultural communication and the resulting systemic shifts should be well known to anyone familiar with the printing press, radio, telephone, and television.

Other disciplines pin different labels on the same concept. In the media studies camp, Sonia Livingstone (2003) describes "media literacy" and examines how its traditional focus on print and audiovisual media has been extended to encompass the Internet and other new media. Media literacy is defined as the ability to access, analyze, evaluate, and create messages across a variety of contexts:

- Access rests on a dynamic and social process, not a one-off act of provision.
- Analytic competencies include an understanding of the agency, categories, technologies, literacy, representations, and audiences for media.
- Evaluation—aesthetic, political, ideological and/or economic (all of which are contested).
- Content creation—the Internet offers increased possibilities for audience content production.

### Community Cocreation

By drawing communities into the consumption and creation of digital content, cultural institutions can take a proactive role in developing new literacy by enabling direct experience of content production and creating environments for community engagement. This initiative is termed "community cocreation" and its implementation is comparatively straightforward: the cultural institution provides ICT infrastructure and training programs, and communities provide original content in the form of narratives, which the community itself produces.

Such community cocreation initiatives are beginning to feature on cultural institutional programs. For example, the State Library of Queensland[3] has provided a mobile multimedia "laboratory" which is made available for community narrative projects in regional areas of Australia. The BBC's "Capture Wales"[4] provides an excellent example of community cocreation: the institution (in this case, the British Broadcasting Corporation) has sponsored a traveling multimedia facility to visit communities and hold workshops in both image manipulation and narrative techniques. This new literacy training has powerful cultural outcomes: the community is empowered to create its own "digital stories," short multimedia narratives constructed from personal photographs and memories. A collection of digital stories can provide a compeling snapshot of a community's cultural identity—the stories are supported by the institution, but created by the community itself. However, media and cultural studies seem to hesitate in their approach to the community cocreation concept. Livingstone (2003, 28) offers a position within media literacy where anchoring content creation may require further research to establish the relation between the reception and production

of content in the new media environment. This includes clarification of the benefits—
to learning, cultural expression, and civic participation—and consideration of the best
means of delivering these benefits. Perhaps it is necessary to look elsewhere for
knowledge to support new literacy programs embedded within community cocreation.
Interaction design is one field which contextualizes the digital creation experience:
Nathan Shedroff (2004) suggests that we consider the meaning of interactivity by envisioning all experiences as inhabiting a "continuum of interactivity," separating passive
traditional media experiences (reading, talking) from interactive new media experiences, the latter being distinguished by:

- The amount of control the audience has over tools, pace, or content.
- The amount of choice this control offers.
- The ability to use the tool to be productive or to create.

If we acknowledge Shedroff's continuum of interactivity, the logical realization
for curatorial practitioners is that the online exhibition should not be a facsimile of
the physical exhibition and neither does the addition of a touchscreen kiosk make a
collection "digital." Shedroff's language—"audience control," "amount of choice,"
"creative tools"—echoes the lexicon of community cocreation. The continuum of
interactivity could provide a simple, yet effective model of how the new literacies can
shift the audience experience from cultural consumption to cultural production.

Although the discussion of community cocreation has focused on tools and training
for new literacy, it is acknowledged that such initiatives may give rise to new social
practices in the same way that the mobile phone has enabled new forms of cultural
communication. It is worth noting that the new literacy training examples mentioned
previously do not attempt to teach evaluation or critical thinking during community
cocreation workshops. So while communities are creating original digital content and
interacting with cultural institutions in new ways, they have yet to contextualize the
new systems to the point where a critical language of cocreative production could provide the skills to analyze what has been created—assuming such skills were missing in
the first place, or even requested.

### Audience Focus

As curatorial practices evolve, curators are providing resources to enable audiences
to engage in cocreation of content. This does not mean that the primary role of the

curator as agent between technology and content, patrimony and program, will cease. Indeed, this role could be strengthened by an audience-focused approach as it will move beyond inclusive policy decisions and provide models of collaboration which allow multiple points of view to coexist. When communities engage with cultural institutions to preserve cultural identity, each party can contribute to the sharing of cultural knowledge and distribution of this knowledge to a wider audience. Cultural institutions can extend the new literacy in this process by providing tools and methods for community cocreation, thus reshaping the process of learning and producing content.

The notion of audience focus has, in some cases, been ignored by some parts of the cultural establishment. This oversight is understandable for institutions which inhabit dense urban clusters offering substantial and diverse audiences. But institutions which represent more distributed cultural constituencies may have to work slightly harder for audience share, and digital community cocreation programs can help the institution to the extent that such programs not only empower ground-up digital cultural creation, they also create new community audiences: imagine the popularity of the museum or gallery that not only empowers the community to create its own digital culture, but also exhibits the work.

This digitally enabled two-way interaction between community and institution is termed Digital Cultural Communication. In Australia, the potential demand for Digital Cultural Communication is high in some segments, notably the secondary education sector. The quality and quantity of the resources available via the country's state library system would grace many curriculum development projects. However, this demand remains largely unrealized: lack of communication between cultural institutions and the education sector means that the significant demand pull that the secondary school audience could exert on cultural institutional "supply" is only beginning to be felt. It is anticipated that a recent formal collaboration between the State Library of Queensland and Education Queensland will start to realize the substantial potential demand within the school system.

The site-specific fixation of curatorship has restricted, to some extent, the strategic deployment of digital programs and it is hoped that community cocreation initiatives will demonstrate to curators how the new literacies can be integrated into the cultural institution's agenda. We need only to look at how the Internet is changing the hitherto site-specific world of personal banking to appreciate the potential of Digital Cultural Communication.

## Remediation: Networks, Space, and Display

### Remediated Networks

One of the key theories which draws technologies, environments, and experiences together is the notion of remediation, defined by Paul Levinson as "the process by which new media technologies improve upon or remedy prior technologies" (1997, 104). Jay Bolter and Richard Grusin prefer to see remediation as the "formal logic by which new media refashion prior media forms" (2000, 59, 273). All three authors confirm that new media does not supersede older media but extends the ongoing process of the reform and refashioning of information. While Bolter and Grusin focus on "immediacy" and "hyperreality" as products of remediated networks, their arguments tend to focus on theoretical aspects rather than a description of how remediation can be considered within a practical experience environment. Acknowledging the medium can only go so far in these cases; as Anders Fagerjørd (2003, 302) reminds us, understanding remediation is as much about different strategies for achieving an unmediated authentic experience. As these authors point out, if "reality" is located in the individual then, by default, it cannot be stable as each individual will interpret reality for themselves in the context that they are in at any given time.

This presents a dilemma for the logic of texts, particularly in genuinely mediated environments such as cultural institutions. As Fagerjørd suggests, to undertake a serious analysis of any medium's text, "a researcher must develop a strategy to find the cultural understanding of a medium's relationship to the real" (2003, 305). Rather than attempt to "seek the real" this discussion of remediation will use specific examples from the cultural sector which demonstrate the remediation of existing networks where the remediated networks, or new physical systems, can operate in tandem with existing networks, thus providing a rich cultural experience for the audience. In so doing, it describes how remediation of narratives and experiences can provide a successful model for cultural institutions. Cultural institutions such as museums and libraries are mediums which deliver messages supported by a network of processes such as collection, registration, publication, display, evaluation, and promotion. These networks can be considered to be remediated through the introduction of new networks, devices, agents, and experiences.

Cultural institutions are a mirror through which institutionalized knowledge is conceptualized in terms of a civic state and defined in global terms. Their appeal is drawn from their ability to engage with social, cultural, and political agendas. The social aspects of cultural institutions include the acts of collecting, exhibiting, and viewing, while the cultural aspects include defining systems of knowledge and linking that knowledge to society at large. Cultural institutions derive political status through acquisition of collections, the management systems that are used to document them, and the public/promotional programs that deliver this knowledge to their audiences.

Just as cultural institutions are civic, social, and political spaces, they are also experience spaces. They present context and content in such a way as to enable audiences to draw meaning from individual experiences. In this way, the cultural institution acts as a continuously remediated environment, positing questions of immediacy and hyperreality, enabling audiences to "make meaning" and draw their notions of reality from access to the remediated network.

For the audience, exhibitions, Web sites, historical sites, theme parks, and education programs interpret the "real" object and communicate through their various media forms. Cultural institutions occupy physical, social, and virtual spaces that are derived from patterns of human occupation and their historic and contextual interaction. Communicative forms, such as exhibitions and Web sites, are social spaces in that they provide places where audiences can gather to navigate and interact with knowledge systems and with each other. The social environments resulting from this interaction help audiences to construct meaning from their experiences. As the social aspects of visiting take many forms, remediated networks present a number of materialities, each of which, while deliberately manufactured, is experienced uniquely each time.

## Cultural Sector Changes

The Web offers an opportunity for audiences to experience, first-hand, the remediated networks of cultural institutions and to interact with narratives, thus creating unique experiences. This interaction transforms the ways in which audiences access and navigate cultural information.

As networked information proliferates, remediated networks provide new programs of cross-institutional collaboration that have the potential to create a shared and trusted cultural heritage network where visitors can follow themes and associations free of the historical collecting patterns of individual institutions (Trant 1998, 107–125). This remediation of network, narrative, and experience makes use of multiple mediums to communicate the values and priorities of the institution.

The remediated networks of cultural institutions are audience focused and revolve around the new literacy required for audience cocreation of content. Remediated networks do not do away with existing systems of curatorial practice but complement existing systems. They offer targeted resources complementary to existing services, for instance, a museum Web site gives a taste of the physical site, while offering in-depth information about the institution that may not be accessible during the physical visit.

So, how can remediated networks serve the cultural institution sector? In the first instance, cultural institutions are concerned with authenticity and quality derived from interpretations of culture from a variety of sources. Remediated networks allow cultural institutions to be seen not as centers of knowledge, but rather as facilitators of networks of reliable, validated information. Jennifer Trant predicted this opportunity in 1998 in describing how the new media landscape presents a place where the traditional roles of author, editor, publisher, distributor, and consumer of information have not only been dramatically altered, but where anyone can become the author or creator of knowledge. Clearly this has become a major issue for cultural institutions, evidenced by the rising number of cocreation programs where communities are brought into the knowledge equation as creators and authors. It is here that Digital Cultural Communication can provide a framework for remediated networks. By establishing systems which support two-way interaction between community and institution, the remediated networks can provide a "reality" that acknowledges and values not only the information coming from a community, but also the community from which it came.

Although the Web has the ability to connect spaces, places, people, and information, in the cultural institution sector it has been used predominately in two discrete groups focusing on services and technology (Fahy 1995, 97–129). The services sector is comprised of services dedicated to the production and dissemination of information products such as database services, libraries, research services, books, journals, or video. Future services in the remediated network can take advantage of audience curiosity by linking information between institutions. Furthermore, by connecting institutions, resources can be pooled to provide literacy training to enable audiences to "make sense" of the increased access to information. The information technology sector produces technology-based goods which enable communities to interact with existing content and to create new content through cocreation programs. Although information has underpinned the work of cultural institutions, the provision of a wide range of information services has yet to be recognized as a major activity (Fahy 1995, 82–96).

Cultural institutions can provide new models for a distributed cultural information space. One of the main reasons for their efficacy results from the chaos and confusion of networked space. Without the traditional visual and spatial vocabulary of communication, audiences are left to their own devices to interpret the "reality" of information they access. As audiences do not have traditional visual and spatial vocabularies for the Web, they cannot rely upon the semiotics of symbols to create the aura of authenticity that has been cultivated by institutions. For instance, early Web models of museum sites used the categorization of objects and their interpretations, the development of autonomous exhibits unrelated to their knowledge and collections, as the basis of the networked space. Many of these sites have now been archived but they showed that the potential of the Web lies, in part, in its ability to link institutions, other collections, and information spaces in general, rather than focusing on a specific museum collection. Early analysis of the Web showed that the experience of connecting a physical visit with a Web visit provided novel interactions which encouraged audiences to return.[5] Our interest is in how we can harness this ability to connect to both physical and virtual spaces as a way of underpinning the new literacy that audiences will develop from their interaction with remediated networks.

## Space and Display

In some of the earliest attempts to describe a virtual aesthetic, Sarah Kenderdine suggested that new media would redefine the treatment of both space and internal relations (1996, 32–42). As institutions were reconsidered, the ways in which museums were used would change, developing deterritorialized models of museum environment.

In the modernist museum paradigm, the geographic address, with its defined real spaces, drew visitors to its doors. These visitors engaged in an interaction with artifact and institution in a personal and physical way. Such engagement led to the definition of cultural experience, providing meaning through authenticity. The connection to reality, with its promise of authenticity endowed the museum with authority. The territory the museum occupied was protected by physical borders, hierarchies of practice, and social/cultural structures derived from its program.

In the deterritorialized space of remediated networks, ascribing meaning to space requires audiences to engage in a type of escape from reality and transport themselves from the physical space within the city to a potential space. Here, the boundaries appear to be more loosely constructed but are, in reality, more tightly controlled by the presenting institution. This model of mediation is the basis from which theme park environments are derived. At the same time, real collections maintain their feeling of substance and authority because of their connection to the deterritorialized environments.

New media technologies allow the convergence of rich content, multimodal communication delivery systems, and the development of new spatial and textual experiences. Visitors are offered an invitation to create the experience of their visit, each visit having the potential to communicate, teach, create, and play with forms. Cultural institutions construct their own narratives through their galleries, and metaphorically through interaction with that space, providing audiences with individual experiences. This interaction is an essential part of the communication between audience and institution.

### Remediated Networks—What We Already Know

While it is tempting to consider the efficacy of remediated networks as a purely contemporary phenomenon, it is worth considering that the remediation of narrative and experience has a long tradition in the physical environment of cultural institutions, particularly museums. Tony Bennett (1995, 195–201) describes how in the modernist museum, exhibitions were "sequences" to tell evolutionary stories, particularly those of the natural sciences. This method, known as "back-telling," was a method of defining a narrative of existence based on the marks left in the environment by particular animals, humans, or activities.

Back-telling provided a performance space where evolutionary narratives could be realized spatially by navigating through a series of rooms, galleries, and objects. Back-telling relied on audiences understanding the museum as a mediator between natural history, anthropology, history, and art.[6] The narratives of the sciences could be described and presented through classification of races, and could be translated through the preservation, collection, and display of objects associated with events in the history of individuals, nations, and races.

Using this method of spatial storytelling, the associated spaces of the museum could correspond with such interpretations by situating galleries specific to a field of study or discipline; for instance, of natural history, mineral exploration, sea life, shipfaring, or prehistoric beasts. Back-telling provided an effective means of communication, for a time, when access to diverse worlds and classes of information was restricted. Viewed in the contemporary context of new media, back-telling can be seen to have provided a remediated network that instructed the public through virtual display, providing an immersive space in which to engage with collections.

Kenderdine (1996, 32–42) suggests that the ideals which underpinned those of the back-telling form were supported by the territorial space of the museum. Galleries were subdivided into paths by circulation systems which represented the social divisions and structures of collected communities. Such paths reflected the hierarchies of privacy and control, which in turn reflected the structure of the institution in a physically diagramed pattern of activities. Such patterns could then be systematically repeated across a number of institutions, thus providing the model for an institutional sequencing of ideals. This network of information spaces provided institutional narratives and supported audience experience by linking knowledge from across cultures and disciplines in a systematic, functional way.

Back-telling as an institutional framework for connecting institution with community is not that far away from the notion of Digital Cultural Communication. Obviously, the modernist museum attempted to control the understanding of narratives and the opportunity for individual experiences by placing limits on the types of behaviors that would be accepted within the building. Even so, the audience was able to construct or make meaning from its physical interaction with knowledge within the institution. Remediated networks, and particularly those that connect institutions with each other, provide a similar functional framework for narratives and experiences to be realized individually.

### Remediated Narratives and Experiences

Another example of remediated networks is the cinema. As one of the closest technologies and representational mediums to new media, cinema presents ways of transcribing practices to discover novel actions, particularly in its manipulation of spatial representation, spatialized narrative, and experience.

In 1996, Michael Wallace asked, "Why go to museums at all?" (107). His peculiarly optimistic view of cultural institutions embraced the notion that given the powerful history of film on the public's perceptions, we could do away with physical environments altogether and deal only with the virtual. Audiences enter cultural institutions with a well-stocked mental film bank and they carry memories of both raw footage and narrative sequences of past and potential future events.

In these situations, audiences cocreate meaning, if not content, by stitching together the images, texts, narratives, and experiences they both bring into the space and are presented with on visiting. In this way, audiences are engaged in a two-way conversation with the institution, a communication exchange that relies as much on the visitors' responses to the space as it does on the content. Hollywood is a master of this two-way interaction, particularly when dealing with narratives that constitute a large component of audiences' historical reality. The plethora of media narratives has already fostered a malleable sense of the past, such that "the mass media is perhaps the single most critical source of popular historical imagination. For many, because cinematic modes of perception seem so real, "movie-past" is the past" (Wallace 1996, 111).

Possibly the most obvious demonstration of Wallace's theory would be the dedication to the blockbuster exhibition. In many instances, such exhibitions are products of deconstructing cinematic process: thus the *Star Trek* exhibition presents the virtual, dematerialized community of interested viewers with the objects of their dedication to deterritorialized space. The special effects exhibition presents the tricks of the trade with reference to the films that made them possible. Such deconstructions of process are the product of mystification of the cinema as a medium of representation.

Each of these archetypal examples provides an environment where the audiences carry maps, places, and events in their heads and the museum presents remediated networks to produce a novel experience. Audiences cocreate meaning through their interaction with these remediated networks, and the experiences they draw from this interaction are as valid and "real" as the objects that inspired them. Klein describes how imaginary maps can replace the authentic place maps, particularly in the examples of well-documented cities such as New York and Los Angeles. Film presents a social imaginary, that is, it uses an environment which virtually represents a real place (New York) but in ways that can only be experienced in the "hyperreal" environment of the cinema. In this way, while audiences cannot know the "real" city, they carry images and memories of it or representations that can create novel experiences. By doing away with reality, the social imaginary induces suspense or a collective memory of a place or event that may never have occurred or existed for the individual, but is equally real as a lived experience.

As the remediated networks supporting the social imaginary are focused on achieving novel experiences, they rely heavily on temporal distortions. For instance, in films based in the Los Angeles western area, the history of the rise and fall of neighborhoods is often replaced with futuristic visions of the city. The way that people interact with these visions is often temporal, that is, gang wars by night, desolate by day, and therefore outside of the useful habitation of the audience. The mechanisms of lighting or lack of lighting are used to represent desolation, violation, and violence (Klein 1997, 16–17).

Just as film has been used to anticipate audience experiences, it has also provided an entirely new space, all-absorbing in depth and movement. Analogous to the remediated network, the filmic imaginary uses a series of devices to shift the viewer from the real to the virtual, each supporting the other. Representations of the actual do, in some instances, replace the lived experience, but as Frederic Jameson proposes, "Representation is both some vague bourgeois conception of reality and also a specific sign system (in the event Hollywood film), and it must now be defamiliarized . . . by a radically different practice of signs" (1991, 123).

Where film can provide a useful example is in the way it can be used to engage audiences in the cocreative environment; active audience interaction creating new meanings as visitors experience objects, spaces, and narratives in unique sequences each time they visit. The cocreative environment does not do away with the historic, social, and political narratives that are embedded within an institution, indeed, the spatial realization of community content validates the cocreative framework. Nor does it deny the knowledge with which audiences enter an institution. Instead, the cocreative environment allows experience to be structured through shared narratives, while new literacy provides the methods to create and share these experiences. Remediated narrative and experience spaces produce visual learning environments based on interpersonal methods of communication and a broader approach to pedagogy. This is why connecting with the secondary sector could provide such rich opportunities: cocreation as a method of audience interaction and community valuesharing. The resulting artifacts produce new types of cultural experiences where the audience is both the reader and the producer, and plays an active role in the remediation of knowledge.

## Conclusion

Exhibition has remained a relatively uncontested site within digital cultural heritage discourse. One reason for this is that while new media technologies have become increasingly important in the rhetoric of museum programs, there is yet to be established a critical literacy for describing the functional link between technology and the experience it affords the audience. Therefore, while interactive and immersive technologies have become commonplace in audience expectations, the way in which cultural interactive experiences are formulated and designed has changed very little. For these to evolve, a framework which allows a two-way interaction between institution and community needs to be developed. Cultural institutions must look at how best to deliver future services that are audience-focused and provide learning opportunities. A new literacy must be devised to deliver interdisciplinary messages and connect content to context. This literacy must privilege audiences in the construction of meaning and address the changing status of audiences in their interaction with the museum as an institution.

This chapter is intended to provide a framework for further development in audience-focused, curatorially managed cocreation programs within the cultural institution sector. It is framed within the notion of Digital Cultural Communication that links communities to institutions through practical, innovative, and technology-derived literacy programs. The aim is to determine the sustainability of convergent and remediated networks, their narratives, and the experiences they afford their communities. In essence, the discussion is consistent with issues of sustainability in the cultural sector and provides a conceptualization for how audiences can "make meaning" with and through the institution in order to create new networks of shared creation and distribution.

## Notes

1. For example, in Thorns 2002, 87.

2. See Schuler 1994.

3. See ⟨www.slq.qld.gov.au⟩ and ⟨www.qldstories.slq.qld.gov.au⟩ (accessed 5 May 2004).

4. See ⟨www.bbc.co.uk/capturewales⟩ (accessed 28 November 2004).

5. See Wallace 1996, 33–54, 101–114, 133–159.

6. Goode, quoted in Bennett 1995, 181.

## References

Bennett, Tony. 1995. *The Birth of the Museum: History, Theory, Politics*. London: Routledge.

Bolter, Jay, and Richard Grusin. 2000. *Remediation: Understanding New Media*. Cambridge, Mass.: MIT Press.

British Library. Available at ⟨www.bl.uk⟩ (accessed 8 October 2004).

Capture Wales. Available at ⟨www.bbc.co.uk/capturewales⟩ (accessed 28 November 2004).

Carroll, Ian. 2002. ABC Digital, "What Has Been Achieved in Year One." Available at ⟨http://www.broadcastpapers.com/broadbiz/ABCDigitalCarroll06.htm⟩ (accessed 19 April 2005).

Fagerjord, Anders. 2003. "Rhetorical Convergence: Studying Web Media." In *Digital Media Revisited: Theoretical and Conceptual Innovation in Digital Domains*, ed. G. Liestol, A. Morrison, and T. Rasmussen, 302–305. Cambridge, Mass.: MIT Press.

Fahy, Anne. 1995. "New Technologies for Museum Communication." In *Museum, Media, Message*, ed. E. Hooper-Greenhill, 82–96. London: Routledge.

Jameson, Frederic. 1991. *Postmodernism, or, The Cultural Logic of Late Capitalism*. Durham, N.C.: Duke University Press.

Kenderdine, Sarah. 1996. "Diving into Shipwrecks: Aquanauts in Cyberspace." *Spectra* 24: 32–42.

Klein, Norman. 1997. *The History of Forgetting: Los Angeles and the Erasure of Memory*. London: Verso.

Leu, D., C. Kinzer, J. Coiro, and D. Cammack. 2004. "Toward a Theory of New Literacies: Emerging from the Internet and Other Information and Communication Technologies." Available at ⟨www.readingonline.org/newliteracies/lit_index.asp?HREF=/newliteracies/leu⟩ (accessed 6 January 2005).

Levinson, Paul. 1997. *The Soft Edge: A Natural History and Future of the Information Revolution*. London: Routledge.

Livingstone, Sonia. 2003. "The Changing Nature and Uses of Media Literacy." Media@lse Electronic Working Papers, London School of Economics, vol. 4.

Nixon, Helen. 2003. "Textual Diversity: Who Needs It?" *English Teaching: Practice and Critique* 2, no. 2: 22–33.

Queensland Stories. Available at ⟨www.qldstories.slq.qld.gov.au⟩ (accessed 26 January 2005).

Schuler, Doug. 1994. "Community Networks: Building a New Participatory Medium." Association for Computing Machinery. *Communications of the ACM* 37, no. 1: 38.

Shedroff, N. 2002. "Information Interaction Design: A Unified Field Theory of Design." Available at ⟨www.nathan.com/thoughts/unified/index.html⟩ (accessed 12 August 2004).

State Library of Queensland. Available at ⟨www.slq.qld.gov.au⟩ (accessed 5 May 2004).

Thomas, Selma. 1998. "Mediated Realities: A Media Perspective." In *The Virtual and the Real: Media in the Museum*, ed. S. Thomas and A. Mintz, 1–18. Washington, D.C.: American Association of Museums.

Thorns, David. 2002. *The Transformation of Cities*. Hampshire, UK: Palgrave Macmillan.

Trant, Jennifer. 1998. "When All You've Got Is 'The Real Thing': Museums and Authenticity in the Networked World." *Archives and Museum Informatics* 12.

Wallace, Michael. 1996. *Mickey Mouse History and Other Essays on American Memory*. Philadelphia: Temple University Press.

# 9 Digital Knowledgescapes: Cultural, Theoretical, Practical, and Usage Issues Facing Museum Collection Databases in a Digital Epoch

Fiona Cameron and Helena Robinson

---

The emergence of the Internet as a ubiquitous global information and communication resource at the end of the twentieth century propelled museums into the digital epoch at a rate unprecedented for institutions that were, in the majority, bastions of traditional collection practices and the empiricist epistemologies that underpinned their foundation.

Museum Web sites have advanced so rapidly that their online presence is now recognized as a distinct genre, with a range of characteristics and user profiles that set them apart from other online entities (Marty and Twidale 2004). Yet, behind the scenes, many museums have struggled to convert their collection documentation into even rudimentary digital collection databases. While many museum Web sites now support online access to their collection databases, the speed of the digitization process has not generally been reciprocated by a review of data quality.[1]

The structure and content of online collection documentation has yet to be questioned more rigorously in both the theoretical and user needs contexts. The challenge to create new collection data models for online collections that are capable of transforming collection documentation tools into effective and sustainable knowledge environments remains. One may also ask how changes resulting from this questioning will reshape museum practices and perhaps lead to unforeseen interpretations of collections.

This chapter draws on the findings of two research projects, *Knowledge Objects*[2] and *Themescaping Virtual Collections*,[3] that explored these issues across the breadth of museological disciplines and online user groups. It provides a synthesis of the data, with a detailed discussion of the theoretical, cultural, and practical implications for museum documentation processes. First, with the findings from *Knowledge Objects*, we address traditional empirical approaches to museum documentation and consider why it is appropriate to reevaluate their legitimacy in the context of the Internet and wider theoretical shifts in the field of museology. This section also explores the ways in

which postmodern and poststructuralist epistemologies and alternative knowledge sys
tems can be incorporated into museum documentation. The aim is to create rich
knowledge environments able to describe, explain, and interpret museum collections
with reference to different historical, disciplinary, cultural contexts and discourses.
Second, utilizing the data from *Themescaping*, we analyze the particular needs of a
range of online museum database users and propose ways in which digital collection
databases can more effectively cater to these different groups. The final and most difficult task in the process of creating a sustainable and feasible online collection documentary form is to reconcile the disparate imperatives of these areas of research. In
particular, it is necessary to consider what changes are required to staff roles in terms
of object acquisition and documentation procedures. In the final part of this chapter,
we suggest viable changes in how museums may think about, construct, and adminis-
ter the collection databases of the future.

## A Changing Understanding of Museum Collection Documentation

Item-level documentation of museum collections has, until recently, been taken for
granted as one of the unchanging givens of museum practice. Other aspects of collection documentation, such as the Australian-based Significance Project and resultant
methodologies for assessing the historical and cultural significance of collections developed in response to recent philosophical debates that question the ability of museum
artifacts to relay unambiguous, concrete information about the past (HCC 2002).
However, the basic descriptive criteria such as "date of manufacture," "maker," and
physical measurements, used by collection managers and curators in documenting
museum objects, have remained largely unaltered. These formulaic descriptions are
rooted in the long-established practices of curatorial disciplines such as art history,
decorative arts, history, science, and technology. Such practices have been perceived
as separate from more "subjective" forms of documentation such as interpretive exhibition text.[4] However, far from being self-evident and unbiased, item-level collection
records represent the primary means by which museums interpret, define, and communicate the significance and heritage value of their objects (Robinson and Cameron
2003). Therefore, what considerations are important in reassessing traditional museum
documentation models in the context of Web-based digital access? More recently, a
social inclusion agenda has been instituted in the UK driven by the new Labour
government. Consequently, museums have been given a greater social mandate and

responsibility to combat social exclusion, disadvantage, and discrimination, and to promote social inclusionist practices in all areas of museum work. In the 2001 report, *Including Museums: Perspectives on Museums, Galleries, and Social Inclusion*, Amanda Wallace, Museums Conservation Manager at the Nottingham City Museums and Galleries, views collections management practices as largely impervious to the challenges presented by the social inclusion agenda. The shaping of collections resources through acquisition and classification procedures is, according to Wallace, neither impartial or objective but rather deeply exclusive (2001, 81). Wallace argues that "much of this exclusivity stems from the historical development of museum specializations—a blinkering and narrow approach to collecting and classification, a lack of meaningful links to discrete collection areas and entrenched attitudes about the value, importance, significance and power of the real thing" (2001, 81). Collection policies structured around distinct collection disciplines, rather than based on a common aim and purpose, minimize the value and limit the meanings of certain objects (2001, 82). Hence, collections management and documentation must also be understood in terms of social purpose:

> Collections management must be understood as a means to an end—not the end in itself. It exists as a process to enable museums to fulfil their broader social remit by maximising the use of collections, giving collections and users a voice and reflecting diverse experiences. (Wallace 2001, 83)

Our research also raised a very important point about the problem of "conceptual fit" between cultural meanings and classification schemes in documentation, highlighting the difficulty of prescribing categories that can be applied universally (Cameron 2000). This was particularly obvious when we investigated issues around access to and the documentation of Maori and Aboriginal collections (Cameron and Kenderdine 2001). One of the main issues for Maori collections is the inadequacy of standard collection documentation, which tends to "place" collections within Eurocentric chronologies, "categories of use and subject." *Taonga* (treasures) have multiple dimensions often incorporating a spiritual dimension, where "boundary making" in a Eurocentric sense raises issues of conceptual/classificatory fit. A Maori curator at Te Papa (Museum of New Zealand) aptly described the dilemma.

> We have problems putting taonga into boxes/classifications. Some are classified under work and industry but they should also be part of a spiritual/religious process. The critical

conceptual base is to consider the relationships between these things and what is the thematic link—more lateral/sideways thinking is required. (Cameron and Kenderine 2001, 31)

By contrast, Aboriginal collections have a different set of problems, one being the lack of provenance information, where cultural meanings were usurped in favor of empirically based documentation processes privileging type and function on entered museum collections during the nineteenth and early twentieth centuries. These now pose constraints on contemporary conceptualization. Considerable effort is required to recover and reconnect items with clan/geographical origins and empower indigenous communities to interpret their own heritage collections.

Even without taking social considerations into account and looking at the issue from a purely technological standpoint, the manner in which digital databases are constructed, and the ways in which digital narratives can work together, enables users to link information in ways previously not possible, thereby calling into question existing documentary structures. The ability to store vast amounts of data, search, retrieve, and distribute information in space, and traverse trajectories of information in new ways sets the scene for a substantial revision of the way information is documented and linked.

Moreover, the need to revisit the fundamental composition of museum documentation resonates with contemporary theoretical discourse. Museological scholar Gaynor Kavanagh (1990, 10) acknowledges that traditional museum documentation practices, founded in empirical rationalism and supporting singular, authoritative historical narratives, need to be reconsidered. Kavanagh argues, albeit in terms of historical research, that:

> The museum databank of objects and a range of other visual material and sound records needs to be understood from a different standpoint, as much more than the raw materials of the historian's craft, but as part, a remainder and a reminder, of cultural expression and social signification where material can have multiple layers of meaning. (1990, 63)

Museum documentation is more than a repository of unadulterated "facts," rather, it constitutes an ideologically and culturally drenched form of text in its own right. While it is impossible to reach a neutral epistemological stance in regard to collections interpretation, one must question whether the understandings of collections privileged by traditional empirical documentation practices have retained relevance or legitimacy in the contemporary postmodern philosophical environment.

The potential to technologically liberate documentation from the empirical model of standardized, linear narrative description to incorporate diverse media and create 3D objects, visualizations, and simulations has major implications for the types of interpretive evidence gathered, recorded, digitized, and created around museum collections. The challenge is to revisit the current epistemological foundations on which documentation is formulated and to consider how diverse cultural and theoretical ideas, for example polysemic interpretive models (ones that recognize the inherent pluralistic meanings of objects), might revise documentation, taking account of these technological potentialities.

### Postmodernism, Poststructuralist Theory, and Collection Databases

Poststructuralist and postmodern paradigms in knowledge creation, as seen through our research and trends in scholarly thinking, fundamentally challenge and undermine traditional concepts about the truth-value of empirically based forms of museum documentation. Although difficult to define as distinct theories, the hallmarks of these approaches signify two broad theoretical positions. First, they reject the fundamental premise of empirical reasoning, namely the existence of objective truth. This notion is instead replaced by the belief in the inherent arbitrariness of totalizing explanations or metanarratives. In this context, any interpretation is seen as a contextually specific construct by a given theorist or author.

Museum theorist and academic, Professor Eilean Hooper-Greenhill, aptly illustrates the shifting nature of knowledge, reason, and the truth-value of particular practices and world views. In her seminal monograph, *Museums and the Shaping of Knowledge*, Hooper-Greenhill explains: "Just as we see the sixteenth-century apothecary's application of pigeons' wings to the patient's chest as both useless and incomprehensible in curing fever, so some of our own everyday actions (and who knows which) will appear equally incomprehensible in the future to others whose knowledge and truth is founded on other structures of rationality" (1992, 10).

Yet, from the mid-nineteenth century, it was the practice of museums to impose a universal ordering structure based on empirical reasoning, to objects in their collections, and these tendencies, evidenced in complex taxonomies and the emphasis on the so-called objective analysis of objects, still persist in a variety of forms in many organizations. Here, a definitive meaning of the past is deemed to lie dormant in material objects and can be exposed through empirical forms of observation, description, and measurement. In the end, one predetermined interpretation of an object emerges usurping all others.

According to this empiricist tradition, museum documentation privileges classification and lengthy descriptions of both physical attributes and the history of an object, rather than so-called "interpretive" text (Robinson and Cameron 2003, 12). Hence, disciplinary classifications, descriptions, and anonymous statements of significance and provenance that are still prevalent in the practices of many museum disciplines have their genesis in the empiricist tradition. Additionally, in contemporary documentation there is an underlying view that object descriptive data, such as measurements, and visual form and meaning and subsequent interpretive text, are distinctly separate fields (2003, 23). That is, the former is an objective entity and the latter, subjective in nature. To summarize, although material collections are no longer given privileged standing as definitive sources of knowledge, legacies still reside in documentation practices.

Like the sixteenth-century apothecary's truth, the empiricist documentary position no longer stands up in a poststructuralist/postmodernist discursive context. Here, the meaning of objects and their significance to the past, to social change, and as expressions of cultural identity is only actualized in the present, and only valid when understood as belonging to the present.

This "new" discursive position and its potential for the epistemological reworking of the role of objects in a museum documentary context was most graphically illustrated in prominent English historian Alun Munslow's critique of social historian Arthur Marwick's book, *The New Nature of History, Knowledge, Evidence and Language* (2005). Here, Munslow rejects what he terms the "reconstructionist" approach to history, whereby historical sources are empirically identified, studied, and assembled to produce an understanding of what "actually" happened in the past (2005). If applied to the museum context, this "reconstructionist" approach views objects as sources of various kinds of historical information, while curators (and others) merely fulfil the role of eliciting this knowledge, rather than creatively contributing to it. The reconstructionist approach can be seen reflected in museum documentation practices that emphasize physical description and other "verifiable" details such as date, maker, technique, style, etc., and one provenance statement of an object (Robinson and Cameron 2003, 20).

In attempting to reposition the role and uses of historical evidence in a context where historical inquiry no longer yields conclusive knowledge of the past, Munslow advocates the acceptance of an "epistemic relativist" approach to historical sources (2003). He argues that epistemic relativism views knowledge of the "real" as derived through our ideas and concepts, including linguistic, spatial, cultural, and ideological compulsions. Munslow goes on to explain: "In acknowledging the sources and the weight of language as an ideologically drenched discourse, epistemic relativists do not deny there is no proximate truth in history but there is more than one way to get at it and, for that matter, represent it" (2003). In terms of museum documentation, this position recognizes the validity of "empirical" records of collection items, but moderates their authority by permitting the inclusion of alternative forms of analysis, documentation (e.g., 3D imaging, sound recordings, oral testimonies, etc.), and a variety of specialist, nonspecialist, and cultural interpretations of object significance.

So, what are the implications for objects as sources of evidence and the future of collection documentation? From the epistemic relativist position, empiricism and other theoretical perspectives become just one of many theories of knowledge in which to place museum collections. Similarly, objects and the "facts" attributed to them are not seen as significant in themselves, but only gain their meaning when placed in the context of an overarching narrative, or "history" (Robinson and Cameron 2003, 21). In other words, the significance of an historical source, or a museum object, is not inbuilt or pre-fixed, but is rather a product of the story that the historian/curator builds around it (2003). As Munslow puts it: "the facts, figures, numbers and record must act as a check on the history we write, but it does not wholly decide the form of our history" (2003). Likewise, for museums, collections act as a form of evidence but do not embody or replicate history itself.

From this perspective, documentation must be viewed in a new way, in that the meaning and significance of a collection item, or its history, must always evade the curator to some extent (2003). Furthermore, objects can be seen as inherently polysemic. That is, they possess the potential to be interpreted in a variety of ways, depending on the nature of their incarnation as museum objects, and are subject to perpetual fluctuations in meaning dependent on various factors, whether cultural, theoretical, disciplinary, institutional, or individual. Hooper-Greenhill draws attention to this when she writes: "a silver teaspoon made in the eighteenth century in Sheffield would be classified as 'Industrial Art' in the Birmingham City Museum, 'Decorative Art' at Stoke-on-Trent, 'Silver' at the Victoria and Albert Museum, and 'Industry' at Kelham Island Museum in Sheffield" (1992, 7). So, an object's meaning, or indeed its classification, is not self-evident or singular, but is imposed on it depending on the position and aims of the museum (Robinson and Cameron 2003, 11).

Taken together, the arguments of Munslow, Hooper-Greenhill, the epistemological stance of poststructuralism and postmodernism, and the principles and practices of emerging social inclusionist agendas suggest that the provision of standard and universal descriptive categories for objects becomes an imposition of an artificial order that fails to acknowledge the polysemy of objects. Systems couched in empiricist ways of thinking hold little regard for emerging user needs and wider social issues. These arguments also call into question the legitimacy of lengthy description as a means of collecting meaningful data about objects, and the authority of the museum to make statements about its collections. In other words, the postmodern epistemological approach questions the validity and longevity of all practices and methodologies (including, inherently, its own) that narrow down the frame of reference for object analysis and interpretation, which happens to be the occupation of all curatorial disciplines.

Digital technologies and information systems offer the potential to promote inclusive practices to a large degree through new discursive, relational possibilities and the ability to store, search, and retrieve vast amounts of data and media. Technically, these capabilities offer the opportunity to emancipate museum objects from narrow and exclusive cultural, disciplinary, and museum-based understandings afforded them through their relegation to particular curatorial areas and institutions. Current documentation, however, has not yet fully engaged these discursive or technological potentialities.

So, how do we bring collections documentation up-to-speed with current thinking, bearing in mind the potential digital technologies have to offer? Here, museums are confronted with a difficult challenge. That is, to provide a conceptual structure and order to disparate collections at documentation level and to acknowledge and utilize previous investments, while simultaneously enabling the inclusion of relevant poststructural and postmodernist paradigms. In other words, museums must devise a way in which the need to provide expert and scholarly information can coexist with an acknowledgment of the fragmentary, arbitrary, and plural nature of object interpretation.

### Integrating User Needs with Digital Museum Collection Databases

Although conceptual imperatives are certainly exerting pressure on museums to rethink their collection documentation processes, these considerations would likely have remained purely academic were it not for the immediate and, arguably, more urgent issue of accountability and the need for museums to meet the needs of the unprecedented number of public users now able to access collection databases online.

In the past, collection databases functioned as internal museum tools for the almost exclusive use of collection managers and curators. Consequently, there was little motivation to alter the standard formats, classifications, and terminologies that characterized collections documentation in paper and, more recently, database forms. By contrast, the presence of collection records on museum Internet sites has opened up database access for a wide range of potential users. Although this move has been touted as the linchpin solution to facilitating expanded public access to museum resources, the exact process for rendering collection databases truly useful and engaging to public users has remained undefined (Cameron and Kenderdine 2001, 31).

Media theorist Lev Manovich argues that each interface has its own grammar of actions, metaphors, and a particular physical interface to knowledge (2001, 73). The first online collection projects, for example, saw traditional museum exhibition and documentation structures directly transposed to the Internet, representing little more than a basic conversion of existing data and methodologies into a digital format. Driven by story metaphors as well as searching interfaces, digital objects are presented via a hierarchical narrative of theme and subtheme.[5] Utilizing traditional museum communication components such as object labels, graphics, and didactic text panels, narratives are presented as a fixed sequence privileging a central theme. Similarly, with most examples, the authority of the museum as author remains intact through prescribed subjects, anonymous narratives, and singular interpretations consistent with the empirical/modernist paradigm of collection organization and interpretation.

More recently, there has been a trend toward more adaptive responses to different usage situations as well as the rethinking of collections data and how it can be put to more productive interpretive ends.[6] Here, authors offer users alternative pathways through collections information while posing greater contextual possibilities around objects through the inclusion of additional multimedia, text-based information, and navigational concepts such as semantic maps. Presenting a more truly postmodern approach to content, navigation, and information retrieval, these concepts empower the user to create pathways through information and incorporate separate modular elements that can be assembled in an almost infinite series of sequences (Sarasan and Donovan 1998).

From an intellectual and scholarly standpoint, most museums and providers of collections management databases have responded to new digital technologies by exploring more effective ways to contextualize collections. For example, the relational capacities of digital technologies have been harnessed in collection automation systems, enabling objects to be contextualized with people, places, events, periods, classification, and multimedia content, with additional fields for the administration and description of collections.[7] Most museum Web sites now contextualize collections according to themes as essays, quantitative data, and digital images (Peacock 2004). Increasingly, tools of navigation and searching employ data visualization models where contextual links and relevance relationships are shown between objects, subjects, and themes, with text-based interpretive essays about the object's context, primary source materials, and media forms such as images, sound, digital video, and 3D object movies.[8] All this, while espousing the historical context of collections—the creation and use of an object and the provision of supporting material such as images and documents— largely fails to include a sustained consideration of user needs, nor does it exploit the inherent plural meanings embedded in collections, although the technical capacity to achieve this exists. Rather than being sourced from databases, much of this information and multimedia content operates as a selection of items curated as an online exhibition via a Web interface.

### The Themescaping Virtual Collections Study

Despite these innovative advances a number of key issues remain. The first, originally raised by Sarasan and Donovan (Cameron 2003b, 325–340), is whether collection databases and exported data presented in the public realm really fulfil the information and pedagogic needs of an emerging community of online users.

This question led to the *Themescaping Virtual Collections* research project in 2000 and 2001, which sought to identify and investigate the motivations and needs of a range of user groups and match these to available and potential searching, browsing, and navigational concepts and content. This entailed profiling user groups, detailing their reasons for accessing online collections, and matching these to current and future access, retrieval, and content requirements, while simultaneously revealing the diversity and breadth of potential uses for these resources. Complementing this, the research also considered future concept and content options, drawing on emerging technical developments and the potential applications for new knowledge paradigms in the creation of innovative knowledge environments, including narrative spaces.

The *Themescaping* research revealed four key user groups. These were curators, collection managers, educators, and nonspecialists. While it is not possible to include a full synopsis of the findings within the context of this chapter, the following provides an overview of the primary concerns, preferences, and similarities between these four groups.

### Curators

Improving research capabilities through quick, reliable access to object records, images, and other forms of collection-based information via intelligent searching and browsing options such as keyword and cross-database searches, multiple indexes and thesauri for subject, function, and classification terms were the primary concerns of curatorial staff (Cameron 2005, 253). Promoting user interests as a searching and organizational principle through personalization as well as fixed sequences, such as question and answer search engines, thematic tours, and collection highlights, was seen as appropriate and useful in expediting public enquiries. These mixed preferences underscore the need, even within the profile of a single-user group, for a range of access points into online collection information. The ability to repackage information to suit the specific needs of a broad range of specialist and nonspecialists, offering both prescribed and non-prescribed navigational opportunities, and incorporating a polysemic notion of object meaning, acknowledges diversity among users as well as the need to present museum collection information according to a range of motivational and intellectual characteristics, beyond that required by professional, in-house users (2005, 255).

### Collection Managers and Registrars

Registrars and collection managers are almost exclusively preoccupied with the tasks of object identification, acquisition, inventory control, and maintaining documentation. Therefore, similar to the curatorial profile, their concerns focus on the increased functionality of search tools such as collection overviews and site maps, graphical representations of browsing pathways, collection percentage statements, the exhibition status of objects, notebooks for storing search results, and language translation (Cameron 2005, 258–259).

### Museum Educators and Teachers

In comparison with curators and collection managers (whose primary concerns are the preservation, documentation, and interpretation of collections), educators and teachers are interested in tools that support a range of learning options along the modernist to postmodern spectrum. That is, from tailored thematic trails and intelligent keyword searches to customizable personalization search engines. Preferences were for traditional hierarchical structures, similar to a book, and timelines and chronologies that emulate familiar learning paradigms, combined with graphically driven interfaces to search, browse, and interpret collections, offering an engaging, interactive experience. Like the two previous profiles, the intellectual characteristics of this range of searching and browsing options aims to predict and recover information according to two conceptual information structures. This includes modernist fixed sequences involving hierarchies of information and linear narratives, where every object has a distinct and well-defined place, existing side-by-side with nonprescribed postmodernist models with multiple relational links between objects and information.

Offering a range of options—the combination of polysemic, object-centered histories, and links to sources and bibliographic information with additional significance hierarchies and narrative-centered themes—was, according to educators, vitally important in supporting project work. These solutions promote two approaches to reading and writing history. The first presents pluralistic solutions encouraging students to engage their own critical faculties on a more sophisticated level, whereas the latter offers a set of knowable "facts."

Virtual and Augmented Reality technologies were also identified as having the potential to deliver detailed, dynamic, immersive experiences in tandem with objects and information in fulfilling learning objectives. "The more realistic the better" was the feedback we received from educators (Cameron and Kenderdine 2001, 30). Scaled 3D simulations of objects, for example, which combine multiple views and object movies have, according to this profile, the potential to allow students to manipulate and explore the form, construction, texture, and use of objects. Similarly, 3D environments, as seen with computer game technology, not only integrate older theatrical/cinematic traditions of illusion with newer ones where the user can freely modify data and select narrative trajectories. They also have the potential to create a contextual ambience around collections and place objects in spatial relationships. A strong pedagogic case was put forward by one participant who stated that "this technique is particularly important for younger children as they need to visually explore the connections between objects, contexts, and stories" (Cameron and Kenderdine 2001, 30). Educators acknowledged the inadequacy of the descriptive power of words as a learning tool, rather emphasizing the need for younger children to engage a range of senses and to experience collections as part of a contextual, immersive narrative.

## Nonspecialist Users

The research of Kravchyna and Hastings (2002) on the informational value of museum Web sites suggests that a large proportion of nonspecialist users access online collections and museum Web sites pre- and post-visiting the physical site. Our findings support this and suggest that potential users value free-choice learning and would access online collections to build on their knowledge base, as a form of entertainment, and to plan visits to museums. Significantly, the belief in the inherent integrity of information a museum Web site has to offer was also a motivating factor in accessing online collections.

Due to the diversity of the profile, multiple skill levels are needed, from simple searching and browsing tools and content for the inexperienced to sophisticated, challenging tools for the initiated, delivering complex rich media to support a range of learning and entertainment needs. Younger members preferred browsing options over specific searches, utilizing 3D space as an exploratory medium along with personalization, theme-generation tools, and "mind maps" with magazine-style interfaces, visual prompts, and interactivity (Cameron and Kenderdine 2001, 30–31). A younger focus group participant put forward a persuasive argument for the use of 3D semantic structures as a tool to enhance the interpretation and communication potential of collections, where "you see relationships and the way they change and interact using the metaphor of 3D space . . ." (2001, 30).

Here, new styles of narrative are created, not only through the exposure of relationships between objects but also through the user's action. This group was less interested in prescribed material, choosing to drive their own pathways through collections and to explore polysemic object-centered narratives with rich streaming media, 3D objects, and visual environments. Solutions such as these attempt to emulate postmodernist principles that are similar to the way our minds normally work, that is, not in a straight line but as a series of networks and associations.

By contrast, older participants preferred traditional exhibition metaphors for structuring information and familiar classifications, such as collection overviews with keyword searches, thematic structures, chronologies/timelines, linear browsing pathways, and searches under known categories.

Clearly the modernist style text is not dead, rather it is emerging in a modified form. It has an important role to play in organizing information according to knowable and easily recoverable information structures. At the same time, online collection users identify the need for a more fluid, malleable structure and the utility of uniquely digital information retrieval tools that will allow them to participate in the creation of multiple narratives and meanings around collections. The next stage of this study, the Australian Research Council-funded *Reconceptualising Heritage Collections* project, currently under investigation, involves a program of research with emerging communities of users combined with interdisciplinary collaborations between educators, curators, and collection managers from museums, and arts, humanities, IT scholars and technicians. The aim is the creation of knowledge-driven uses and novel ways of imparting information and experiences via polysemic, multidisciplinary, and multimedia knowledge environments.[9] This will lead to a greater understanding of the cultural, social, educational, and potential economic returns of digital assets.

### Emerging Trends and Recommendations for Changes in Approaches to Museum Collection Databases Online

The Internet and the digital media it supports are advancing with ever-increasing complexity and unforetold speed. Yet, our research into the theoretical and user imperatives being brought to bear on collection databases strongly indicates that museums must take a long hard look before they leap into the digital realm, with all the resources (both financial and personnel) that this entails.

Generally, digital technologies create a greater emphasis on action and psychological engagement. Spatial representations of data allow new navigational possibilities and relationships between objects to be seen via a 3D metaphor enabling new polysemic knowledge structures to be realized. Reading of flat text will give way to narration, exploration, and highly integrated user-centered interpretive spaces. A recent report by Digicult on the future of digital heritage space suggests that developments in artificial intelligence will allow for more context-sensitive search and retrieval, effective computing through advanced profiling, learning and adaptive systems, speech recognition, and cognitive vision (Geser and Pereira 2004, 36). Personalization will advance to produce more sophisticated user-tuned interaction capabilities and adaptive user interfaces tailored to the individual (2004). In these ways, digital technology challenges the notion of a singular, fixed, homogenous, and authoritative museum voice and lends greater legitimacy to multiple interpretations of objects and the collections where they reside. However, recent research, including our own, also suggests that, in terms of departing from traditional museum documentation methods and embracing everything that digital media has to offer, we should be wary of throwing the baby out with the bath water. Both theoretically and in the context of key user groups, it is clear that effective online museum collection databases will have to present knowledge in a variety of formats, across the modernist to postmodernist spectrum. With a view to reconciling this apparent dichotomy, the final section of this paper offers recommendations for specific areas of museum practice, policies, staff roles, and information-asset-management that warrant attention, providing a catalyst for new directions in the development of collection databases online.

## Museums, Authority, and New User Relationships

Postmodern theoretical shifts and the need to include alternative knowledge systems in collections documentation compel museums to adopt a polysemic approach to the collation and presentation of collection information. Our research has shown, however, that the potentially excessive fragmentary and subjective nature of the information that may emerge out of this process can be intimidating to users. Many users do not want to take full responsibility for the interpretive process and continue to look to the museum to provide trustworthy, authoritative, and meaningful scholarly information. Hence, there is a need to provide solutions to integrate these two apparently conflicting approaches.

One solution to integrating these seemingly incompatible imperatives would entail the curator acting as an expert subject specialist by authoring information and prototyping experiences, or as facilitator by selecting and defining all possible links and trajectories. Knowledge-making on the part of the user would be limited to the selection and recombination of predetermined elements or paths chosen from a menu/catalog/database designed by the curator/designer, rather than the creation of a whole new interpretive framework. Here, the user is conceived as a spatial wanderer, traversing information and freely selecting trajectories and viewpoints, rather than a "passive," directed observer. The user is also conceived as an individual whose underlying thoughts and desires are identified through devices such as user profile interfaces. Although the use of new database narratives is not as directed as linear and hierarchical ones, this is not to say that hyperlinking cannot be used to seduce the user through a careful arrangement of arguments and counterarguments. In terms of authority, users still identify with the curator's mental picture with regard to collections information, a position fundamental to modernist interpretation. According to our user research, authoritative text remains acceptable only if the author is acknowledged.

## Computer Ontologies and Collection Worldviews

Our investigations suggest that two distinct and culturally specific collection world views emerge due to the computer's ontology and theoretical poststructuralist/ postmodernist and modernist structures. The first, in the form of 3D information spaces, presents a model of the world governed by relational connections between things. Here chronologies, historical time, and privileged narratives give way to a flattened interpretive structure, where various viewpoints or sources can be juxtaposed in space, each seen as equally valid (Manovich 2001, 78). The latter, modernist structure continues to act as an alternative to the above. Here the world continues to be organized into hierarchies of information, where one privileged narrative offers a truth statement about the meaning and significance of a particular object or collection ensuring that the museum's interpretive authority remains intact. Chronology also remains the foundational and organizational structure.

However, both conceptual structures can be combined and qualified by each other through the combination of elements that belong to each distinct field. Modernist text can be qualified by hyperlinking to other sources and the use of visible authoring, whereas postmodernist views can be moderated by the presence of authoritative texts to offer some certainty in the interpretive process. Providing a number of ways into information allows for cross-referencing between two conceptual structures, the prescribed modernist and the self-guided poststructuralist/postmodernist form, each operating in a parallel universe.

## Promoting Plurality of Meaning

While many curators are aware of the polysemic contexts of objects, appreciate the merits of other disciplinary and culturally specific interpretive frameworks, and experiment with integrating themes in exhibitions, this approach has been slow to be applied to documentation. In many cases, collection documentation is still structured according to the traditionally rigid disciplinary frameworks specific to curatorial fields such as art history, social history, science, decorative arts, and anthropology.

This discrepancy could be resolved through collaborations between departments/ curators to discuss potential meanings and the significance of selected objects, and the writing of joint significance statements to expose a range of meanings and opinions across disciplinary areas. Alternatively, should the intellectual and logistic challenges of a coauthored approach prove too difficult, museums could consider adding a comments section to object records, open to contributions by curators not directly involved with the object's research and acquisition. A regular bulletin could advertise upcoming or current acquisitions to curators across departments, allowing for specialists in different fields to contribute to an object's documentation. Potentially, information could be added to digital object records by public users via the Internet. This process would allow disciplinary and cultural frameworks to be pulled together while expanding the potential meanings of objects.

However, an important prerequisite to achieving these aims includes, fundamentally, a greater emphasis in museums on encouraging awareness among museum staff of current theoretical and cultural issues, and their repercussions on museum practice. Museum managers need to support a more dynamic culture of intellectualism and facilitate greater dialogue between museology theoreticians, practitioners, the education sector, and the communities from which collections originate. Collections acquisition and documentation should not be seen as a means to an end, but rather as the starting point where value exists in relation to wider social, cultural, and civic goals. Indeed, this reflects a shift from a predominant interest in collections management to wider collections use. This conceptual and attitudinal change will require institutions to acknowledge the need for realistic time allocation to professional education in the busy schedules of curatorial staff.

Moreover, in order to take inclusive practices into account, there needs to be a consideration of bigger issues relating to the way in which decisions about collecting and documentation are made. A wider consultation process is desirable in the recording and research of collections; one that takes into account the information needs and aspirations of the original owners and wider specialist audiences. For whose benefit is collection documentation carried out? Does it have meaning and relevance outside particular disciplines? Is it based on public need and clear outcomes? (Wallace 2001, 83). Further democratizing and regularly reviewing museum processes of value- and meaning-production has the potential to reframe the way many collections are developed and challenge restrictive assumptions about significance and value.

## Polysemic Documentation and a Shared Interpretive Approach

A new polysemic knowledge model invites the opportunity to create categories and associated linkages through documentation that could potentially connect objects with a whole range of cultural, social, historical, technological, artistic, and disciplinary contexts. For example, links to related resources such as primary source materials, bibliographies, Web sites, and current exhibitions, and a provision for user interpretations, could broaden meaning-making options. Furthermore, establishing relationships between objects on the basis of extended fields and additional ontologies could form the basis of retrieval methods. However, developing concepts about how this metadata is to be handled and standardized will continue to be an important issue.

When considering the development of polysemic multimedia knowledge environments the vision of a future semantic Web becomes compelling and possible. Here, semantically related information is envisioned as interrelated islands of knowledge-representation retrieved by software agents that have the ability to understand and reason over information, and filter, select, and retrieve Web resources according to user profiles. Such a level of interconnectivity not only has the potential to reduce information overload but also retrieves a network of highly relational information, multimedia, and objects across domains that are interdisciplinary and intercultural, thus enhancing polysemic interpretations.

Making resources self-describing, retrievable, and presentable based on semantic frameworks, ontologies, and controlled vocabularies will be a challenge. To achieve this, according to the Digicult report, requires the sector along with other domains to construct new conceptual frameworks with strong reference ontologies (for example CIDOC Conceptual Reference) and semantic tags to facilitate the integration, mediation, and interchange of heterogeneous cultural heritage information to form a coherent global resource (Crofts 2003). Once ontological harmonization is achieved and more sophisticated access, retrieval, and browsing systems, including querying systems that relate descriptive, conceptual, and categorical data together are in place, it has the potential to support new paths for the meaningful exploration of resources and networked knowledge-sharing (Geser and Pereira 2004, 21, 36).

### New Approaches to Registration—Extending Thesauri, Nomenclatures, and Glossaries

Our research revealed that problems of conceptual structure and naming within existing thesauri and nomenclatures need to be dealt with for documentation purposes. Ensuring greater intellectual access to collections and enriched data means extending existing thesauri, nomenclatures, and glossaries to reflect the plural meanings of objects. Most basically, this needs to occur during the documentation process in order to create a greater range of search and naming options, thus contributing to and expanding interpretive options.

Currently, institutions require the user to accept the institutional way of organizing information and the meanings ascribed to objects. There is, however, a strong case for the integration into documentation of alternative classification systems that acknowledge indigenous knowledge models.

## Use of Non-text-based Information

A trend away from descriptions and text-based provenanced information demonstrated through user research and existing online collections suggests in the future non-text-based evidence such as video, audio, and 3D simulations will become predominant in the support of cinematic, dynamic, and more immersive experiences in documentation (Cameron 2003, 24). Experience-based and affective virtual reality and visual knowledge-rendering will partially replace other means of knowledge rendering, particularly textual, over the next ten years. VR systems will also be able to connect more effectively to knowledge bases to explore alternative interpretations (Geser and Pereira 2004, 47). Here, we need to consider how objects can be documented at acquisition, especially those deemed most significant, in ways that capture a range of information. For example, this could include the digital recording of significance through comments by makers, users, and donors, the documentation and recording of the pre-museum contexts of objects, and the digital rendering of objects in 3D and object movies.

## Drawing on Wider Museum Information Assets to Best Advantage

In Kevin Donovan's formative paper on the future of collections databases, he convincingly argued for the building of knowledge bases drawing together the wider information assets of a museum into a centralized resource (1997). Another important task is to consider how existing museum data can be directed towards more productive ends in the documentation and interpretive process. Research files, documents, recorded interviews, graphics, audio/visuals, publications, interactives, 3D objects, and educational materials are all currently held in separate files and collections. In the future, all these components could be indexed, managed, and delivered by database tools while contributing to polysemic interpretive models such as relational thematic linkages, multimedia presentations, rich research resources, and multimodal/augmented and real life visits, including context-aware location visits, peer-to-peer, and social network applications (Geser and Pereira 2004, 30). Moreover, in the next ten to fifteen years the development of smart tags—devices to read and connect to databases—will enable the retrieval of this range of collections information in a variety of contexts and modalities (2004, 12). Nevertheless, the challenge of delivering multimedia within the context of constantly evolving digital platforms, as well as digital preservation and migration into new formats, will continue to have major implications for the ways in which collections information can be preserved, accessed, configured, and interpreted in the future.

Additionally, the emergence of more creative content environments, in particular advanced 3D visualizations and multimodal sensory experiences to which this material will contribute, are expected to become cheaper and more persuasive within the next twenty years (2004, 52). According to Digicult, key areas for research and development include information-rich representations of heritage objects, buildings, sites, simulations, and game-like action spaces as well as more sophisticated 3D modeling of objects to a high degree of realism (2004, 30). In a collections context, this, along with the museum's information assets, could potentially lead to the creation of complex cultural interpretations such as highly detailed and dynamic visualizations and navigation environments that have strong popular and pedagogic appeal. These, however, will only have value in a museum context if connected to scholarly-based information, alternative/cultural interpretations and supporting documentation. In any case, as Scali and Tariffi argued, the rethinking of collection management and multimedia delivery systems to provide effective access to such multimedia content will have major implications on how collections information can be configured and elucidated (2001). In addition, tools for expressive scene design, scripting, dynamic rendering, and seamless integration of animated 3D components in augmented/virtual reality environments will become important for collections documentation and in configuring and delivering experiences (Geser and Pereira 2004, 16).

With a universe of applications for new digital technologies opening up for museums, as well as the need to effectively draw together existing information resources, museums will need to consider the creation of new staff roles responsible for the digitization and linking of related data. These new information brokers will be responsible for identifying documentary sources and creating relationships between data in previously unrelated fields or disparate media categories.

## Linking Practice to Contemporary Knowledge-making

A shift from the predominant use of highly prescribed authored information, text-based descriptions, and significance statements to a greater inclusion of interpretive materials around selected significant objects will involve new curatorial roles. Curators may become more involved in bringing together and linking forms of evidence. Here the curator, in conjunction with educators, has the potential in a collections context to become an experience broker. That is, to enable the creation of a range of user contexts and preferences, drawing the user's attention to a specific object, its relations to others, and suggesting routes through information based on specific profiles. These experiences could be captured, stored, and reused in a range of knowledge and intelligent networked environments to support activities beyond Web applications, from smart mobile devices to personal digital assistants (Geser and Pereira 2004, 30). Likewise, the tasks of collection managers may witness a greater emphasis on creating and linking digital resources. Nevertheless, as our research suggests, many users will continue to look to museum curators and collection managers to provide authoritative scholarly information in the form of authored significance statements, narrative-centered histories, and chronological frameworks.

One answer to the problem of workload is to institute a program of documentation where objects deemed as significant, albeit within the museum's frame of reference, get priority standing. Objects further down the hierarchy could be recorded according to a more simplistic polysemic template, where hyperlinks in documentation are used as the primary tool in the creation of relationships and links to sources. At the same time, consideration needs to be given to how significance statements can be written in a compelling and engaging way for lay audiences, without limiting their ability to participate in the cycle of meaning-making.

In order to achieve these aims, curators must learn to think reflexively about the limited lifespan and legitimacy of collection information and their interpretations of objects, while periodic revision and archiving needs to be instituted as part of the documentation process. In addition, museums could endeavor to expose the epistemological, disciplinary, and cultural frameworks in which objects are interpreted through disclaimers, authored text, and linked curatorial essays about disciplinary contexts, the nature of museum significance assessment procedures, and so on. Indeed, this approach to curatorial consciousness is not foreign to many practicing curators. One participant in the science and technology focus group discussion, as part of the *Knowledge Objects* project, commented that "as a curator, I would define myself as a 'knowledge broker:' we have to be aware of notions of pluralism and acknowledge the lack of singular authority of the museum or curator" (Robinson and Cameron 2003, 26).

## Conclusion: Collection Databases in Light of New Information-capture and Delivery Technologies

Perhaps museums have been reluctant to leave the comfort zone of the traditional documentation practices that have underpinned their culture and activities for so many decades. Yet, the Internet and digital technologies are radically different media for communication than any other previously utilized by museums. The technology impels us to rethink basic museum practices because, when incarnated within it, they are already fundamentally changed.

From a contemporary theoretical standpoint, digital technologies and the Internet potentially represent the ultimate postmodern media. After all, these tools are culturally created and informed by current epistemologies. The technical possibilities they offer resonate with the intellectual characteristics of the poststructuralist text, where information can be organized, manipulated, segmented, reworked, and delivered in modular and multifarious ways (Manovich 2001, 131). For example, the ability to retrieve and juxtapose collection data in a variety of ways will enable alternative and sometimes mutually contradictory object interpretations to appear.

Faced with this epistemological uncertainty, it is not surprising that many potential users of online collection documentation, including some museum professionals, still prefer to have the option of following prescribed, museum-authored pathways through the data. Increasingly, the imperative on museums will be to develop a high degree of flexibility and adaptability in terms of policies, procedures, and staff training, in order to enable them to absorb evolving technologies, reassess their cultural role and authority and, importantly, to continually evaluate and respond to the changing expectations of the audiences/users of the resources they provide. Moreover, institutions need to envision themselves as part of a larger collaborative and cross-sectorial network rather than as singular digitization projects, in order to create open systems that can fully participate in the vision of the semantic Web, enable relationship mapping, and produce high-quality metadata that can be brought into a global knowledge sphere. Indeed, this may prove a challenge for organizations that are traditionally insular, bureaucratic, and have relied heavily on established procedures and protocols. Additionally, research is required on the social function of information and knowledge creation, experience profiling, and in "selling" cultural heritage content to a broader consumer market. As

Donald Sanders, President, Institute for the Visualization of History, United States, stated in the recent Digicult report, "The Future Digital Heritage Space:"

> Certainly global access to all the wonderful things that we will be creating is a noble goal, but unless we can demonstrate that millions will demand our content, demand new content and demand new functionality from our content (as from the downloadable music revolution . . .) we will remain a quaint curiosity. Likewise, unless our worlds become more sophisticated, globally accessible, easily navigable, and linked to multiple databases we will not gain attention, the funding nor the credibility we need to build large libraries. (Geser and Pereira 2004, 53)

Our research suggested that nonspecialists and educators are currently largely unaware of the benefits digital access to cultural heritage resources might provide (Cameron and Kenderdine 2001). But, most importantly, in a time of rapid change in technologies, there is a need for advanced platforms, tools, and services to support long-term preservation and for migration technology to enable the transfer of data into new relationship-encoding formats for the perpetual availability of collections information (Geser and Pereira 2004, 53).

The reality is that museums, which in the majority are paid for with public dollars, have no choice but to remain accessible and inspiring to their audiences in order to remain viable. In the foreseeable future, it appears that the only solution for museums is to commit to putting the key issues of digital collection documentation, some of which have been outlined in this paper, into practice through innovative projects in what may turn out to be, with constantly shifting goal posts, an ongoing yet potentially exciting and liberating experiment.

### Notes

1. This problem first raised in Thomas and Mintz 1998, 2.

2. *Knowledge Objects* was an Australian Research Council–funded Sesqui grant with partners the University of Sydney History Department and the Powerhouse Museum.

3. *Themescaping Virtual Collections* was an Australian Research Council–funded project with partners the University of Sydney History Department, the Powerhouse Museum, and Vernon Systems Ltd.

4. See Robinson and Cameron 2003, 325–340, and Cameron 2005, 243–259.

5. Examples include: American Strategy, at ⟨http://www.americanstrategy.org/home.html⟩ and National Maritime Museum, Collections Online, at ⟨http://www.nmm.ac.uk/

collections⟩ (both accessed 13 April 2005); The Paul J. Getty Museum collections, at ⟨http://www.getty.edu/art/collections⟩ (accessed 20 April 2005); and the Museum of English Rural Life collections, at ⟨http://www.ruralhistory.org/interface.index.html⟩ (accessed 20 April 2005).

6. Examples include the EMP Digital Collections, Experience Music Project, at ⟨http://www.emplive.com⟩; Hypermuseum, ⟨http://www.HyperMuseum.com⟩; Revealing Things, Smithsonian National Museum of American History, ⟨http://www.si.edu/revealingthings⟩; and History Wired, Smithsonian National Museum of American History, ⟨http://historywired.si.edu/index.html⟩.

7. KE Software, EMu, KE Software, at ⟨http://www.kesoftware.com/emu/index.html⟩; Willoughby Associates, Multi mimsy, at ⟨http://www.willo.com/mimsy_xg/default.asp⟩; and Vernon Systems, Collection Vernon Systems Ltd., at ⟨http://www.vernonsystems.com⟩ (accessed 13 April 2005).

8. Best practice models include National Museum of American History "History Wired," National Museum of American History, at ⟨http://historywired.si.edu/index.html#⟩; Powerhouse Museum "Behind the Scenes," at ⟨http://projects.powerhousemuseum.com/virtmus/⟩; and National Museum of Australia History browser, at ⟨http://www.nma.gov.au⟩ (all accessed 20 April 2005).

9. The ARC project *Reconceptualising Heritage Collections* is a three-year project, commencing in 2005 with partners the University of Sydney, University of Technology Sydney, and the Powerhouse Museum. These issues were raised as key areas for research in Geser and Pereira 2004.

## *References*

American Strategy. 1998. Available at ⟨http://www.americanstrategy.org/home.html⟩.

Besser, Howard. 1997. "Integrating Collections Management Information into Online Exhibits: The World Wide Web as a Facilitator for Linking 2 Separate Processes." *Museums and the Web 1997 Conference*, 16–19 March, Los Angeles, Calif. Available at ⟨http://www.archimuse.com/mw97/speak/besser.htm⟩ (accessed 13 April 2005).

Besser, Howard. 1998. "The Transformation of the Museum and the Way It's Perceived." In *The Wired Museum Emerging Technology and Changing Paradigms*, ed. K. Garmil-Jones, 153–170. Washington, D.C.: American Association of Museums.

Cameron, Fiona R. 2000. "Shaping Maori Histories and Identities: Collecting and Exhibiting Maori material culture at the Auckland and Canterbury Museums, 1850s to 1920s." Ph.D. thesis, Massey University, New Zealand.

Cameron, Fiona R. 2002. "Wired Collections: The Next Generation." *International Journal of Museum Management and Curatorship* 19, no. 3: 309–315.

Cameron, Fiona R. 2003a. "The Next Generation: Knowledge Environments and Digital Collections." *Museums and the Web 2003 Conference*, 19–22 March, Charlotte, North Carolina. Available at ⟨http://www.archimuse.com/mw2003/papers/cameron/cameron.html⟩ (accessed 13 April 2005).

Cameron, Fiona R. 2003b. "Digital Futures I: Museum Collections, Users, Information Needs, and the Cultural Construction of Knowledge." *Curator* 46, no. 3, July: 325–340.

Cameron, Fiona R. 2005. "Digital Futures II: Museum Collections, Documentation and Shifting Knowledge Paradigms." *Collections: A Journal for Museum and Archive Professionals* 1, no. 3: 243–259.

Cameron, F. R., and S. Kenderdine. 2003. "Themescaping Virtual Collections: Accessing and Interpreting Museum Collections Online." Unpub. ms. University of Sydney and the Powerhouse Museum.

Crofts, Nick, Martin Doerr, Tony Gill, Stephen Stead, and Matthew Stiff, eds. 2003. "Definition of the CIDOC Conceptual Reference Model produced by the ICOM/CIDOC Documentation Standards Group, continued by the CIDOC CRM Special Interest Group, Version 3.4.9, 30 November 2003." Available at ⟨http://cidoc.ics.forth.gr/⟩.

Donovan, Kevin. 1997. "The Best of the Intentions: Public Access, the Web & the Evolution of Museum Automation." *Museums and the Web 1997 Conference* 16–19 March, Los Angeles, Calif. Available at ⟨http://www.archimuse.com/mw97/speak/donovan.htm⟩ (accessed 13 April 2005).

EMP Digital Collections, Experience Music Project. 2001. Available at ⟨http://www.emplive.com⟩ (accessed 13 April 2005).

Geser, Guntram, and John Pereira. 2004. "The Future Digital Heritage Space: An Expedition Report." *Digicult* Thematic Issue (December). Available at ⟨http://www.digicult.info/pages/Themiss.php⟩ (accessed 13 April 2005).

Hein, H. S. 2000. *The Museum in Transition: A Philosophical Perspective.* Washington, D.C.: Smithsonian Institution Press.

Heritage Collections Council. 2002. *Significance: A Guide to Assessing the Significance of Cultural Heritage Objects and Collections.* Canberra: Commonwealth of Australia.

History Wired, Smithsonian National Museum of American History. 2002. Available at ⟨http://historywired.si.edu/index.html⟩ (accessed 13 April 2005).

Hooper-Greenhill, Eilean. 1992. *Museums and the Shaping of Knowledge.* London: Routledge.

Hypermuseum, Starlab, Semantics Technology and Applications Research Laboratory, Belgium. 2001. Available at ⟨http://www.HyperMuseum.com⟩.

J. Paul Getty Museum collections, The J. Paul Getty Museum. Available at ⟨http://www.getty.edu/art/collections⟩ (accessed 13 April 2005).

Kavanagh, Gaynor. 1990. *History Curatorship*. Leicester and London: Leicester University Press. KE Software, EMu, KE Software. Available at ⟨http://www.kesoftware.com/emu/index.html⟩ (accessed 20 April 2005).

Lyotard, Jean-François. 1994. *The Postmodern Condition: A Report on Knowledge*. Trans. Geoff Bennington and Brian Massumi, Minneapolis: University of Minnesota Press.

Kravchyna, V., and S. K. Hastings. 2002. "Informational Value of Museum Web Sites." *First Monday* 7, no. 2 (February). Available at ⟨http://www.firstmonday.org/issues/issue7_2/kravchyna/index.html⟩ (accessed 13 April 2005).

Manovich, Lev. 2001. *The Language of New Media*. Cambridge, Mass.: MIT Press.

Marty, Paul F., and Michael B. Twidale. 2004. "Lost in Gallery Space: A Conceptual Framework for Analysing the Usability Flaws of Museum Web Sites." *First Monday* 9, no. 9 (September). Available at ⟨http://www.firstmonday.org.issue9_9/marty/index.html⟩ (accessed 20 April 2005).

Marwick, A. 2001. *The New Nature of History, Knowledge, Evidence and Language*. London: Palgrave.

McLuhan, Marshall. 1964. *Understanding Media: The Extensions of Man*. New York: McGraw-Hill.

Metropolitian Museum of Art collections, Metropolitian Museum of Art. Available at ⟨http://www.metmuseum.org/works_of_Art⟩ (accessed 13 April 2005).

Munslow, Alun. 2001. "Book Reviews, Institute of Historical Research." *Reviews in History link, Discourse Section*. Available at ⟨http://www.history.ac.uk/reviews/discourse/index.html⟩ (accessed 20 April 2005).

Museum of English Rural Life collections, Museum of English Rural Life. Available at ⟨http://www.ruralhistory.org/index.html⟩ (accessed 20 April 2005).

National Maritime Museum, Collections Online, National Maritime Museum. Available at ⟨http://www.nmm.ac.uk/collections⟩ (accessed 13 April 2005).

Peacock, Darren, John Doolan, and Derek Ellis. 2004. "Searching for Meaning, Not Just Records." *Museums and the Web 2004 Conference*, March 31–April 3, Arlington, Virginia/Washington, D.C. Available at ⟨http://www.archimuse.com/mw2004/abstracts/prg_250000705.html⟩ (accessed 13 April 2005).

Powerhouse Museum. 2001. "Behind the Scenes," Powerhouse Museum. Available at ⟨http://projects.powerhousemuseum.com/virtmus/⟩ (accessed 13 April 2005).

Revealing Things. 1998. Smithsonian National Museum of American History. Available at ⟨http://www.si.edu/revealingthings⟩ (accessed 13 April 2005).

Robinson, H., and F. R. Cameron. 2003. "Knowledge Objects: Multidisciplinary Approaches in Museum Collections Documentation." Unpub. ms. University of Sydney.

Rosenau, Pauline. 1992. *Postmodernism and the Social Sciences: Insights, Inroads, and Intrusions.* Princeton, N.J.: Princeton University Press.

Sarasan, Lenore, and Kevin Donovan. 1998. "The Next Step in Museum Automation: Staging Encounters with Remarkable Things (the Capture, Management, Distribution and Presentation of Cultural Knowledge On-Line." *Occasional Papers on the Value and Use of Museum Information*, Willoughby Associates. Available from ⟨http://www.willo.com/servicesold/article_nextstep .html⟩ (accessed 13 April 2005).

Scali, Gabriele, and Flavio Tariffi. 2001. "Bridging the Collection Management System Multimedia Exhibition Divide: A New Architecture for Modular Museum Systems." *ICHIM Conference*, Milan, 3–7 September. Available at ⟨http://www.archimuse.com/ichim2001/abstracts/ prg_115000625.html⟩ (accessed 13 April 2005).

Seely Brown, J., and P. Duguid. 2001. The *Social Life of Information*. Boston: Harvard Business School, 2000 Stuer, P., R. Meersman, and S. De Bruyne. "The HyperMuseum Theme Generator System: Ontology-based Internet Support for the Actual Use of Digital Museum Data for Teaching and Presentation." *Museums and the Web 2001 Conference* 14–17 March, Seattle. Available at ⟨http://www.archimuse.com/mw2001/papers/stuer/stuer.html⟩ (accessed 13 April 2005).

Thomas, Selma, and Ann Mintz. 1998. *The Virtual and the Real: Media in Museums*. Washington, D.C.: American Association of Museums.

Wallace, Amanda. 2001. "Collections Management and Inclusion." In *Including Museums: Perspectives on Museums, Galleries and Social Inclusion*, ed. J. Dodd and R. Sandell, 80–83. Leicester: University of Leicester, Department of Museum Studies.

Willoughby Associates. 2002. "Multi Mimsy." Willoughby Associates. Available at ⟨http:// www.willo.com/mimsy_xg/default.asp⟩ and via Vernon Systems, Collection Vernon Systems Ltd. Available at ⟨http://www.vernonsystems.com⟩ (accessed 13 April 2005).

# 10 Art Is Redeemed, Mystery Is Gone: The Documentation of Contemporary Art

Harald Kraemer

"You can be a museum. Or you can be modern. But you can't be both." For more than three years this statement by genius Gertrude Stein was our motto. Written on a piece of paper, fixed on our wall, this wisdom was a stimulus and a wake-up call for us fools, who in spite of warning sought to explore the dark secrets of the documentation of contemporary art. The main aim of this project,[1] undertaken in Cologne, was to make a practical contribution to the development of strategies and structures for a methodology and for the documentation of modern and contemporary art. Furthermore it was to demonstrate, by means of the specific traits of modern art, the necessity of new documentation procedures and digital technologies. Of central importance are the multimedia electronic technologies, with which traditional static methods of aesthetic documentation can be significantly extended. The diverse prerequisites and specific demands of contemporary art involve a changed methodology of analytic access and documentation. Thus the task of the subproject reads: In what respect could a renewed methodology of documentation question contemporary art meaningfully? And what roles might interactive possibilities of the digital and multimedia technology play here?

Meanwhile, the project remains unfinished under circumstances of budget redistribution, and the questions have changed since I started to reflect about documentation in the early 90s (1994, 93–99). How can we document our own experiences for future generations and not only the facts? How can we visualize the invisible mystery and timeless charm of art objects? And, reflecting the different levels of realities, how can we empower users to rely on their own critical perception? Time will show. This chapter discusses the complexities of contemporary art documentation and some of its relevant aspects. After a short summary about the state of art, I will address a number of problems with the perspectives.[2] It is not my aim to give an overview of the various current theories, the postmodernism discussions about media-archaeology, iconic turn, performative turn, etcetera, etcetera. Considering some case studies, my text will simply provide my perspective on the current discussion.

### State of the Art

A deluge of new information is produced daily by the art trade, art criticism, art magazines, artists, scientists, universities, museums, and the machinery of exhibitions. On one hand, there's the discursive occupation of art, on the other hand, we find the dandy-like discourse celebrating itself. The discourse, the *mise en scène*, the documentation, the communication, are transdisciplinary, interdisciplinary, intermedia, and hypermedia forming a multifunctional network of information, media, and communication structures (Zembylas 1997). This is one side. On the other side we find the provocations of contemporary art. Recent art is open, transient, interdisciplinary, multimedia-based, processual, discursive, and dependent on concept and context, as well as increasingly aimed at interactivity with the recipient. Because contemporary art is so diverse it is in need of documentation in a wider sense, in case it should at some point be subjected to scientific questions and tests of authenticity by those who were not present at its conception and presentation. It is symptomatic that descriptions and visualizations of an eventful work of art seldom give a concrete and characteristic presentation of it. With the traditional methods of documentation, which are still valid as the basis of scientific art research, modern works of art are insufficiently recorded. One of the fundamental differences between modern works of art and traditional works of art is the transient process (Ketnath 1999, 369–372). Documentation of the transient nature of modern works of art (e.g. performance, actionism, kinetic objects) can only be archived by means of process-related media. Time and space-dependent, concept-related works of art need to be reconstructed in their wholly functional context. Interactive media art, for instance, at times needs supplementary descriptions and rudimentary analytic penetration when constructing neutral documentation, otherwise these dimensions would be lost and could hardly be reconstructed. Digital and multimedia technologies provide the necessary capabilities for documenting contemporary works of art. On the one hand, there are huge capacities for storing text, especially image and film data; on the other hand, modern database management systems as well as hypermedia links and metadata offer new possibilities of accessibility. These capabilities seem custom-designed to contribute to the creation of Digital Collections.[3] After testing various text and image database management systems, as well as thesauri and rule types used in museums of modern and contemporary art, it seems that the current systems—structured for the traditional documentation of individual traditional works of art—often lag behind the demands made by contemporary art.[4] The experiments to try to extend database systems used for traditional works of art to incorporate contemporary art have failed. The existing text and image database management systems are mostly unfit for contemporary artworks. Structured for the traditional documentation of individual traditional works of art they are mainly descriptive, their strength being the accessibility of iconography. But faced with the challenge of universality and richness of artistic expression, verbal descriptions reach their limits. How do you describe a room-related color painting by David Tremlett? Or institution-related installations by Gerwald Rockenschaub, which confronted the visitors with open windows? Or the falsification of a museum tour produced by an artist who plays the role of a museum guide like Christian Philipp Müller? Or the moments of shock on discovering living asylum-seekers sitting in Santiago Sierra's cardboard boxes in Berlin? Or the interactive dramas and masterpieces of Net art shown in the SFMOMA exhibition 010101 and on their Web site? In the classical definition of museology, documentation has the principal duty to register and catalog the objects of the collection in their totality.[5] But documentation is a "category" and part of a process (Fluegal 2001, 19–29).

It is also the basis, the aim, and the result of research. Anne Bénichou (Department of Visual Arts, University of Ottawa), in her research on the documentation of contemporary art for the Canadian Heritage Information Network, talks about the "delicate complexity" of documenting contemporary art and suggests using conventional methods and reduced data fields to quantitatively solve the problems (1994). Describing the fast progress of communication and multimedia technology used by artists in the last ten years, John S. Weber of SFMOMA speaks of "the beginnings of a tectonic shift" (2001, 15–23). Nicholas Serota, Director of the Tate Gallery, in his book, *Experience or Interpretation* (1996), discusses the process of transformation taking place in art museums unleashed by unconventional pieces of art. Austrian media artist Peter Weibel describes this process in museums as a "phase out model," which will be followed by a placeless digital discourse (1998, 22–27). These four positions outline the challenges confronting art museums.[6]

In this sense the documentation of contemporary art requires rethinking and the implementation of strategic procedures. Documentation has to become a category and a strategy, which must be used in an active way by the researcher, curator, registrar, but also by the artist, and user. In search of the historical truth and the reconstruction of original relationships, documentation in a new sense will be an integral part of the work of art. The artist-registrar relationship must become an equal partnership which enables the development of diverse strategies for the optimal documentation of works of art. The instruments of digital and multimedia technology require specific art efforts. The changing role of documentation poses an entirely new set of problems and perspectives. These will have consequences for the questions asked by later generations of researchers and, due to the increasing presence of digital media appliances in museums, will also change the viewing habits of visitors to the museums.

## *Problem of the Quality of Information*

The value of information disappears. As there is no time to reflect on information and to discern its relevance, we are producing endlessly and registering everything. We are captured by the "tarantula"[7] of information technology. The accumulation of information with the aim of comprehensiveness but without a sense of the whole may be imposing, but it leads us nowhere. Technological developments are such that artists, registrars, curators, researchers, and viewers alike get carried away. The requirements and questions at stake are by no means clear. The computer has created new forms of work. It simplifies some working processes and administers a large quantity of information, but it gives no information about values, about the relevance or the quality of information. Digital data makes no distinction between Duchamp, Duchamp's urinoir, the photographer of Duchamp's urinoir, Duchamp's own words about his urinoir, and the art historian gibberish about Duchamp's urinoir. So who is responsible for the weighting of information? What can be asked, what is actually asked? How much of all this is dead knowledge? What knowledge must be retained? How does information differ in substance? In an exchange of ideas about the challenge of understanding art, Austrian painter Robert Lettner suggested that:

> The original art work is what it is. Only the interpretation is multifaceted. There is always some reaction: new ideas, new fantasies. Surely, a piece of information only triggers sensory effects if associations can be formed with it or on the basis of it. So works of art that flicker across the screen are to be perceived more as instant art, whereas an ink drawing still has something mysterious about it. In most cases, though, people want the information only in order to be able to experience these mysteries. (1998, 15–23)

## Problem of Research

Looked at more closely, the crisis around keywords and terminology is actually a crisis of research, or, to be more precise, of the progressive separation of exhibition-oriented research in museums from problem-oriented research at universities (Boersch–Supan 1993, 51–68). Object-oriented basic research that would lead to inventories and stock catalogs, so badly needed, are often relegated to the realm of fiction. To educate the sector about different views of research it is necessary to reform and reorganize teaching methods and the branches of study. In German-speaking countries you can count the university courses for museology and museums informatics on one hand (Scheffel 1997). Instead of painstaking, in-depth investigations that could potentially result in knowledge, comparisons are amassed, and artworks, selected almost at random, are used to support or embellish wild theories. Simple statements and comparisons are mistaken for research. Simple descriptions and the copying of press releases are also mistaken for art criticism. The helplessness or rather impotence of art science, museum education, and art criticism in dealing with contemporary art is daily illustrated in exhibition catalogs and articles in art reviews. Restrictions in classification and a narrow terminology lead to an impoverishment of the language. The computer forces theoretical diversity into narrow formulas. Abstract works of art rarely require precise descriptions. Rather, terms for composition, material, technique, and, most importantly, subjective impressions and knowledge based on experience are more useful. What we need is a dynamic terminology, not artificial, not encyclopedic, but work-immanent and growing out of the artistic concept like liquid architecture. The terminology needs terms extracted from texts, interviews, and concepts of the artists themselves and a handful of articles by art critics, art historians, and art connoisseurs. We need quality by selection and not quality by quantity, as the numberless miniarchives—documentation as a self-fulfilling prophecy—collecting the same press releases and invitation cards have shown us every day.

## Problem of Reality

As researchers' predilections and questions change with time, the relation between art and reality and the artist as a person has changed. The question of the medium is rarely of interest nowadays. Art is a process that can produce a specific relation to reality, i.e. the viewer is given an opportunity to experience artistic realities and to experience reality artistically. Traditional categories are no longer valid, even though they continue to exist. Other special fields, such as genetic engineering, cybernetics, artificial intelligence, cyberspace, as well as ecology, sociology, and politics have been introduced.[8] While media artists provide a stimulating impetus and influence technological development, the museum—especially the museum of contemporary art—develops more and more into a keeper of meaningless fragments incorporating artistic conceptions of all sorts. Performance and action relics, spatial installations taken out of their original location and context, videos, computer animations and conceptions to be translated into reality by the viewer, a photograph left behind after an installation in a public space—all of these are caught and misrepresented by the outmoded concerns of the museum environment.

## Problem of the Future of the Museum

The apparent timelessness of classical ancient works of art is confronted by the transience of contemporary works. Works of contemporary art, and the texts that document them, are changing, often twice a week. Some of them are timeless, all are transitory. Provocative works pass away. After fulfilling their guiding function, most contemporary works of art go into storage. Mostly we're producing ephemeral texts, catalogs, and multimedia products. Two months after publishing nobody needs them. Museums are like theaters or movies with daily moving programs and fast-changing events. The traditional understanding of chronology disappears. Technological development and the media combine past, present, and future. The world becomes a museum and art criticism is partly responsible for that (Jeudy 1987). The art leaves the studio. There are no boundaries. Anyone becomes an artist. Or a curator. Or a critic.[9] For Tate Gallery Director Nicolas Serota "the best museums of the future will seek to promote different modes and levels of interpretation by subtle juxtapositions of experience" (1996, 54). This is right, but nowadays, one of the main tasks of the museum is to critically deal with the possibilities of electronic reproductions (Reifenrath 1997, 27–40). The meaning and importance of museums are changing in the age of digital revolution. Notwithstanding its present main function as an arthouse cinema, tourist attraction, and boutique, the museum is still an institution of enlightenment in the classic sense, a school of all senses, and, now, more than ever, it is obliged to guide the visitor towards critical viewing and experience.[10] As Serota points out: "Our aim must be to generate a condition in which visitors can experience a sense of discovery in looking at particular paintings, sculptures, or installations in a particular room at a particular moment, rather than find themselves standing on the conveyor belt of history" (1996, 55). The museum of the twenty-first century—according to discussions arising from a conference about the changing meaning of art museums, held in Bonn, 1996[11]—will have to be a constructive counterpart to the deluge of reproductive media images, and it will also have to consider itself an interactive transmitter actively influencing the opening of electronic elbow spaces and the creation of new visual codes. But even though technology and the related industry seem to work for education, the focus is still on play. Will multimedia technology be able to fulfill the requirements of the museums of the future?

## Problem of the Change in the Meaning of Documentation

The term "documentation" is not only used differently in the field of contemporary art. Museology, conservation, art science, and the art trade are all in the process of redefining the process (Fluegal 2001, 19–29). But with regard to recent art the consequences are complex and have a lasting effect. In the documentation used by the restorer, for example, there exists a patchwork of divergent opinions. Part of the strategy of the conservational staff as it is shown in the INCCA[12] (International Network for the Conservation of Contemporary Art) project is to interview living artists; but the restorers are mainly interested in questions of conservation, so that these interviews have little benefit for documentation generated by art science, museology, or the art trade. It's just one example from many to show the juxtaposition and the noncooperation between different committees working in the same field. In his paper, "The Classroom of the Future," Michael Greenhalgh, from the Australian National University in Canberra, outlined a clear view of the problems: "Perhaps we were too optimistic about the speed of progress; perhaps insufficient attention was paid to the costs of infrastructure. Certainly there has been less cooperation than would have been either efficient or healthy: people still reinvent the wheel on their own—and this, ironically, in the age of the greatest cooperative vehicle yet seen, namely the Web" (Greenhalgh 2002). So it's absolutely necessary to foster cooperation and communication with other committees working in the same field. The definition or redefinition of documentation—or whatever the name in future will be—has to take place within a community. To stimulate, coordinate, and lead a discussion about these points will be one of the main challenges of museology in cultural studies in the future. Due to the possibilities of digital and multimedia technology, the changing styles of contemporary art, and the change of semantic import of the art museums, the "means of documentation" (Bretwieser 1999) changes in such a way that it becomes more a strategy for communication with different user groups. Successful documentation of a work of art, dependent on time and space, contributes considerably to a continuation of high-quality artistic presence in the art world, exhibitions, and museum research. This is also an indication of why documentation has a new, important position. New criteria must be discussed and found because documentation by means of new technologies will be part of the interpretation of recent works of art.[13]

## Perspectives

On June 6, 1856, after his first experience of rail travel, French painter Eugène Delacroix wrote, "Seeing means nothing anymore. You arrive to depart" (1997). Behind these simple words one detects a warning, that technology will change our perception fundamentally and that we're going to lose the ability to reflect on different levels of reality. Nearly 150 years later—shortly after the labor pains of the digital revolution—we're in the eye of the hurricane and it offers a chance to construct new worlds of perception using our current cultural knowledge and the technical possibilities offered by cyberspace, multimedia, and interactivity. Documentation is changing from a passive retrospective to an actively participating form for archiving contemporary works of art. Of central importance will be the multimedia and electronic technologies with which traditional static methods of aesthetic documentation can be significantly extended. The diverse prerequisites and specific demands of contemporary art involve a changed methodology of analytic access and documentation. And which role do the interactive possibilities of the digital and multimedia technology play here?

## Documentation as a Work of Art

In some cases documentation as a result of an artistic action is the work of art. Then it is necessary to show the difference between the documentation as a work of art and documentation in a museological sense, i.e. the documentation of the documentation of a work of art.

Example 1: In April, 1994 the American artist Andrea Fraser undertook a project with the Generali Foundation in Vienna (Fraser 1995). In accordance with her preliminary prospectus for corporations, the project was conceived as a service to be provided in two phases, the first defined as interpretive and the second as interventionary. The results of this project have been published in an exhibition and as a report. Shall we put this report into the library or in a showcase?

Example 2: For a retrospective show of her work the French media artist Marie-Jo Lafontaine documented her video installations from 1979 to 1999 in the form of a CD-ROM (1999). It is structured according to video installations, topics, interviews, chronology, and a list of her exhibition shows. The interactive network of this CD-ROM—composed of images, film stills, video sequences, talks, texts, and music—is the best strategy for documenting the multimedia aspects and the intercontextuality of her work. Via a Web-link the visitor can access the actual information on Lafontaine's Web-archive. So this approach is a perfect solution for art historians as well as curators, art dealers, and collectors of her work.

### Somewhere between Construction and Reconstruction

The reconstruction of the past through documentation, like the famous Merzbau of Kurt Schwitters (2005), the construction of spaces of experience like static or permanent installation, and the construction over a specific period of time, such as a performance, become specific items in the documentation of recent art works. Freezing the moment and documenting the cosmos of the whole is the challenge for this strategy of documentation. Anna Oppermann's installation "Umarmungen, Unerklaerliches und eine Gedichtzeile von R.M.R." is a long-dated conversion of annotated fragments of newspapers, poems, and texts as well as her own drawings.[14] The art historian Carmen Wedemeyer, and Martin Warnke, the godfather of German cultural informatics, have designed in HTML a hypermedia image-text-archive which allows navigating in the micro- and macrocosmos of Anna Oppermann. The whole installation is divided into different groups and each group has several pieces (figure 10.1). An index has enabled the reconstruction of this complex installation and so it's possible to retrieve the comments of the artist herself as well as those of the art historian.

### Communication

Communication is an integral part of documentation. There is a need to document the artist's own interpretations, the methodological strategies of art science and art criticism, and the behavior of recent recipients of the information. This means that the artist's own opinion is a powerful tool for the interpretation of a work of art as well

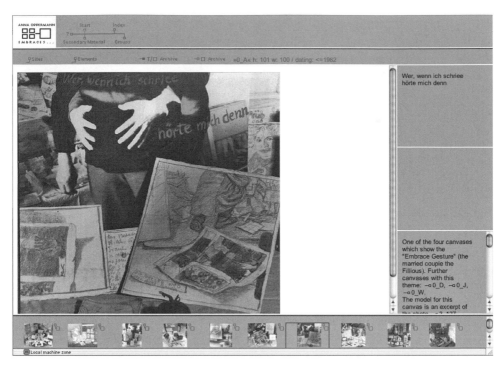

Figure 10.1   *Anna Oppermann, Umarmungen, Unerklaerliches und eine Gedichtzeile von R.M.R.*, hypermedia archive, CD-ROM. © 1998 Universitaet Lueneburg, Stroemfeld Verlag, Frankfurt/Main and Edition Lebeer Hossmann, Hamburg—Bruxelles.

as the statements of visitors regarding the work. To communicate the artist's own statements, the following strategies are usual.

*Example 1*   In the CD-ROM "Improvisation Technologies" the choreographer and dancer William Forsythe (1999) provides insights into his dance philosophy and methodology. This brilliant example of "oral history" combines theory and practical experience by using the artist as a teacher and interpreter.

*Example 2*   The artist presents the information in the form of a "film trailer." This is an effective way of presenting the artist's statements. But using "short cuts" is also a trick to create a personal atmosphere. Other examples of this technique are the collector couple Harry and Moo Anderson,[15] or Roy R. Neuberger.[16] Both CD-ROMs bear the signature of Peter Samis, the Program Manager for Interactive Educational Technologies at SFMOMA.

*Example 3*    The strategy of using "hypertext" structures has been mentioned with the example of Anna Oppermann.

Registrars and art historians need to be involved in the process so that they can document the essentials. It's not helpful to archive every press release or article but it's absolutely necessary to document the soul of the work of art. The meaning of a work of art changes over time. The questions we have and the answers we get show that a work of art isn't a dead object, but part of an ongoing conversation in an evolving society.[17] A main part of the documentation of contemporary art should be the communication between the different target groups on one side and the different meanings and methodologies on the other. It's essential to establish communication between these groups because the work of art is part of an existing discussion. Larry Friedlaender, codirector of the Stanford Learning Lab, has shown that these problems are not art-specific, but essential for the common transfer of knowledge.

"How are we to analyze the structure of such a hybrid experience, and how do we design this kind of complex space? How do we integrate media-rich environments with people-rich ones and make them human, warm, conducive to learning? How do we organize these experiences for the user so they make sense without robbing them of their inherently rich and spontaneous quality?" (Friedlaender 1999).

At the CHArt Conference 2001 in London, Sylvia Lahav, Mac Campeanu, and Jean Kerrigan gave an impressive insight into the making of the online distance-learning course "An Introduction to Tate Modern." In cooperation with Tate Modern and City Literary Institute in London the team focused the online course on the Landscape suite at the museum:

> The landscape suite would also allow us to engage online students in a more personal way, to ask questions about their own environment, where they lived, their surroundings etc. Our concept was that the course might attract the elderly, the homebound, those in hospital or those in prisons and by using the first lesson to invite our students to describe their own surroundings we could make a personal and important connection. (Lahav 2002)

And this was the right way, as all the reactions of the twenty online students have shown. "We had incredible answers, not just one-liners, but amazingly insightful, perceptive, philosophical discussions about the notion of museums, space, landscape and then, of course, each one of our chosen artists." The communication with the online visitor brings a warm atmosphere and intensity in the exchange of ideas. This dialogue with users and their experiences is what is required when communicating about art, especially contemporary art. But "tutoring/facilitating an online course is incredibly time consuming. Tutors/facilitators need to handle their students sensitively but firmly. A great deal of intellectual and factual information is needed to answer student responses" (2002).

In the project Virtual Transfer Musée Suisse,[18] we find other strategies to construct or reconstruct history. The Virtual Transfer is not a digital collection, not a portal, not a virtual museum, but a vision of interaction and communication with the museum visitors and users.[19] It aims to be a strategy for direct communication in five languages (German, French, Italian, Romansh, and English) and will work from Spring 2005 onwards as a sort of "online agency" for the Swiss National Museums. Unlike the flood of information available from the Web, the Virtual Transfer offers the opportunity to rediscover the charm and charisma of objects in the collections, their history, and their impact on the viewer. Places and stories that can be explored interactively, a choice selection of objects, highly personalized forms of address, and successful dramatic scenes presented in multimedia form, all stimulate and inspire users, and in this way activate their own creativity. The Virtual Transfer offers a labyrinth of experiences. The objects of the museum collections become the narrators and the stories they have to tell are part of our own life. These encounters with interactivity allow the viewer to activate his/her own creativity. The story of the invention of the historic "Notzimmer"—knockdown furniture consisting of four chairs, two beds, one table, one wardrobe, dishes, and cutlery in case of need—is told by the daughter of Mauritius Ehrlich and visualized by a newsreel from 1945 and two trick films (figure 10.2). Thinking about his own fate, Ehrlich had to leave Vienna in 1938 as a refugee and he constructed the "Notzimmer" to help others. The interview with his daughter ends with an appeal for help. With this personal background the user will get more information about the Swiss asylum policy in the years from 1933 to 1945 as well as Web links to the refugee department of the UN.[20] Or a second example: the fictive eyewitness on Memento Mori. Five hundred and fifty years after his death in 1456 Rudolf von Ringoltingen, the former mayor of the city of Bern in Switzerland, talks to a contemporary audience about why he has spent a lot of money to shorten his time in purgatory.

## Navigation

The accumulation of information with the aim of comprehensiveness may be imposing, but it leads us nowhere. The communication between the different target groups on

Figure 10.2   *Notzimmer, Virtual Transfer Musée Suisse*, CD-ROM. © 2003 Swiss National Museum, Zurich and Transfusionen, Zurich—Vienna—Berlin.

one side and the different meanings and methodologies on the other needs navigation. Only a guided navigation, supported by interactive technology, brings the information to different target groups and allows understanding of different meanings and functions of a work of art, as well as individual interpretations to show the colorful richness (Kraemer 2001, 199–228). The following four examples show different strategies for this challenge.

*Example 1*   As an outgrowth of its ongoing research in the use of educational multimedia to contextualize and enhance the art-viewing experience, John S. Weber and Peter Samis of the San Francisco Museum of Modern Art have realized several interactive interferences in the permanent collection (2001). "Points of Departure, I and II: Connecting with Contemporary Art of the SFMOMA" (March 23–October 28, 2001; November 17, 2001–June 9, 2002) takes off from a number of premises that collectively prototype working methods and delivery platforms for a museum of the future, for example, six galleries arranged according to themes, and multimedia educational resources delivered directly into the galleries via iPAQ, Pocket PC, PDAs, and smart tables with touchscreen interfaces. Specifically developed for each gallery these tools have been tailored to each theme and include every work in the show. In front of a painting of Gerhard Richter the user can choose between different short films about the painter. It's really impressive to see the artist's own studio, to follow the process, listen to his voice, and get information about how to make a painting given by Gerhard Richter himself. The painting is explained by the producer, i.e. clear explanations and no art historian jargon.

*Example 2*   The CD-ROM, "Art and Industry," produced in 2000 for an exhibition about the founding of the Museum of Applied Arts in Vienna is a good example of the relation between content and navigation.[21] The navigation-line of the main story is in chronological order following the four chapters and the main events of the story with the possibility to increase the knowledge via sublines (e.g. chapter 3: main line: "Vienna World Fair 1873"—subline: "Center vs. suburb" (i.e. the upper circles vs. the working class)—main line: "The Emperors pavilion"—subline: "Fascination Orientalism"). The second main part about the pluralism of styles and historicism is differently navigated. Concerning the democratic pantheon of the different styles (e.g. neogothic, neorenaissance, japonism, orientalism), the navigation shows them all, so that no style comes into prominence (figure 10.3). So the user learns through the navigation that you can understand history linearly and chronologically in a single-line order, following event by event, as well as through a complex network of coexisting reciprocal influences.

*Example 3*   On the occasion of an exhibition the visitor might find not far from Picasso's masterpiece "La Vie" (1903) the interactive display "Exploring Picasso's La Vie," conceived by Holly Witchey and Leonard Steinbach, (The Cleveland Museum of Art), and produced by Cognitive Applications.[22] Presented at the Museums and the Web Conference 2002 in Boston, Holly Witchey explained the intention of the project: "Our goal was to create an interactive which would involve a number of visitors in a gallery. We wanted something more than a kiosk in a corner. Exploring Picasso's La Vie not only spurred visitors to go back to the beginning of the exhibition and look again at the actual work of art, it proved to be a successful multigenerational and multimodal learning experience. The interaction design is clearly arranged, the navigation plain and straight, the amount of information is manageable. There is no overflow of information. The "less is more" strategy of the famous American architect, Mies van der Rohe (1986), is applied. But simply the best is the appearance of the screen, with its ability to swivel from a horizontal angle of ninety degrees to an upright position

GESCHICHTE  BIOGRAPHIEN  **HISTORISMUS**  ANFANG  ENDE

»Ist die Verwendung von
Vorlagenwerken nicht ein
pädagogischer Unfug?«
(JOSEPH AUGUST LUX, 1905)

Historismus
Stilpluralismus
Neogotik
Neorenaissance
Neobarock
Neorokoko
Neoklassizismus
2. Biedermeier
Orientalismus
Japonismus
Purismus
Volkstümlich
Vorlagenwerke

HISTORISMUS  **STILPLURALISMUS:**  VORLAGENWERKE
NEOGOTIK NEORENAISSANCE  NEOBAROCK  NEOROKOKO  NEOKLASSIZISMUS
2.BIEDERMEIER  ORIENTALISMUS  JAPONISMUS  PURISMUS  VOLKSTÜMLICH

Figure 10.3  *Kunst und Industrie—Die Anfaenge des Museums fuer angewandte Kunst in Wien*, kiosk system and CD-ROM. © 2000 MAK—Austrian Museum of Applied Arts, Vienna and Transfusionen, Zurich—Vienna—Berlin.

(figure 10.4). The exceptionally large screen, with its proximity to Picasso's original painting, ensures a high level of audience attendance.

*Example 4*   "Vienna Walk" is the prototype of an interactive movie produced in 1998 by the Austrian media company Science Wonder Productions.[23] Shot on 16 mm film, Michael Perin-Wogenburg and his team realized a dynamic encyclopedia about the city of Vienna with "Hypervideoengine," a tool combining the possibilities of digital film, hypermedia, and the Web. "Vienna Walk," presented on the 1999 Museum and the Web Conference in New Orleans, was more than a demo; it was a futuristic vision of real interactivity and intelligent knowledge transfer for the field of cultural heritage-tourism-e-commerce. The contents: A virtual institution has planned to get a encyclopedia of the values (economic, cultural, and energy) of a European capital city. Three agents (Tatjana, Pauline, and Tomo) have to fulfill their missions. They have

Figure 10.4 *Exploring Picasso's La Vie*, kiosk system. © 2002 The Cleveland Museum of Art and Cognitive Applications, London.

ROOM: BAROQUE. ROCOCO. CLASSICISM    CURATOR: CHRISTIAN WITT-DÖRRING

WERNER DEPAULI-SCHIMANOVICH-
GOETTIG_SCIENTIST
MAILTO    www.schimanovich.at

**museum of applied art**

Figure 10.5    *Vienna Walk*. © 1998 Science Wonder Productions, Vienna.

twenty-four hours and they bring the user—their accomplice from the real world—to specific locations in Vienna. With interactive crossings you can leave the story (follow a virtual skater to meet other people) or follow a subline (listen to Pauline discussing the purpose of art with a famous scientist at the museum (figure 10.5); if you don't agree, send him an e-mail), shop for diamonds with Tatjana and pay online, or peruse the actual program of the opera and buy tickets on the Web site while you are interacting with the film. Vienna Walk can show us that the relationship between navigation and information has changed. Intelligent navigation leads us through the world of knowledge and is part of the content at the same time.

## Borderland

At the borderland of documentation we are faced with a lot of challenges and questions. A lot of the existing museum Web sites are mainly online brochures designed using the traditional methods of print layout. But a Web site is a process, an action plan, and not a self-contained print product. To understand the users as people seeking to integrate information and to lead the visitor to real interactivity—a dialogue and not a monologue—will be one of the main goals of future work. Meanwhile, a lot of museums are defining new positions on the Web. The relaunch of the Australian Museums Online—a "metacenter for museums online collections"—has taken one stand in the museum Web-world.[24] Recently the term "virtual museum" has become indistinct and means different things like digital collection, cyber-museum, online museum, virtual space, interactive installation, but also information gateway or simply portal (Schweibenz 2001). It is on the basis of this multiuse of the term "virtual museum" that the Swiss project has been named Virtual Transfer. This name stands for an integrative laboratory, a medium with challenges in interface and interaction open for common developments. These recent works of art have specific criteria like the acceleration of the explosion of art terms such as Net art,[25] and the interdisciplinary expansion of its contents. Will the fusion of real and virtual works of art in real and virtual museum spaces create a new kind of museum in an integrative or performative sense? At the Museum and the Web Conference 2002, in Boston, Pia Vigh, Director of CultureNet Denmark, gave an overview of the problems of documenting and preserving Net art.[26] In her conclusion she remarked that Net art is not thrust upon museums, but museums collecting contemporary art want Net art and feel bound to collect it. "If museums have to take Net art seriously, they have to start with the already established competencies and viable forums outside the museums. Museums that wish to cover Net art should join these forums. For the sake of the reputation of museums in the Net art environment, it is essential that they do not appear to be parasites."[27] Another topic at the borderland is the transformation of a unique opus to a series of staged events, and the mutation from the unique original signed by the artist to digital reproductions and copies. Will the crossover of original and digital reproductions lead us to a brave new world of images and movies creating a semantic language of organic and inorganic aesthetics, as the Austrian artist, Robert Lettner, has shown in his digital paintings? In his exhibition entitled "Pictures on Magical Geometry," Lettner has presented in the Viennese Seccession the results of the amalgamation of an organic process (i.e. drawing, which is dependent on the psyche and artistic abilities) with an inorganic medium, the computer (figure 10.6). Furthermore, we have to ask how can we document the use of interactive artworks undeviatingly self-creating, like Dieter Kiessling's masterpiece "Continue."[28] In this interactive binary from the 1997 drama of the German artist you have the freedom of selecting "quit" (white field)

Figure 10.6    Robert Lettner, *Digital Painting*. © 2002 Robert Lettner, Vienna.

and "continue" (black field). If you choose "continue" the two fields change to 4 fields, then 8, 16, 32, 64, 128, 256, 512, 1024, 2048 fields, and so on. After a while—from one moment to the next—the screen is filled with black and white dots. Your free-will decision has been hijacked by the system. The interactivity will become purely accidental and the idea of free will is no longer applicable. The helpless user quickly learns the truth and the real possibilities of interactive systems. Interactivity—multiple choices of digital one way streets—can also be a catalyst for creative action. "Continue" contains both, the invitation to participate creatively and the confrontation with an inflexible system. The first means a joyful collaboration, the second calls for the user to be inconsistent with the system. Real interactivity grows out of the actuality of action. It is productivity that leads to the creation of reality and the meaning of perception.

## Fragments and Figments of Knowledge

Each item of information has its raison d'être in terms of learning. It makes little sense to feed more and more data into the computer, creating a collection of dead material, useless classifications, and senseless information. Documentation is interpretation. This means that registrars, scientists, restorers, artists, and art critics are responsible for the documentation of art. Therefore, it is necessary for a museum or an archive to not only understand itself as an information provider, but to act as an information broker.

Information is the capital of the knowledge bases named museums, archives, libraries, and art trade. The boundaries between them will collapse; digital collections will combine and create information with a long time value. Museums legitimize and constitute values, but the museum is nothing in particular; it's just one place where we can experience art. Recently, the museum is transforming from a sanctuary to a production center. Transparency and mobility become the significant criteria of documentation, reacting to the requirements of the users as well as the producers. Thus documentation, which is now to be applied in nonstandard and immanent manner (as appropriate to works of art), changes from a passive retrospective to an actively participating form for archiving contemporary works of art. Documentation becomes a medium, which must be used actively by the scientist, curator, registrar, but also by the artist and user, for the search of historical truth and the reconstruction of the original relationships. In this manner, it can become an integral part of the work of art. Contemporary art requires rethinking and a strategic procedure in its documentation. The artist turns into an active partner of the registrar and the scientist. Together, they must develop diverse strategies for the optimal documentation of the artist's work. The instruments of digital and multimedia technology require specific art efforts. This cannot be carried out by the present database management systems or, at best, only minimally. The changing role of documentation—which on the one hand will have consequences for the questions asked by later generations of researchers and on the other hand, due to the increasing presence of digital media appliances in museums, will also change the viewing habits of visitors to the museums—poses an entirely new set of problems and questions. These points must be understood as an undergraduate course for a methodology and as a big challenge for museology as science. They need to be discussed thoroughly and then the possible structures and strategies for handling future measures of documentation by digital technologies should be put together in the form of recommendations and guidelines. Most important will be the discussion and definition of different technical means, forms of photographic and (hyper)textual assimilation, and the applicability of advanced software-technological possibilities, for example, pattern recognition, color, and texture analysis, etc. All of these measures will contribute to guarantee the future of fundamental scientific art research.

But the museum is not an archive for storing unlimited information produced by the art critics or art historians for the science of art. The museum is not a democratic forum for all artists. To collect all information is misguided and without relevance for the future. We have to accept the demise of certain aspects of our work. This should be one of the considerations of a museum's policy. It is important to remember this when creating multimedia products for the future. Concentrate on the essentials and transfer this knowledge in the form of really good stories. The museum makes an important contribution to humanity. We all have the duty to lead our visitors and users to a critical seeing and understanding of art and reality. Critical seeing is distinctive seeing. Thus, interactive museum guides and digitalized masterpieces are ubiquitous today. Multimedia is displacing our beloved postcards, those auratic stand-ins for original works of art. "In the future, art lovers will be able to enjoy artworks in their own home, instead of museums and galleries, as art is increasingly published on CD-ROM. Libraries and museums could function as transmission centers which interested persons will no longer visit, they will only log onto them. Hence, the museum no longer possesses originals in the classic sense, but only administers sets of data" (Jacobs 1994, 29). This was written in 1994, and now it is a reality. In recent years the museum has changed from a content pool to a content provider. In years to come, the museum will change from a content provider to an information broker. The museum harbors the untold wealth of collected life experiences. But to communicate this richness to the visitor, especially the user, a guided navigation is necessary. Interactivity is the multiple choices of digital one-way streets, and in most cases interactivity is more interpassive than interactive. The secret is to understand that real interactivity grows out of the actuality of an action. It is the productivity that leads us to the creation of reality and the meaning of perception. The method of navigation is not responsible for my choice. My preferences, my associations, and my own history lead me through the contents, a many-layered dramaturgy which allows access to different levels for different user groups. Not all of these contents are for everyone, but everybody has the chance to find something. The navigation should work with the content and this means: Using interactivity in documentation is the way to control the creativity of your user. The challenge for documentation from the point of view of contemporary art will be to construct spaces of experience and define strategies to support communication, navigation, and the process of creativity. Additionally, it will be necessary to document the behavior of the recipient as a coauthor of the artist and document the interactivity and the change of meaning of the work of art perceived by different viewers at different times. To navigate this flow will be one of the gigantic provocations of our future. But my attempt ends as it has started—with a recognition of my limits as art historian: "On the day when I know all the emblems," Kublai Khan asked Marco Polo, "shall I be able to possess my empire, at last?" And the Venetian answered: "Sire, do not believe it. On that day you will be an emblem among emblems."[29]

## Notes

1. Under the title "Methodology for the Documentation of Contemporary Art," a research project was realized, 1999–2001, at the interdisciplinary Kulturwissenschaftliches Forschungskolleg (SFB/FK 427) of the German Universities of Aachen, Bonn, and Cologne. This project (no. B4) was initiated by professors Hubertus Kohle and Harald Kraemer. The other members of the team in the subproject B4 were Nicole Birtsch, Kathrin Lucht, Martina Nied, Simone Schmickl, and Christina Hemsley. Some of the following aspects are based on the discussions of the ICOM/CIDOC Contemporary Art Working Group, founded in 1994 at the CIDOC conference in Washington, DC. A part of the text has been published under the title "Fragments and Figments of Knowledge: The Documentation of Contemporary Art," *CHArt London 2001 Conference Proceedings* vol. 4 (2002), available at ⟨http://www.chart.ac.uk⟩ (accessed 14 April 2005).

2. I am very grateful to Sarah Kenderdine and Fiona Cameron for inviting me to pursue these questions of documentation and contemporary art in their deserving publication. For the support of my research I am grateful to Hubertus Kohle (Ludwig-Maximilians-Universitaet, Munich) and the board of directors of the Kulturwissenschaftliche Forschungskolleg (SFB/FK 427) of the University of Cologne for their generosity. I would like to thank my colleagues Maria José Ferreira Moniz Pereira (C. Gulbenkian Foundation, Lisboa), Larry Friedlaender (Stanford University), Hartmut John (RAMA, Brauweiler), Norbert Kanter (Transfusionen, Berlin), Sarah Kenderdine (Museum Victoria), Vincent de Keijzer (Gemeentemuseum Den Haag), Robert Lettner (University of Applied Arts, Vienna), Frances Lloyd-Baynes (V&A, London), Xavier Perrot (Ecole du Louvre, Paris), Michael Perin-Wogenburg (Science Wonder Productions), Peter Samis (SFMOMA), Angela Spinazze (Chicago), Sirkka Valanto (The Finnish National Gallery, Helsinki), Kim H. Veltman (Maastricht McLuhan Institute), Oliver Vicars-Harris (London), Friso Visser (Den Haag), Holly Witchey (Cleveland Museum of Art), and Patricia Young (CHIN, Ottawa) for their suggestions.

3. Regarding Digital Collection, see Kraemer 2001, 162–179.

4. Neither ICONCLASS, the Art and Architecture Thesaurus, the German rule type MIDAS, nor the data-field catalog of the German Museum Association take the complex needs of contemporary art into consideration. For ICONCLASS and MIDAS, see Linz 1996.

5. See Waidacher 1999, and chapter 7 in Herbst and Levykin 1988, 146–169.

6. Ambitious endeavors to alter the situation have taken the form of, for example, conferences and initiatives, such as Corzo 1999, Hummelen and Sillé 1999, Vektor: European Contemporary Art Archive, at ⟨http://www.vektor.at/⟩ (accessed 14 April 2005), and Kraemer and John 1998.

7. *Tarantula*, directed by Jack Arnold, United States (Universal, 1955).

8. See San Francisco Museum of Modern Art 2001, Klotz 1996, Roetzer 1993, Matysik 2000.

9. The "Kunstkritikmaschine"—realized by daphne-art-works—is a software using fragments of real art reviews, combining them to create new art criticism. See ⟨http://kunstkritik.de.vu⟩ (accessed 14 April 2005).

10. See Raphael 1989, Imdahl 1981, and Belting and Gohr 1996.

11. See Kraemer and John 1998, especially Steiner, Spiegl, and Siepmann.

12. See Weyer and Heydenreich 1999, 385–388.

13. From "Documentation of Contemporary Art: Perspectives and Methodologies." Paper presented at Meeting of the Contemporary Art Working Group of ICOM/CIDOC, London, 6–10 September 1999.

14. See Wedemeyer 1998; for R.M.R., see The Rainer Maria Rilke Archive, at ⟨http://www.geocities.com/Paris/LeftBank/4027/⟩ (accessed 14 April 2005).

15. See San Francisco Museum of Modern Art 2000.

16. See Chronicle Books/Eden Interactive 1994.

17. See Schulze 2001 and Heidkamp 2001.

18. See ⟨http://www.virtualtransfer.com⟩ and ⟨http://www.musee-suisse.ch/vtms⟩.

19. See Jaggi and Kraemer 2004, 8, and Kraemer 2004.

20. UNHCR, The UN Refugee Agency, at ⟨http://www.unhcr.ch/cgi-bin/texis/vtx/home⟩ (14 April 2005).

21. See Noever 2000.

22. Information at ⟨http://www.clevelandart.org/exhibcef/picassoas/html/9476994.html⟩ and ⟨http://www.cogapp.com/home/clevelandPicasso.html⟩ (both accessed 14 April 2005).

23. Vienna Walk is a CD-ROM based on Soerenson Quicktime and one of the first DVD-ROMs on MPEG2 with a length of more than two hours. See Science Wonder Productions 1998; more information at ⟨http://www.perin-wogenburg.com⟩ (accessed 14 April 2005), and Kraemer 1998, 13–15.

24. About the relaunch of the AMOL, see Kenderdine 1999.

25. Possible strategies for documenting Net art and media art are part of the project "Database of Virtual Art" of the Humboldt-University Berlin. See Grau 2001, 135–140.

26. Information about Net art institutions is available at ⟨http://n2art.nu⟩, ⟨http://www.kulturnet.dk⟩, ⟨http://rhizome.org⟩, and ⟨http://netzspannung.org⟩ (all accessed 14 April 2005). See also Vigh 2002.

27. Ibid.

28. See Kiessling 1997; see note 25, and Kraemer 2001, 214–215.

29. See Calvino 1978, 22–23, with many thanks to Peter Samis.

## References

Albers, Josef. 1975. *Interaction of Color*. New Haven: Yale University Press.

*American Visions. 20th Century Art from the Roy R. Neuberger Collection*. 1994. San Francisco: Eden Intercative, CD-ROM.

*The Anderson Collection: Art as Experiment / Art as Experience.* 2000. Ed. San Francisco Museum of Modern Art. San Francisco: SFMOMA, CD-ROM. Also available at ⟨http://www.sfmoma.org/anderson/⟩ (accessed 15 April 2005).

Belting, Hans, and Siegfried Gohr. 1996. *Die Frage nach dem Kunstwerk unter den heutigen Bildern.* Karlsruhe: Hochschule für Gestaltung.

Bénichou, Anne. 1994. *Normes de Documentation en Art Contemporain.* Ottawa: Ministre des Approvisionnenments et Services.

Boersch-Supan, Helmut. 1993. *Kunstmuseen in der Krise. Chancen, Gefaehrdungen, Aufgaben in mageren Jahren.* Munich: Deutscher Kunstverlag.

Breitwieser, Sabine. 1999. *Sammlung Archiv Kommunikation: Bedingungen heute—Ueberlegungen fuer morgen.* Ed. Generali Foundation, Vienna: Generali Foundation and Cologne: Verlag der Buchhandlung Walther Koenig.

Calvino, Italo. 1978. *Invisible Cities.* San Diego: Harvest.

Carroll, Lewis. 1984. *Alice's Adventures in Wonderland.* London: Basingstoke, Macmillan Children's Books.

Chronicle Books/Eden Interactive, ed. 1994. *American Visions: 20th Century Art from the Roy R. Neuberger Collection.* CD-ROM. San Francisco: Chronicle Books/Eden Interactive.

Corzo, Miguel Angel, ed. 1999. *Mortality Immortality? The Legacy of 20th Century Art.* Los Angeles: The Getty Conservation Institute.

Delacroix, Eugène. 1997. *Eine Auswahl aus den Tagebuechern.* Ed. H. Platschek. Frankfurt/Main: Insel.

Dickinson, Emily. 1924. *The Complete Poems of Emily Dickinson.* Boston: Little, Brown.

Erdman, David V., ed. 1982. *The Complete Poetry & Prose of William Blake.* Berkeley: University of California Press.

Fluegel, Katharina. 2001. "Dokumentation als museale Kategorie." *Rundbrief Fotografie* Sonderheft 6: 19–29.

Forsythe, William. 1999. *Improvisation Technologies: A Toll for the Analytical Dance Eye.* CD-ROM. Karlsruhe: Zentrum für Kunst und Medientechnologie.

Fraser, Andrea. 1995. *Report: EA Generali Foundation.* Vienna: EA Generali Foundation.

Friedlaender, Larry. 1999. "Keeping the Virtual Social." In *Museums and the Web*, conference proceedings, ed. D. Bearman and J. Trant. New Orleans, 11–14 March, CD-ROM.

Grau, Oliver. 2001. "Datenbank der Virtuellen Kunst." In *EVA Berlin*, conference proceedings, ed. J. Hemsley and G. Stanke, 135–140. Berlin: GfaI.

Greenhalgh, Michael. 2002. "The Classroom of the Future." In *Digital Art History—A Subject in Transition: Opportunities and Problems*, CHArt Conference, London, 27–28 November 2001, ed. A. Bentkowska, T. Cashen, and J. Sunderland. Available at ⟨http://www.chart.ac.uk/⟩ (accessed 15 April 2005).

Heidkamp, Philipp. 2001. "Menschen, Medien und Moeglichkeiten: Perspektiven im Interface- und Interaktionsdesign." In *Museums and the Internet*, conference proceedings. Hagen: Historisches Centrum. Available at ⟨http://www.mai-tagung.de/FachDez/Kultur/Unsichtbar/Maitagung/Maitagung+2001/beitraege.htm⟩ (accessed 15 April 2005).

Herbst, Wolfgang, and K. G. Levykin. 1988. *Museologie*. Berlin: VEB.

Hummelen, Ijsbrand, and Dionne Sillé, eds. 1999. *Modern Art: Who Cares? An Interdisciplinary Research Project and an International Symposium on the Conservation of Modern and Contemporary Art*. Amsterdam: The Foundation for the Conservation of Modern Art and the Netherlandish Institute for Cultural Heritage. Available at ⟨http://www.incca.org/⟩ (accessed 15 April 2005).

Imdahl, Max. 1981. *Bildautonmie und Wirklichkeit: Zur theoretischen Begruendung moderner Malerei*. Mittenwald: Maeander Kunstverlag.

Jacobs, Leo. 1994. "Interaktive kun stwellen." *Screen Multimedia* 6: 22–31.

Jaggi, Konrad and Harald Kraemer. 2004. "The Virtual Transfer Musée Suisse." *ICOM News* 57, no. 3: 8.

Jeudy, Henri Pierre. 1987. *Die Welt als Museum*. Berlin: Merve-Verlag.

Kenderdine, Sarah. 1999. "Inside the Meta-center: A Cabinet of Wonder." In *Museums and the Web*, conference proceedings, New Orleans, 11–14 March, ed. D. Bearman and J. Trant. CD-ROM.

Ketnath, Artur. 1999. "How to Conserve Motion." In *Modern Art: Who Cares? An Interdisciplinary Research Project and an International Symposium on the Conservation of Modern and Contemporary Art*, ed. I. Hummelen, D. Sillé, 369–372. Amsterdam: The Foundation for the Conservation of Modern Art and the Netherlandish Institute for Cultural Heritage. Available at ⟨http://www.incca.org/⟩ (accessed 15 April 2005).

Kiessling, Dieter. 1997. "Continue." *Artintact*. Karlsruhe: Zentrum fuer Kunst und Medientechnologie, vol. 4, CD-ROM.

Klotz, Heinrich ed. 1996. *Perspektiven der Medienkunst*. Karlsruhe: Edition ZKM.

Kraemer, Harald. 1994. "Believe Your Eyes and Get the Picture: Artworks and Museums in the Age of Electronic Communication." In *AURA: The Reality of the Artwork between Autonomy, Reproduction and Context*, ed. Wiener Secession, exhibition catalogue, 93–99. Vienna: Wiener Secession.

Kraemer, Harald. 1998. "Vienna Walk: Ueber den Prototyp eines Interaktiven Films." *Museen im Rheinland* 1: 13–15.

Kraemer, Harald. 2001a. *Museumsinformatik und Digitale Sammlung*. Wien: WUV.

Kraemer, Harald. 2001b. "CD-ROM und Digitaler Film: Interaktivitaet als Strategie der Wissensvermittlung." In *Euphorie Digital? Aspekte der Wissensvermittlung in Kunst, Kultur und Technologie*, ed. C. Gemmeke, H. John, and H. Kraemer, 199–228. Bielefeld: Transcript Verlag.

Kraemer, Harald. 2002. "Fragments and Figments of Knowledge: The Documentation of Contemporary Art." In *Digital Art History—A Subject in Transition: Opportunities and Problems*, CHArt Conference, London, 27–28 November, ed. A. Bentkowska, T. Cashen, and J. Sunderland. Available at ⟨http://www.chart.ac.uk/⟩ (accessed 15 April 2005).

Kraemer, Harald. 2004. "Erlebnisse zwischen Vision und Realitaet: Virtueller Transfer Musée Suisse und Museum Schloss Kyburg." In *Museums and the Internet*, conference proceedings. Bonn: Rheinisches Landesmuseum. Available at ⟨http://www.lvr.de/FachDez/Kultur/Unsichtbar/Maitagung/Maitagung+2004/beitraege.htm⟩ (accessed 15 April 2005).

Kraemer, Harald, and Harmut John, eds. 1998. *Zum Bedeutungswandel der Kunstmuseen: Positionen und Visionen zu Inszenierung, Dokumentation, Vermittlung*. Nuremberg: Verlag für moderne Kunst. Available at ⟨http://www.kulturnet.dk/⟩ (accessed 15 April 2005).

*Kunstkritikmaschine*. 2000. Realized by daphne-art-works, Dresden: Hochschule fuer Bildende Kuenste. Available at ⟨http://kunstkritik.de.vu/⟩ (accessed 15 April 2005).

*Kunst und Industrie: Die Anfaenge des Museums fuer angewandte Kunst in Wien*. 2000. Edited and produced by MAK—Austrian Museum of Applied Arts and Die lockere Gesellschaft—Transfusionen. Vienna: MAK—Austrian Museum of Applied Arts and Transfusionen, CD-ROM.

Lafontaine, Marie-Jo. 1999. *Installations Vidéos 1979–1999*. Paris: Réunion des Musées de France. CD-ROM. Also available at ⟨http://www.marie-jo-lafontaine.com⟩ (accessed 14 April 2005).

Lahav, Silvia, Mac Campeanu, and Jean Kerrigan. 2002. "An Introduction to Tate Modern." In *Digital Art History—A Subject in Transition: Opportunities and Problems*, CHArt Conference, London, 27–28 November, ed. A. Bentkowska, T. Cashen, and J. Sunderland. Available at ⟨http://www.chart.ac.uk/⟩ (accessed 15 April 2005).

Lettner, Robert, and Harald Kraemer. 1998. "Art Is Redeemed, Mystery Is Gone: Conversations with Robert." In *Robert Lettner: Bilder zur magischen Geometrie*, ed. Wiener Secession, 19 November 1998–17 January 1999, exhibition catalogue, 15–23. Wien: Wiener Secession.

Linz, Barbara. 1996. *Ansaetze zur Inhaltserschließung abstrakter Kunstwerke und Moeglichkeiten der Einbindung in MIDAS*. Potsdam: University of Applied Sciences, diploma, 11 October. *Netzspannung*. Available at ⟨http://netzspannung.org/⟩ (accessed 15 April 2005).

Matysik, Reiner, ed. 2000. *Zukuenftige Lebensformen*. Berlin: Vice Versa Verlag.

Mies van der Rohe, Ludwig. 1986. *Less Is More*. Zurich: Waser Verlag.

Noever, Peter, ed. 2000. *Kunst und Industrie: Die Anfaenge des Museums fuer angewandte Kunst in Wien*. Vienna: MAK Austrian Museum of Applied Arts. Also available on CD-ROM.

Raphael, Max. 1989. *Wie will ein Kunstwerk gesehen sein?* Frankfurt/Main: Suhrkamp Verlag.

Reifenrath, André. 1997. "Relation und Realitaet: Von den Problemen der Informationsabbildung in Elektronischen Systemen." In *Kunstgeschichte Digital*, ed. H. Kohle, 27–40. Berlin: Reimer Verlag.

*Rhizome*. Available at ⟨http://rhizome.org/⟩ (accessed 15 April 2005).

Roetzer, Florian. 1993. "Das neue Bild der Welt: Wissenschaft und Aesthetik." *Kunstforum* 124, no. 11/12: 69–235.

Samis, Peter. 2001. "Points of Departure: Integrating Technology into the Galleries of Tomorrow." In *International Cultural Heritage Informatics Meeting*, conference proceedings, Milan. Available at ⟨http://www.archimuse.com/ichim2001/abstracts/prg_115000640.html⟩ (accessed 15 April 2005).

San Francisco Museum of Modern Art, ed. 2000. *The Anderson Art Collection: Art as Experiment/ Art as Experience*. San Francisco: SFMOMA.

San Francisco Museum of Modern Art, ed. 2001. *010101: Art in Technological Times*. San Francisco: SFMOMA.

Scheffel, Regine. 1997. "Positionspapier zu Taetigkeitsbereich und Berufsbild in der Museumsdokumentation." *Mitteilungen und Berichte aus dem Institut fuer Museumskunde Berlin* 10.

Schneede, Uwe M., and Martin Warnke, eds. 2004. *Anna Oppermann in der Hamburger Kunsthalle*. CD-ROM. Hamburg: Kunsthalle.

Schulze, Claudia. 2001. *Multimedia in Museen: Standpunkte und Perspektiven Interaktiver Digitaler Systeme im Ausstellungsbereich*. Wiesbaden: Deutscher Universitaets-Verlag.

Schweibenz, Werner. 2001. "Das virtuelle Museum. Ueberlegungen zum Begriff und Wesen des Museums im Internet." In *Museums and the Internet*, Conference Proceedings, Hagen: Historisches Centrum, avaiable at ⟨http://www.mai-tagung.de/FachDez/Kultur/Unsichtbar/ Maitagung/Maitagung+2001/beitraege.htm⟩ (accessed 20 April 2005).

Schwitters, Kurt. *Merzbau*. Available at ⟨http://www.merzbau.org/Schwitters.html⟩ (accessed 20 April 2005).

Science Wonder Productions, ed. 1998. *Vienna Walk Demo*. CD-ROM/DVD-ROM. Vienna: Science Wonder Productions.

Serota, Nicholas. 1996. *Experience or Interpretation: The Dilemma of Museums of Modern Art*. London: Thames and Hudson.

Siepmann, Eckhard. 1998. "Madame Sosostris erschaut die Zukunft des Kunstmuseums." In *Zum Bedeutungswandel der Kunstmuseen: Positionen und Visionen zu Inszenierung, Dokumentation, Vermittlung*, ed. H. Kraemer and H. John, 178–183. Nuremberg: Verlag für moderne Kunst.

Spiegl, Andreas. 1998. "Das Museum: Auf seinem Weg vom Ort zum Raum, vom Archiv zum Pool und zurueck." In *Zum Bedeutungswandel der Kunstmuseen: Positionen und Visionen zu Inszenierung, Dokumentation, Vermittlung*, ed. H. Kraemer and H. John, 28–34. Nuremberg: Verlag für moderne Kunst.

Steiner, Barbara. 1998. "Selbstinszenierung: Das kuenftige Kunstmuseum—ein Erlebnispark?" In *Zum Bedeutungswandel der Kunstmuseen. Positionen und Visionen zu Inszenierung, Dokumentation, Vermittlung*, ed. H. Kraemer and H. John, 35–41. Nuremberg: Verlag für moderne Kunst.

Chuang Tzu. 1965. *The Way of Chuang Tzu*. Ed. Thomas Merton. New York: New Directions Publishing Corporation.

UNHCR, The UN Refugee Agency. Available at ⟨http://www.unhcr.ch/cgi-bin/texis/vtx/home⟩ (accessed 14 April 2005).

*Vektor—European Contemporary Art Archive*. Available at ⟨http://www.vektor.at/⟩ (accessed 15 April 2005).

*Vienna Walk Demo*. 1998. Edited and produced by Science Wonder Productions. Vienna: Science Wonder Productions, CD-ROM. Also available online at ⟨http://www.perinwogenburg.com/⟩.

Vigh, Pia. 2002. "Hacking Culture." In *Museums and the Web*, ed. D. Bearman and J. Trant. Conference Proceedings, Boston, 18–20 April, CD-ROM.

*Virtual Transfer Musée Suisse*. 2003. Edited and produced by Swiss National Museum and Transfusionen. Zurich: Swiss National Museum and Transfusionen, CD-ROM. Also available at ⟨http://www.virtualtransfer.com/resp. http://www.musee-suisse.ch/vtms⟩ (accessed 15 April 2005).

Waidacher, Friedrich. 1999. *Handbuch der Allgemeinen Museologie*, 3d ed. Wien: Boehlau.

Weber, John S. 2001. "Beyond the Saturation Point: The Zeitgeist in the Machine." In *010101. Art in Technological Times*, exhibition catalogue. San Francisco: San Francisco Museum of Modern Art, 15–23.

Wedemeyer, Carmen, ed. 1998. *Anna Oppermann: Umarmungen, Unerklaerliches und eine Gedichtzeile von R.M.R.* Frankfurt/Main: Stroemfeld/Roter Stern, CD-ROM.

Weibel, Peter. 1998. "Zur Zukunft des Kunstmuseums." In *Zum Bedeutungswandel der Kunstmuseen. Positionen und Visionen zu Inszenierung, Dokumentation, Vermittlung*, ed. H. Kraemer and H. John, 22–27. Nuremberg: Verlag für moderne Kunst.

Weyer, Cornelia, and Gunnar Heydenreich. 1999. "From Questionnaires to a Checklist for Dialogues." In *Modern Art: Who Cares? An Interdisciplinary Research Project and an International Symposium on the Conservation of Modern and Contemporary Art*, ed. I. Hummelen and D. Sillé. Amsterdam: The Foundation for the Conservation of Modern Art and the Netherlandish Institue for Cultural Heritage, 385–388. Available at ⟨http://www.incca.org/⟩.

Zappa, Frank. 1979. *Joe's Garage*. EMI.

Zembylas, Tasos. 1997. *Kunst oder Nichtkunst: Ueber Bedingungen und Instanzen Aesthetischer Beurteilung*. Vienna: WUV.

# 11 Cultural Information Standards—Political Territory and Rich Rewards

Ingrid Mason

Cultural information standards provide the infrastructure for the collection and preservation of, and access to, digital cultural heritage. Information standards that support the collection and preservation of digital cultural heritage determine what is collected (selection policies and guidelines) and kept (preservation policies and methods). Information standards that support access to digital cultural heritage operate on two levels: intellectual (learning about a collection item through its description) and physical (interacting with a collection item). Virtual access facilitates intellectual and physical access through making information about a cultural item and the actual item (born or digitized) available via the Internet, thus—digital cultural heritage. Virtual access provides an immediacy and proximity to cultural heritage. The benefit of being "closer" to cultural heritage, or making it more accessible, is the increased opportunity to relate to one's own culture (nationalism) or other cultures (globalism). Virtual access extends access to cultural resources for educational, entertainment, or commercial purposes.

Cultural knowledge spaces are constructed with information shaped and influenced by sociopolitical forces.[1] Cultural information standards that facilitate and extend access to cultural heritage are sociopolitical in nature. In a virtual cultural information-sharing and knowledge-enabling environment the development of new cultural information standards therefore revisits and reconsiders cultural traditions, social practices and strata, economic disparities, and political dynamics. Cultural resources are coordinated and territories mapped as the cultural information standards are established. Collaboratively developed and maintained cultural information standards stabilize the environment for cultural collection and access. The consistency provided by new cultural information standards enables new social and cultural practices (associated with digital cultural heritage) to evolve and flourish. Digital cultural heritage comes at a cost, that is, the process of standardization requires significant international goodwill, interest, and investment in people, time, and technology, by those that are participating in developing the standards.

Relevant questions are: What are cultural information standards and what is their relationship to digital cultural heritage? What drives the development of cultural information standards? What are the sociopolitical forces that influence cultural information standards? How do sociopolitical forces impact upon digital cultural heritage? What are the benefits of virtual access to digital cultural heritage?

## Cultural Information Standards

What are cultural information standards and what is their relationship to digital cultural heritage? Cultural information standards provide the foundations for the collection and access to cultural heritage. Cultural information standards are curatorial (keeping practices), semantic (terminology and documentation), and technical (Web design and communication) in nature. These information standards guide curatorial interpretation, provide intellectual structure, and technically enable resource discovery and retrieval of cultural information and artifacts.

Curatorial standards guide the interpretation of social phenomenon and this is represented through the collection of cultural heritage. Collection and preservation policies have been rewritten and incorporate new approaches to the selection of digital cultural heritage. The selection of digital cultural heritage is problematic as the technical standards that enable its preservation are still rudimentary and experimental. In addition to this, the ability to extend the collection of cultural heritage and include digital material often means a reprioritization of limited resources. Curatorial skills to understand the construction of digital cultural heritage and interpret the embedded cultural information and knowledge require a technical literacy, knowledge of new cultural information standards, and production design trends.

Semantic standards coordinate intellectual access to cultural information through controlled vocabularies, documentation practices, and data structures (records and markup languages). The development, maintenance, and use of controlled vocabulary such as thesauri, subject headings, and Dublin Core (DC)[2] metadata element sets aids with standardizing resource discovery across multiple information sources and diverse sociopolitical contexts. Examples of controlled vocabulary are the Art and Architecture Thesaurus (AAT),[3] ERIC thesaurus (education),[4] Library of Congress (LCSH),[5] and medical subject headings (MESH).[6] Examples of metadata element sets are the

Australian (AGLS)[7] or New Zealand (NZGLS)[8] Government Locator Services, and Education Network Australia (EdNA)[9] metadata standard. Cultural documentation practices, that is, archival, artifactual, and bibliographic description practices, differ according to the nature of the collection and the cultural repository type. These description practices use different methods, terminology, data structures, and levels to provide intellectual access. Broadly speaking, the description methods differ in that provenance and context have primacy in archival and artifactual description, whereas traditions and standards of publishing have primacy in bibliographic description. Examples of standardized data structures for collection description are ISAD(G) (archival),[10] SPECTRUM (artifactual),[11] and MARC (bibliographic).[12] The International Committee for Documentation of the International Council of Museums (ICOM-CIDOC) Conceptual Reference Model (CRM) offers the opportunity to map cultural heritage information across diverse repositories.[13]

Technical standards aid with selecting, preserving, and accessing digital cultural heritage, for example, markup languages have emerged with the advent of the Internet to structure documents and facilitate access to online resources. Markup languages incorporate a data structure (or metadata) and presentation styles. There are three types of metadata used in digital cultural heritage: administrative, descriptive, and technical. Examples of markup languages in digital cultural heritage for descriptive metadata are Encoded Archival Description (EAD)[14] and Textual Encoding Initiative (TEI).[15] An example of a markup language that incorporates technical metadata is Metadata for Images (MIX)[16] in XML. Soon a standard for administration and preservation metadata will emerge from the Preservation Metadata: Implementation Strategies (PREMIS) project to structure preservation metadata.[17] Technical standards also coordinate physical storage and access to cultural information and artifacts through transfer protocols for database searching (Z39.50, SOAP).[18] A markup language that incorporates all three types of metadata (administrative, descriptive, and technical) is Metadata Encoding and Transmission Standard (METS).[19] Digital preservation (digitization, emulation, normalization, and migration) is possible with new technical standards, and supports long-term access to digital cultural heritage. Guides such as CURL Exemplars in Digital Archives (CEDARS) or NINCH provide the principles to guide preservation practices.[20] The majority of digital cultural heritage comprises few common file formats (html, jpeg, tif, doc) and there are efforts to establish cultural information standards that can support the long-term preservation of a multitude of file formats by building up information about them such as the PRONOM File Format Registry.[21]

## Trust and Collaboration

What are the drivers for the development of cultural information standards? Major international collaborative ventures speak volumes of the shared interests and benefits of digital cultural heritage and the desire to broaden access. For example: the Metadata Encoding Transmission Standard (METS)[22] program to implement metadata standards to enable cross-sector unionization of cultural information for international resource discovery and retrieval of digital cultural heritage; the Digital Preservation Coalition (DPC)[23] to develop the tools and resources to enable the preservation of the Internet, to name just two. Technical and semantic standards that ensure the information architecture and technical infrastructure of cultural institutions can support local communities and also enable interaction and collaboration with other cultural users and agencies.

As cultural agencies develop their digital archives to preserve digital cultural heritage there is significant professional communication, on both formal and informal levels, to share information and contribute peer review. This collective professional effort enables cultural agencies (and the private sector companies contracted to assist with that development) to work together to create robust information system design and technical infrastructure. Common cultural information practices and collegial openness and liaison provide a major support for the development of information standards, from the highly conceptual to the procedural, for the preservation of digital cultural heritage.[24] The united need for digital preservation has driven this phenomenal international effort. Cultural agencies have identified the highly conceptual Open Archival Information system (OAIS) reference model as a central point of reference that will inform how they develop digital archives, for example the e-Depot developed by the Koninklijke Bibliotheek (National Library of the Netherlands).[25] The OAIS reference model developed by the Consultative Committee for Space Data Systems was the result of a considerable collaboration of national space agencies and is publicly available.[26] This reference model is on its way to becoming an international standard for digital archiving.[27]

The development and maintenance of cultural information standards is heavily reliant on trust and collaboration. In the knowledge management and systems discourse, Maija-Leena Huotari and Mirja Iivonen define the overlapping features of trust as follows: "(1) Trust is based on expectations and interactions. (2) Trust is manifested in people's behavioral patterns. (3) Trust makes a difference" (2004, 8). Huotari and Iivonen describe the effect trust has on collaboration: "Trust has an effect on collaboration and the development of the structural dimension because it produces more interactions between the interdependent members of an organization or network. Therefore, the enhancement of trust in collaboration is crucial for knowledge creation" (2004, 15). Mutual trust, interest, and benefit drive effective collaboration (which drives a sharing of expertise and investment). These behaviors are observable in the information standard-setting for digital cultural heritage. In an environment of trust, rich information flows freely, and more knowledge is enabled. The technical, semantic, and communications standards (developed to assist with the collation and unionization of cultural information) are influenced by sociopolitical factors. So too are the information system and infrastructural standards (developed to assist with the preservation of digital cultural heritage) influenced by sociopolitical factors. If the sociopolitical environment in which cultural information standards are developed is one of openness (which it seems to be), with high levels of trust and collaboration, this bodes very well for the future of digital cultural heritage.

### Sociopolitical Forces

What are the sociopolitical forces that influence cultural information standards? The social nature of information, its relationship to data and knowledge, and its ability to be shaped, organized, and well-managed is what enables social connectivity (Brown and Duguid 2000). Social connectivity stimulates old relationships and creates new ones, expanding opportunities for new ideas, and building efficient and reliable networks of trust.[28] If the social relationships are not stabilized by consistency, respect, trust, or consideration, the information does not flow well or it is a constrained flow. Social connectivity is predicated on understanding and being cognizant of your community's resources, abilities, interests, and needs. That social awareness then needs to be integrated into the information standards developed for digital cultural heritage to enable cultural knowledge. Challenging traditional cultural engagement and proximity in the configuration of new knowledge spaces means a move away from that which has become intellectually rarefied (and calcified), and inhibiting of cultural connection to intellectually accessible, flexible, and engaging cultural connection. A considerable amount of attention in the larger national and international digital cultural heritage development programs is on the competitive nature and risk-management aspects of digital cultural heritage.[29] Who got there first? What commercial advantage is there in this? What happens if it is not economically feasible? Very little of the discussion examines the sociopolitical forces, social benefits, and what impact this will have culturally if power is unevenly held, or if it fails to impact. The sociopolitical forces that promoted an increased collection of cultural heritage and extended exposure to cultural information—for education and edification—emerged in the Victorian era to encourage the working classes into cultural institutions as a means of "improving" these class members (Bennett 1992, 1–8).

Why create digital cultural heritage and make it widely available? What social and economic impact is widening access going to have? What are the collective or particular benefits in having extended access? What is the social and economic value of digital cultural heritage? Aside from the fundamental need to consolidate and coordinate the collection and preservation of digital cultural heritage there is very little in the way of commentary that investigates the sociopolitical implications of widening access to cultural heritage and the sociopolitical drivers behind keeping it. Perhaps these social values are so embedded in cultural practice they seem self-evident. The fact that social and economic forces work rather at odds with each other is cause for tension and a situation that is ripe for exposure and debate. It pays to know what sociopolitical forces are at play when coursing into any new territory.

The investigation into what happens when people and digital technology collide and collude was being discussed in the area of cybernetics and technological innovations, and has emerged more recently as the concept of "cyberspace" in interdisciplinary discourse.[30] The philosophical shift from a physical to a virtual arrangement of collection, preservation, and access to cultural heritage revisits old and new social and cultural practices and assumptions, and sociopolitical power structures. What is socially acceptable or commonplace in the physical world can become enormously complex to recreate in the virtual. What has been inhibiting in the physical world can become enormously simple to enable in the virtual. It is vital that the construction of cultural information standards are guided by the understanding and interpretation of sociopolitical purpose in giving shape to the form and function of digital cultural heritage. A constant application of effort to anticipate the less-empowered is required to counter inequity. Openness in cultural information standard-setting permits broad and rich participation in design. This openness (or "federal thinking") in approach will provide robust information architecture to build up, preserve, and provide access to digital cultural heritage.

Browsing the shelves of a library or drifting through a gallery is a socially embedded process in the physical world. The ability to design and reconstruct, digitally, over mere decades what has slowly been developed over hundreds of years is an enormous leap forward in social and cultural expression and interaction. If there is any discussion of commercial interest it is really in terms of human computer interaction and the ability to design technology that enables this social behavior—this is one of the biggest challenges to the library profession (OCLC 2005). User-driven enabling technology is readily observable in consumer and leisure sectors, e.g. supermarkets and gaming. The same technology is beginning to be experimented with and translated for use in cultural institutions and stimulating new practices, e.g. wireless, 3D, and digital preservation.[31]

This advance in cultural computing relies upon a strong grasp of the information management practices that support the wide and socially intuitive flow of cultural information (Brown and Duguid 2000). The success and impact of digital cultural heritage in terms of function and service delivery is overwhelmingly reliant upon integrating socially driven information practices and reliability into its design. The notion of "stickiness" in Web use analysis is an acknowledgement that the ability to tap into information needs and the virtual "zeitgeist" when creating digital social space is crucial to understanding all digital relationships, i.e. economic, social, and cultural. Cultural knowledge spaces require similar attention to notions of structuring, offering, and response with regard to handling information and relationship building. Far more than "online transactions," cultural information standards form the spatial orientation for digital cultural heritage explorers.

## Sociopolitical Models

How do those sociopolitical forces impact upon digital cultural heritage? Information standards are information systems. Information systems influence and are affected by the sociopolitical context in which they are implemented. Tom Davenport and Larry Prusak classify five major political models that reflect cultural and governance structures of organizations and impact upon the management and sharing of information within the organizations (1992, 53–65). These political structures are technocratic utopianism, anarchy, feudalism, monarchy, and federalism. These conceptual structures have been widely used in the information field to analyze the nature and success (or failure) of information system implementation (in organizations) and examine how readily information is shared. Davenport and Prusak caution: "One reason the stakes are so high in information politics is that more than information is at stake. In order to arrive at a common definition of information requirements, organizations (sic) must often address not just the information they use, but the business (sic) practices and processes that generate the information" (1992, 54).

Davenport and Prusak find federalism as a governance structure is more effective in terms of "information quality, efficiency, commonality, and access" (1992, 55). Federalism is a more successful means for managing information because it "treats politics as a necessary and legitimate activity by which people with different interests work out among themselves a collective purpose and a means for achieving it" (1992, 59). Political power is exerted differently in the other four political models, that is, political power is wielded (feudalism), unmediated (anarchy), ignored (technical utopianism), or vested in individuals (monarchy), and the ability to effectively manage information is reduced and the information flow is restricted. Information flows, and is shared, in all of these situations, but how much of it flows, how this occurs, and to whom it flows is determined by sociopolitical forces. The imbalance in political power in the governance structures of the other four political models makes developing and sharing cultural information standards less possible. If cultural information standards are not collaboratively developed and shared, then information is not collectively accessible and it is less feasible to satisfy these information needs readily and equitably.

Debates raging about the digital divide have a bearing here in terms of whose cultural information needs are being served by the development of cultural information standards that support digital cultural heritage. The digital divide is "a social/political issue referring to the socioeconomic gap between communities that have access to computers and the Internet and those who do not. The term also refers to gaps that exist between groups regarding their ability to use ICTs (Information and Communications Technologies) effectively, due to differing levels of literacy and technical skills, as well as the gap between those groups that have access to quality, useful digital content and those that do not. The term became popular among concerned parties, such as scholars, policy makers, and advocacy groups, in the late 1990s."[32] It is also worth considering the sociopolitical forces that actively inhibit or override cultural exploration or expression counter to that of those holding the political (and economic) power.

The collection of and access to cultural heritage is primarily aimed at serving the cultural information needs of local or immediate communities. It is important to develop cultural information standards that are unique, reflect that proximity, and where the larger, dominant commonalities are no longer of benefit, or indeed of relevance. For example, the development of a Dublin Core metadata element set to reflect the unique nature of Maori language sets the balance between federalism on a larger and smaller collective scale linguistically in New Zealand (DCMI 2005). The fifteen elements of Dublin Core are a basic means of allowing diverse metadata sources to be pooled together for international information retrieval. For enriched searching, metadata element sets are constructed following international standards permit information retrieval on a more granular and localized level. As a result, cultural artifacts and information held in cultural institutions around the world are accessible generically and specifically. For example, in a multitude of repository collections around the world there are descriptions of a stone adze or weapon. However, a description of a stone adze or weapon with specific terminology referring to composition, such as greenstone, pounamu, nephrite, or jade, begins to focus the search and reflect different provenance or local terminology. Even in this example, the cultural communities that might hold these types of artifacts, that have cultural or social links, may use disparate languages, have varying levels of access to technology and resources, and possess conflicting attitudes to sharing cultural information. The ability to make those links means negotiating cultural information standards that are acceptable (and feasible) for generic and specific information retrieval.

Davenport and Prusak promote federalism as a means of managing information and establishing standards for cultures that celebrate empowerment and widespread participation. Davenport and Prusak acknowledge that this end is harder to achieve and takes more time but it provides a stronger basis for knowledge enabling. Standardization has an organizing effect upon cultural information, in so far as it provides a reliable infrastructure for cultural information exchange and exploration. If the collective purpose in creating and providing access to digital cultural heritage is to empower and share its richness, then negotiating and sharing the development of cultural information standards is a means of achieving this. Standardizing cultural information permits consistency in information structuring and sharing, at different levels (e.g., international, national, or regional); in different contexts (e.g., scientific, educational, or cultural); different language groups (e.g., European, Asian, or Middle Eastern); or different description standards (e.g., archival, artifactual, or bibliographic).

The five political models presented by Davenport and Prusak are a simple means to highlight what happens when power is, or is not, distributed evenly when implementing information systems. The translation and sensing skills of cultural information professionals channeled into developing information standards can be tested against them. Clearly, social and economic imperatives drive investment in digital cultural heritage. The question then is: how are these imperatives influencing the development of digital cultural heritage? The purpose in establishing information standards is to augment the collection and preservation of, and access to, cultural heritage. The expansion of cultural heritage into the digital realm can be seen as cultural expression: regionalism, nationalism, or globalization with the attendant reinforcements and biases. The political influence in the global context provokes some sober thinking when reflecting on the production of information standards to construct international cultural knowledge spaces and the need for good diplomatic relations. Political and economic realities influence the information design of digital cultural heritage projects and there is pause for thought about what impact this is having on the collective cultural knowledge space. Strategically, the overwhelming presence of developed nations and the predominance of English as the language of the Internet has already had a colonizing and homogenizing effect.[33]

National and/or large cultural institutions worldwide, mostly those in developed countries, are allocating vast sums of public money in cultural digital preservation projects and digitization.[34] The first ventures and lessons of digital development have occurred and the integration of new ventures into current operations is occurring. Collection information and the development of interactives and online exhibitions require the coordination of information standards within and without cultural repositories. The interest in cross-cultural resource discovery requires the coordination of information standards across cultural repositories (archives, galleries, libraries, museums) with diverse information practices, collections, and communities. The interest in cross-national cultural repository resource discovery requires the coordination of information standards across nations with diverse political and cultural relationships (trade, ethnicity, and religion). The interest in cross-cultural research and communication, facilitated by access to cultural information, is achievable when information standards make it possible for resource discovery across shared knowledge spaces through collaboration, diplomacy, trust, and mutual benefit.

All this requires a consolidation of practice within and across cultural information professions, sectors, and nations. This diplomatic exercise does indicate that the social and economic agreements and protocols associated with traffic and trade within and between nations: air, ocean, or land, are most definitely being replicated in the development of information standards to support cultural traffic and exchange in the institutional, sectoral, regional, national, and international knowledge spaces for digital cultural heritage (Veltman 2004). Where the coordination of cultural information across the Tasman Sea between Australia and New Zealand seems feasible there are considerable challenges in cultural information structuring, exchange, and unionization between Pakistan and India, North and South Korea, the Republic of Ireland and Northern Ireland. The opportunity to realize commonality in language, history, and heritage in information terms, that is, standard-setting, takes the discussion of cultural knowledge spaces into some very highly charged places. If information by its very character is social, then any structuring of it reflects the nature of cultural relationships and social exchange—challenging and beneficial or overwhelming and overpowering— political in the extreme!

## Sociopolitical Outcomes

What are the benefits of virtual access to digital cultural heritage? The benefits of virtual access to digital cultural heritage are that it: reinforces the physical presence of cultural heritage, i.e., an Internet presence; maintains social relevance with the social shift toward an Information Society;[35] extends the opportunity for interaction, i.e., 24/7 business hours of the Internet; extends access within and without the cultural organization, e.g., through interactives, virtual exhibitions, handheld guides, Web sites, online exhibitions; extends the opportunity for inter-organizational collabora- tion, e.g., unionized cultural description information and federated access to digital collections; and it reinterprets and establishes a digitally interactive relationship to cul- tural heritage.

The maintenance or improvement of people's interaction and satisfaction with ac- cess to cultural heritage is predominantly what motivates cultural practitioners in extending their skills, developing technologies, sharing insights, and seeking funding for digital cultural heritage. The reassessment of information standards and cultural practices in the volatility of social change exposes them to uniting and disuniting, advantaging and disadvantaging, political and economic interests. In the process of institutional and cultural transformation, the social fabric can be rent, or rendered more robust. The social and economic impact of digital cultural heritage and digital cultural connectedness is *yet* to be fully realized or, in fact, understood. This shift to digital practice could be constrained by economic imperatives and undermined by technocratic imperialism and the inability of cultural institutions to rigorously assess (pragmatically, with social ideals intact) and use sustainable and appropriate business models.[36] The risk in establishing cultural information standards with fiscal drivers is the commodification, reduction, and diminishment of the potential for digital cultural heritage to flourish and have widely exploitable social and economic value. Little of the political envisioning that relates to public sector digital development (e-government) speaks explicitly of the social motivations of the public sector—that is, public service and social good as they create public infrastructure and thus public knowledge space.[37]

There is some discussion of the digital divide; however, there is little recognition of the social needs for and benefits arising from digital cultural heritage that will form part of that public knowledge space.

The cost benefits of long-term investment in information infrastructures and standards for digital cultural heritage are being examined. To achieve the vision of digitally facilitated or augmented access to cultural heritage there may be a reprioritization of resources in other areas of cultural heritage. Considerable sums of public money have been invested in seeding projects. The next step is to ascertain how, what, and where these ventures should be situated or where monies can be gained to maintain and continue to develop them.[38] Where digital cultural heritage projects commenced with a combined social and economic imperative, they continue in that vein, e.g., Scottish Cultural Resources Access Network (SCRAN), Art Museum Image Consortium (AMICO), and ARTstor.[39] Where they have had a solely social imperative, e.g., Picture Australia, Australian Museums & Galleries On-Line (AMOL), Matapihi—Open the Window (NZ), and Online Archive of California (OAC),[40] they continue in their outreach or social connective role and form part of the new digital cultural infrastructure. It seems there is a "back to reality" in focus and a bridging between new and old strategic and financial concerns within the larger national and international development programs.[41] The requirement to be consistent with social mandate and inclusive of new concerns and cultural information needs means a serious review of collecting, preservation, and access in the cultural sector. This reprioritization is to take advantage of the more widely available and exploitable value of digital cultural heritage.

## Summary

The proficiency and efficiency with which standardizing and managing cultural information is achievable requires due care, and potentially renders it a prosaic exercise. Information standards are technical and intellectual in nature and facilitate the discovery process (information search and retrieval) thereby fulfilling information needs and providing access. Information practices, enabled by information standards, are social in nature, and are part of the metaphysical engagement with information as a rich social and cultural resource. The intellectual and social facility integrated into information standards of digital cultural heritage is realized in what cultural information practices it enables and the embedded cultural knowledge which becomes absorbed. If digital cultural heritage is a collective good, then potential social and economic value is the outcome. By its very social nature, physical and virtual cultural heritage will always be influenced by politics and guided by social forces. Therefore, any political agenda that sets in motion and funds information design and standardization in the cultural sector comes with mixed imperatives. The goodwill shown by larger and more economically powerful nations and cultural institutions in leading the development of information standards and sharing insights is considerable. The investment reflects an understanding of the value of digital cultural heritage and the openness reflects an understanding of the value of collaborative effort.

Cultural knowledge spaces are about cultural territory, and in many ways, the information standards developed to enable interaction are as much arid technical and intellectual concepts as they are gravid with social and political forces in influencing information practices. It pays to bear in mind the potent nature of those forces when confronting issues of cultural power broking, (appropriation, isolation, alienation, or sublimation) in the digital realm. Openness in information standards (technical, semantic, telecommunications) facilitates access, and social and economic value will arise from that.[42] Cultural practitioners must transfer the social awareness, integral to their roles as cultural information creators and knowledge enablers, into their shaping of information standards for digital cultural heritage. If cultural information standards are the rules of engagement and signposts for cultural exchange and they facilitate access, then it will be because social and cultural practices and trust are embedded in the design of the virtual terrain in which digital cultural heritage is kept and made available. As long as cultural information professionals, in the developed world particularly, understand their role as social interpreters and cultural brokers in this pioneering and mapping exercise, there is an opportunity for wider cultural congress.

### Notes

1. See Davenport, Eccles, and Prusak 1992; Bourdieu and Wacquant 1992; and Baron, Field, and Schuller 2000.

2. Dublin Core Metadata Initiative.

3. The Getty, "Art & Architecture Thesaurus Online."

4. Education Resources Information Center (ERIC), "Thesaurus."

5. Library of Congress, "Tools for Authority Control, Subject Headings."

6. United States, National Library of Medicine, National Institutes of Health, "Medical Subject Headings."

7. National Archives of Australia, "Record Keeping, Government Online, AGLS."

8. NZ E-government, "NZGLS Metadata Standard."

9. Education Network Australia, "EdNA Metadata Standard."

10. International Council on Archives, "ISAD(G) General International Standard Archival Description."

11. Museums Documentation Association, "SPECTRUM: The UK Museum Documentation Standard."

12. International Federation of Library Associations and Institutions (IFLA), "Universal Bibliographic Control and International MARC Core Programme."

13. See International Council of Museums, "The CIDOC Conceptual Reference Model," and Gill 2004.

14. Library of Congress, "Encoded Archival Description."

15. Text Encoding Initiative, "What Is the TEI Consortium?"

16. Library of Congress, "Metadata for Images in XML Standard (MIX)."

17. Online Computer Library Center, "PREMIS."

18. World Wide Web Consortium, "Extensible Markup Language (XML)."

19. Library of Congress, "METS Official Web Site."

20. See CURL Exemplars in Digital Archives, "Cedars Guide to Digital Preservation Strategies" and National Initiative for a Networked Cultural Heritage, "The NINCH Guide to Good Practice in the Digital Representation and Management of Cultural Heritage Materials."

21. National Archives, "PRONOM, the File Format Registry."

22. Library of Congress, "METS Official Web Site."

23. See Digital Preservation Coalition.

24. Koninklijke Bibliotheek 2002.

25. Koninklijke Bibliotheek, "The History of the e-Depot."

26. National Aeronautics and Space Administration, "Reference Model for an Open Archival Information System."

27. National Aeronautics and Space Administration, "US Efforts towards ISO Archiving Standards: Overview."

28. See DigiCULT, "Thematic Issue 5: Virtual Communities and Collaboration in the Heritage Sector"; Putnam 2002; and Fine 2001.

29. See DigiCULT, "Technology Watch Brief 8: Technologies and New Socioeconomic Business Models [draft]"; Joint Information Systems Committee, "Funding Opportunities," "Information Environment, Sustainability Study," "Portal—User Requirements and Sustainability, Study Proposal," and "Strategic Activities: Information Environment"; MINERVA, "Coordinating Digitisation in Europe"; Council on Library and Information Resources, "Building and Sustaining Digital Collections"; and the National Digital Information Infrastructure and Preservation Program Web site at ⟨http://www .digitalpreservation.gov/⟩.

30. See Spiller 2002, Stille 2002, and Rheingold 1991.

31. See Koninklijke Bibliotheek, "E-depot and Digital Preservation, Expert Centre"; the VROOM Web site, at ⟨http://www.vroom.org.au⟩; and Spinazze et al. 2004.

32. Wikipedia definition: ⟨http://en.wikipedia.org/wiki/Digital_divide⟩ (accessed 3 January 2005).

33. Australian Broadcasting Corporation, "Lingua Franca."

34. See The Stationery Office, "Irreplaceable Records on the Human Cost of War to be Preserved"; JISC, "Digitisation Programme"; NDIIPP; National Library of New Zealand, "Media Releases"; and National Archives of Australia, "Digitising Records for Improved Accessibility."

35. See the World Summit on the Information Society Web site, at ⟨http://www.itu.int/ wsis/⟩.

36. DigiCULT, "Technological Landscapes for Tomorrow's Cultural Economy."

37. Commonwealth Network of Information Technology for Development Foundation, "Country Profiles of E-governance, 2002"; and Doczi 2000.

38. See DigiCULT, "Technological Landscapes for Tomorrow's Cultural Economy"; JISC, "Funding Opportunities"; and the National Science Foundation Web site, at ⟨http:// www.nsf.gov⟩.

39. See the Scottish Cultural Resources Access Network Web site, at ⟨http://www.scran.ac .uk⟩; Art Museum Image Consortium, at ⟨http://www.amico.org⟩; and ARTstor, at ⟨http://www.artstor.org/info/⟩.

40. Picture Australia, at ⟨http://www.pictureaustralia.org⟩; Australian Museums & Galleries Online, at ⟨http://www.amol.org.au⟩; Matapihi—Open the Window, at ⟨http://ndf

.natlib.govt.nz/about/matapihi.htm⟩; and Online Archive of California, at ⟨http://www
.oac.cdlib.org⟩.

41. See DigiCULT, "Technological Landscapes for Tomorrow's Cultural Economy"; JISC,
"Funding Opportunities"; and the National Science Foundation Web site.

42. See Baron, Field, and Schuller 2000; Bourdieu and Wacquant 1992; Coleman 1988 and
1990; and Fine 2001.

## References

AMOL (Australian Museums & Galleries On-Line). Available at ⟨http://www.amol.org.au⟩
(accessed 7 July 2004).

Art Museum Image Consortium (AMICO). Available at ⟨http://www.amico.org⟩ (accessed 7
July 2004).

ARTstor. Available at ⟨http://www.artstor.org/info/⟩ (accessed 13 February 2005).

Australian Broadcasting Corporation (ABC). 2002. "Lingua Franca 19/01/2002, Language and
the Internet (transcript)." Available at ⟨http://www.abc.net.au/rn/arts/ling/stories/s416337
.htm⟩ (accessed 13 February 2005).

Baron, Stephen, Tom Field, and John Schuller. 2000. *Social Capital: Critical Perspectives*. Oxford: Oxford University Press.

Bennett, Tony. 1992. "Museums, Government, Culture." *Sites* 25: 1–8.

Bourdieu, Pierre, and Loïc J. D. Wacquant. 1992. *An Invitation to Reflexive Sociology*. Chicago:
Chicago University Press.

Brown, John Seely, and Paul Duguid. 2000. *The Social Life of Information*. Boston: Harvard
Business School Press.

Coleman, James S. 1988. "Social Capital in the Creation of Human Capital." *American Journal
of Sociology* 94 (Suppl): 95–120.

Coleman, James S. 1990. *Foundations of Social Theory*. Cambridge, Mass.: The Belknap Press.

Commonwealth Network of Information Technology for Development Foundation (COMNET-
IT). 2002. "Country Profiles of E-governance." Available at ⟨http://www.unesco.org/
webworld/news/2002/e-governance.rtf⟩ (accessed 25 February 2005).

Council on Library and Information Resources (CLIR). 2001. "Building and Sustaining Digital
Collections: Models for Libraries and Museums." Available at ⟨http://www.clir.org/pubs/
reports/pub100/pub100.pdf⟩ (accessed 20 May 2004).

CURL Exemplars in Digital Archives (Cedars). 2002. "Cedars Guide to: Digital Preservation Strategies." Available at ⟨http://www.leeds.ac.uk/cedars/guideto/dpstrategies/⟩ (accessed 25 January 2005).

Davenport, Thomas H., Robert G. Eccles, and Laurence Prusak. 1992. "Information Politics." *Sloan Management Review* (Fall): 53–65.

DigiCULT. 2001. "Technological Landscapes for Tomorrow's Cultural Economy, Unlocking the Value of Cultural Heritage." Available at ⟨http://www.digicult.info/pages/report.php⟩ (accessed 25 February 2005).

DigiCULT. 2003. "Technology Watch Brief 8: Technologies and New Socio-economic Business Models [draft]." Available at ⟨http://www.digicult.info⟩ (accessed 20 May 2004).

DigiCULT. 2004. "Thematic Issue 5: Virtual Communities and Collaboration in the Heritage Sector." Available at ⟨http://www.digicult.info/downloads/html/1075714044/1075714044.html⟩ (accessed 25 February 2005).

Digital Preservation Coalition (DPC). Available at ⟨http://www.dpconline.org/⟩ (accessed 3 January 2005).

Doczi, Marianne. 2000. "Information and Communication Technologies and Economic Inclusion: Addressing the Social and Economic Implications of Limited E-literacy and Access to Information and Communication Technologies," March. Available at ⟨http://www.med.govt.nz/pbt/infotech/ictinclusion/⟩ (accessed 25 February 2005).

Dublin Core Metadata Implementation (DCMI). n.d. "Localization and Internationalization Working Group." Available at ⟨http://dublincore.org/groups/languages/⟩ (accessed 3 January 2005).

Dublin Core Metadata Initiative. Available at ⟨http://dublincore.org/⟩ (accessed 13 February 2005).

Education Network Australia. n.d. "EdNA Metadata Standard." Available at ⟨http://www.edna.edu.au/metadata⟩ (accessed 13 February 2005).

Education Resources Information Center (ERIC). "Thesaurus." Available at ⟨http://eric.ed.gov/ERICWebPortal/Home.portal?_nfpb=true&_pageLabel=Thesaurus&_nfls=false⟩ (accessed 13 February 2005).

Fine, Ben. 2001. *Social Capital Versus Social Theory*. London: Routledge.

The Getty. "Art & Architecture Thesaurus Online." Available at ⟨http://www.getty.edu/research/conducting_research/vocabularies/aat/⟩ (accessed 13 February 2005).

Gill, Tony. 2004. "Building Semantic Bridges Between Museums, Libraries and Archives: the CIDOC Conceptual Reference Model." *First Monday*, May. Available at ⟨http://www.firstmonday.org/issues/issue9_5/gill/index.html⟩ (accessed 13 February 2005).

Huotari, Maija-Leena and Mirja Iivonen. 2004. "Managing Knowledge-based Organizations Through Trust." In *Trust in Knowledge Management and Systems in Organizations*, ed. M.-L. Huotari and M. Iivonen, 1–29. Hershey: Idea Group.

International Council on Archives. 1999. "ISAD(G) General International Standard Archival Description." Available at ⟨http://www.ica.org/biblio/cds/isad_g_2e.pdf⟩ (accessed 13 February 2005).

International Council of Museums. 2003. "The CIDOC Conceptual Reference Model." Available at ⟨http://cidoc.ics.forth.gr/⟩ (accessed 13 February 2005).

International Federation of Library Associations and Institutions (IFLA). "Universal Bibliographic Control and International MARC Core Programme." Available at ⟨http://www.ifla.org/VI/3/p1996-1/unimarc.htm⟩ (accessed 13 February 2005).

Joint Information Systems Committee (JISC). n.d. "Digitisation Programme." Available at ⟨http://www.jisc.ac.uk/index.cfm?name=programme_digitisation⟩ (accessed 17 July 2004).

Joint Information Systems Committee (JISC). "Funding Opportunities." Available at ⟨http://www.jisc.ac.uk/index.cfm?name=funding⟩ (accessed 3 June 2004).

Joint Information Systems Committee (JISC). "Information Environment, Sustainability Study: Common Services & Digital Infrastructure, Study Proposal." Available at ⟨http://www.jisc.ac.uk/index.cfm?name=funding_iesustainability⟩ (accessed 3 June 2004).

Joint Information Systems Committee (JISC). n.d. "Portal—User Requirements and Sustainability, Study Proposal." Available at ⟨http://www.jisc.ac.uk/index.cfm?name=funding_portals⟩ (accessed 2 June 2004).

Joint Information Systems Committee (JISC). n.d. "Strategic Activities: Information Environment." Available at ⟨http://www.jisc.ac.uk/index.cfm?name=about_info_env⟩ (accessed 3 June 2004).

Koninklijke Bibliotheek (National Library of the Netherlands). n.d. "The History of the E-depot." Available at ⟨http://www.kb.nl/dnp/e-depot/dm/geschiedenis-en.html⟩ (accessed 19 February 2005).

Koninklijke Bibliotheek (National Library of the Netherlands). 2002. "Workshop: Digital Preservation, Technology & Policy, 13 December 2002." Available at ⟨http://www.kb.nl/hrd/dd/dd_links_en_publicaties/workshop2002/presentations.html⟩ (accessed 19 February 2005).

Koninklijke Bibliotheek. "E-depot and Digital Preservation, Expert Centre." Available at ⟨http://www.kb.nl/kb/resources/frameset_kenniscentrum-en.html⟩ (accessed 3 June 2004).

Library of Congress. "Encoded Archival Description (EAD)." Available at ⟨http://www.loc.gov/ead/⟩ (accessed 13 February 2005).

Library of Congress. n.d. "Metadata For Images in XML Standard (MIX)." Available at ⟨http://www.loc.gov/standards/mix/⟩ (accessed 13 February 2005).

Library of Congress. "METS (Metadata Encoding and Transmission Standard), Official Web Site." Available at ⟨http://www.loc.gov/standards/mets⟩ (accessed 3 July 2004).

Library of Congress. n.d. "Tools for Authority Control, Subject Headings." Available at ⟨http://www.loc.gov/cds/lcsh.html#lcsh20⟩ (accessed 13 February 2005).

Matapihi—Open the Window. n.d. Available at ⟨http://ndf.natlib.govt.nz/about/matapihi.htm⟩ (accessed 17 July 2004).

MINERVA, Progress Report of the National Representatives Group. n.d. "Coordinating Digitisation in Europe." Available at ⟨http://www.minervaeurope.org/publications/globalreport.htm⟩ (accessed 2 June 2004).

Museums Documentation Association. n.d. "SPECTRUM: The UK Museum Documentation Standard." Available at ⟨http://www.mda.org.uk/spectrum.htm⟩ (accessed 13 February 2005).

National Aeronautics and Space Administration (NASA)—Consultative Committee for Space Data Systems. 2002. "Reference Model for an Open Archival Information System (OAIS)." Available at ⟨http://ccsds.org/publications/archive/650xObl.pdf⟩ (accessed 19 February 2005).

National Aeronautics and Space Administration (NASA). n.d. "US Efforts Towards ISO Archiving Standards—Overview." Available at ⟨http://ssdoo.gsfc.nasa.gov/nost/isoas/us/overview.html⟩ (accessed 19 February 2005).

National Archives of Australia (NAA). 2002. "Digitising Records for Improved Accessibility." Available at ⟨http://www.naa.gov.au/publications/corporate_publications/digitising_thing.pdf⟩ (accessed 25 February 2005).

National Archives of Australia. n.d. "Record Keeping, Government Online, AGLS." Available at ⟨http://www.naa.gov.au/recordkeeping/gov_online/agls/summary.html⟩ (accessed 13 February 2005).

National Archives [of the UK]. n.d. "PRONOM, the File Format Registry." Available at ⟨http://www.nationalarchives.gov.uk/pronom/⟩ (accessed 3 January 2004).

National Digital Information Infrastructure and Preservation Program (NDIIPP). Available at ⟨http://www.digitalpreservation.gov/⟩ (accessed 17 July 2004).

National Initiative for a Networked Cultural Heritage (NINCH). 2002. "The NINCH Guide to Good Practice in the Digital Representation and Management of Cultural Heritage Materials." Available at ⟨http://www.nyu.edu/its/humanities/ninchguide/index.html⟩ (accessed 2 June 2004).

National Library of New Zealand (NLNZ). 2004. "Media Releases: National Library to Capture New Zealand's Digital Heritage, 30 May 2004." Available at ⟨http://www.natlib.govt.nz/bin/media/pr?item=1085885702⟩ (accessed 17 July 2004).

National Science Foundation (NSF). Available at ⟨http://www.nsf.gov⟩ (accessed 3 June 2004).

NZ E-government. "NZGLS Metadata Standard." Available at ⟨http://www.e-government.govt.nz/nzgls/standard/index.asp⟩ (accessed 13 February 2005).

Online Archive of California (OAC). Available at ⟨http://www.oac.cdlib.org⟩ (accessed 3 July 2004).

Online Computer Library Center (OCLC). n.d. "PREMIS (PREservation Metadata Implementation Strategies)." Available at ⟨http://www.oclc.org/research/projects/pmwg/⟩ (accessed 3 January 2005).

Online Computer Library Center (OCLC). 2004. "The 2003 Environmental Scan: A Report to the OCLC Members." Available at ⟨http://www.oclc.org/reports/escan/⟩ (accessed 25 February 2005).

Picture Australia. Available at ⟨http://www.pictureaustralia.org⟩ (accessed 3 July 2004).

Putnam, Robert D., ed. 2002. *Democracies in Flux: The Evolution of Social Capital in Contemporary Society*. Oxford: Oxford University Press.

Rheingold, Howard. 1991. *Virtual Reality*. New York: Summit Books.

Scottish Cultural Resources Access Network (SCRAN). Available at ⟨http://www.scran.ac.uk⟩ (accessed 7 July 2004).

Spiller, Neil, ed. 2002. *Cyber Reader: Critical Writings for the Digital Era*. London: Phaidon.

Spinazze, Angela, Geri Gay, Michael Stefanone, and Emily Posner. 2002. "Understanding Visitor Expectations and Museums as Mobile Computing Environments: A Report on Handhelds in the Museum Landscape. Handscape Symposium Final Report, Computer Interchange of Museum Information Consortium (CIMI)." Available at ⟨http://www.cimi.org/handscape/hs_symposium_0602_final.html⟩ (accessed 3 June 2004).

The Stationary Office (TSO). n.d. "Irreplaceable Records on the Human Cost of War to Be Preserved." Available at ⟨http://www.tso.co.uk/latestinformation/site.asp?FO=1142929&DI=520917⟩ (accessed 3 July 2004).

Stille, Alexander. 2002. *The Future of the Past*. New York: Farrar Straus & Giroux.

Text Encoding Initiative. n.d. "What Is the TEI Consortium?." Available at ⟨http://www.tei-c.org/⟩ (accessed 13 February 2005).

United States, National Library of Medicine, National Institutes of Health. "Medical Subject Headings." Available at ⟨http://www.nlm.nih.gov/mesh/meshhome.html⟩ (accessed 13 February 2005).

Veltman, Kim H. 2004. "Towards a Semantic Web for Culture." *Journal of Digital Information* 4, no. 4. Available at ⟨http://jodi.ecs.soton.ac.uk/Articles/v04/i04/Veltman/⟩ (accessed 2 June 2004).

VROOM (Virtual Room). Available at ⟨http://www.vroom.org.au⟩ (accessed 3 June 2004).

World Summit on the Information Society (WSIS). Available at ⟨http://www.itu.int/wsis/⟩ (accessed 19 February 2005).

World Wide Web Consortium (W3C). "Extensible Markup Language (XML)." Available at ⟨http://www.w3.org/XML⟩ (accessed 3 January 2004).

# 12 Finding a Future for Digital Cultural Heritage Resources Using Contextual Information Frameworks

Gavan McCarthy

The imperative to manage accurate and comprehensive information to meet a variety of needs has long been acknowledged by the cultural heritage community. Some of this information, generally in the form of objects and associated records, is recognized as necessary for the management or curation of cultural heritage objects today, but the community is also aware that some or much of this information will be required to ensure that valuable objects continue to be effectively managed over the long term. In practice, this deceptively simple requirement has been difficult to implement.

Other areas of society, for example the radioactive waste community, have been developing their thinking about these issues under the rubric of sociotechnical sustainability. The influence of the Brundtland Commission in 1987 is most noticeable in this area. It stated that "sustainable development is development that meets the needs of the present without compromising the ability of future generations to meet their own needs."[1] These principles of sustainability could equally be applied to the long-term management of digital cultural heritage resources and may well provide the conceptual framework in which we can conceive a path forward. However, the issues surrounding the successful transfer of digital objects or indeed information about digital objects to future generations have proven to be resiliently problematic. In practice, the rate of technological change or perhaps from this perspective the rapid rate of redundancy remains a major stumbling block and therefore indicates that there are fundamental conceptual issues that need to be addressed.

Information needs to be preserved in such a way that a future society can use it with confidence. In other words, the information objects or records must be readily comprehensible by a future audience, not just people inside a given community, but all others with some sort of interest or concern. However, the experience gained thus far in the archives and museum world, as well as in the wider community, suggests that existing systems and practices are still ill-equipped to meet information needs over the long term.

During the past thirty or forty years there have been some notable examples of the human race investing considerable resources to obtain nonreproducible information, (such as space exploration), only to discover years later that it cannot be accessed. Although this is often a result of insufficient planning for its long-term management, there appear to be two major reasons why these failures have occurred:

*Media redundancy*   where physical changes in either the medium or the supporting technology have rendered the data unreadable; and

*Epistemic failure*   where there has been inadequate preservation of the information necessary to explain the structure and meaning of the data (known in some fields as metadata).

Media redundancy, epitomized in the digital world by short lifespans and machine dependency, has been a major preoccupation of those dealing with the curation of digital resources.[2] It has been recognized as a natural and inevitable process that applies to all media and is an issue that must always be addressed when considering long-term information transfer. However, it is the second point—perhaps the most critical—which has only recently started to be addressed by the cultural heritage community. It is the need to sustain our knowledge of the existence of the elements that comprise the complex sociotechnical framework which is at issue. The nature of the technology and cultural heritage objects themselves has been the focus of exhaustive study by scientists, technologists, historians, curators, and archivists. The knowledge generated by these studies is typically condensed and abstracted in professional publications and managed through the library system. But what of all the other elements of the sociotechnical framework—the politics, the intra- and interorganizational relationships, the day-to-day running of a museum or archive, the standards framework, the technology providers, the implicit know-how that enables things to work, the human to human interactions—how is the knowledge of these sustained over time?

Where information has survived and remained accessible for many years it has often been the result of fortuitous circumstances rather than well-planned and resourced processes. The information required to manage the long-term viability of digital cultural heritage resources is sufficiently important that the present generation cannot afford to ignore information preservation failures of the past. It must also be prepared to invest the necessary resources now, not just to remedy the inadequacies of past practice but also to prepare systems for the future.

Epistemic failure typically occurs when information is recorded and preserved in isolation and exists, therefore, out of context. Information has limited use, for example, if its provenance is unknown, its significance is unclear, and the creator cannot be consulted to explain semantic ambiguities and ontological subtleties. Relatively simple systems, including catalogs, designed to preserve resources, can be deployed, but if they are not meshed within a broader contextual information framework there is a significant risk that they will not meet society's needs, either now or in the future. The limitation of catalog thinking has been a theme addressed by leading thinkers in library informatics and is reflected in the work of Carl Lagoze (2000).

It is argued in this chapter that the systematic preservation of contextual information is currently the most likely means by which we can mitigate epistemic failure. The chapter highlights the fact that the creation and subsequent management of digital cultural heritage resources gives rise to a considerable amount of information and knowledge; often held by one or two individuals. As a rule, it is hoped that adequate information is captured in the records generated by the heritage community, for the heritage community, for current purposes. However, it cannot be said that the knowledge necessary to understand that information by outsiders to that community, and by future generations, is being effectively preserved.

## Past Thinking and New Ideas

### A View from the Archives

In the cultural heritage community, in particular the archives area, much work has been done over many generations to tackle the question of effective information transfer through time. In the 1960s the Australian archival community made a substantive move away from the catalog mentality that dominated thinking internationally about archival management and description. The systems analysis, undertaken by Scott (1966) and others, of what actually was going on in real life, particularly in relation to government records, suggests that a network approach may be much more effective. In practice this meant separating but relating the description of materials from the description of the people and organizations responsible for their creation and management. What they set up was a means by which they could start to map the dynamic, ever-changing complex sociopolitical environment in which Australian government records were created. This simple but seminal change in thinking has proved to be remarkably well-suited to the digital world. This approach to archival management has now permeated the international archival community and forms the foundation for the two standards promulgated by the International Council on Archives, the International Standard Archival Authority Records for Corporate Bodies, Persons and Families (ISAAR[CPF]) (2004), and ISAD(G) General International Standard Archival Description (2000).

However, probably due to the limits of the technology at the time, with a few notable exceptions, the thinking stagnated as to how this descriptive conceptual framework might be used for the linking and sharing of common knowledge (McCarthy 2000). As a result, silos of unconnected knowledge banks of archival materials, some with elegantly maintained internal contextual information architectures, became standard.

In the museum world, a similar development in thinking occurred that became embodied in the work of Martin Doerr and others (Crofts 2003).This resulted in the development by the International Council of Museums, International Committee on Documentation, of a conceptual reference model known as the CIDOC CRM.[3] The emphasis of their work was on the provision of definitions and a formal structure for describing the implicit and explicit concepts and relationships used in cultural heritage documentation. The development of the Functional Requirements for Bibliographic Records by the International Federation of Library Associations and Institutions also revealed a commonality of issues and a fundamentally similar solution in terms of informatics.[4]

*Sociotechnical Complexity*

The last decade has seen a blossoming of studies looking at the nature of complexity from a variety of perspectives drawing on the humanities, social sciences, and mathematics. Helga Nowotny, currently Chair of the European Research Advisory Board (EURAB) and based in Switzerland, has been tackling such issues as the need for socially robust knowledge in evolving complex environments (1999, 12–16). She draws on the work of Niklas Luhmann, one of Germany's foremost figures in social theory, who notes that complexity is inherent in social systems—it is "the unobserved wilderness of what happens simultaneously" (Luhmann 2000). She continues: "Luhmann's reference to the ultimately unfathomable complexity of the world—that which happens simultaneously—implies also its ultimate uncontrollability. While this definition of complexity has an elegance that differs from that used in the natural sciences, it has the advantage of leaving space for the invention of social mechanisms of coping, aimed at reducing its otherwise unbearable degree of uncontrollability. All human societies have therefore invented means of coping with uncertainty and ways of reducing complexity" (Nowotny 2003, 66–78). Once the broad concept of contextual information frameworks is understood it becomes clear that they have long since been key mechanisms humans have used to help make sense of the world around them and the culture they create. Indeed, it could be argued that they are intuitively and essentially human.

In mathematics and physics, studies of the nature of open complex networks really got underway in the mid 1990s. In 2002 Albert-László Barabási, a leader in this field, published a major work that explored the issue of "how everything is connected to everything else and what it means for science, business and everyday life" (2002). The surprising and ubiquitous properties of these complex networks were shown to share mathematical foundations such as power laws but also to have links with the now popular idea of the small-world effect. Barabási noted: "Real networks are not static, as all graph theoretical models were until recently. Instead, growth plays a key role in shaping their topology. . . . There is a hierarchy of hubs that keep these networks together, a heavily connected node closely followed by several less connected ones, trailed by dozens of even smaller nodes. No central node sits in the middle . . . controlling and monitoring every link and node. There is no single node whose removal could break [a network]."

This work provided an intellectual and conceptual milieu in which new strategies could be conceived to deal with the intransigent problems of information transfer to future generations and the viability of digital cultural heritage resources. The nature of the cultural heritage community, in itself a complex sociotechnical network, and the materials it preserves, would appear ideal for this type of treatment.

## Threats to Effective Knowledge Transfer

As time goes on, locating, accessing, and understanding information becomes more difficult in an inadequate or poorly managed system. The inevitable loss of implicit or contextual information is particularly significant and this is exacerbated by the fact that skilled and knowledgeable staff tend to be more mobile these days, moving on to other projects. For a cultural heritage community at the dawn of a new series of challenges arising from the creation and curation of digital cultural heritage resources, it is more important than ever to recognize the existence of specialist knowledge and to prevent its loss. It is worth recording that there have been some extraordinary examples of human cultures, notably in Canada and Australia, where knowledge has been sustained in communities for many thousands of years (Harris 1997). The relative social and environmental stability that enabled this sustainability of knowledge is not a feature of contemporary Western society.

## Physical Threats

The loss of knowledge can be extreme, total, immediate, and irreversible (catastrophic loss); for example, destruction of a records archive or museum store by fire. Alternately, the loss can be limited, selective, and relatively slow (graceful degradation); for example, the natural and inevitable disintegration of the physical medium. The severity of knowledge loss may be reduced or indeed eliminated by planning and implementing countermeasures. These may include making multiple copies of records or digital objects in a variety of different media, migrating records to new media, and the regular sampling of records to assess the state of the media. The preoccupation with the media life as the primary issue has led to the development of new media such as laser-etched silicon carbide tiles with remarkable properties of endurance and resistance to environmental factors (Sugiyama 2003).

## Change and the Dispersal of Knowledge

Work undertaken by staff of the Australian Science and Technology Heritage Centre[5] since 1985, in a range of science, technology, and broad cultural settings has revealed that one of the most serious and likely sources of knowledge loss is institutional or organizational change. Change, often characterized as a matrix of simultaneous and sequential events, can occur at a personal level when a particular expert retires taking with them valuable implicit knowledge. At the other extreme, a country might become a victim of some form of major disruption such as war or terrorist assault resulting in the destruction of material and professional intellect.

An everyday example of how knowledge can be lost through change is the increasing use of short-term contract staff. In order to spread costs, reduce timescales, and improve efficiency, many organizations use third-party experts who meticulously build up a knowledge base which is subsequently lost when the contract ends. Thus, the accumulated knowledge becomes increasingly dispersed and disconnected. If this knowledge dispersal is then combined with organizational change, the likelihood of losing the information increases dramatically.

## Responsibility

Although information at all levels faces these risks, it is the critical contextual information, not systematically managed, that is most at risk of loss. Organizations creating and managing digital resources must therefore identify the full range of risks and implement effective mitigating or countermeasures. Generic approaches can be problematic as the risks tend to be context-dependent, varying from organization to organization and from one country to another. Any top-tier policy should recognize the different risks and allow custodians to develop systems best suited to counteract local threats.

## A Contextual Information Framework

If captured and documented, contextual knowledge that has been accumulated over the years should enable subsequent generations to understand the significance of preserved digital cultural heritage resources. This could be used to form the nucleus of a continually evolving sustainable knowledge base. If systematically documented to locate its position in both time and space, this contextual information, by its very nature, will map the changes in the sociotechnical environment. Through access to this comprehensive, reliable, and accurate knowledge each generation should have the necessary confidence to make informed judgments and decisions about further curation and management. Conversely, if no action is taken, valuable implicit knowledge will be lost and may ultimately render any associated records and other information objects incomprehensible and the resource itself unusable.

Therefore, a possible strategy for preserving knowledge about digital cultural heritage resources is through the development and maintenance of a distributed contextual information framework. This framework would be designed to link together sources of knowledge that may be of value to a future society. A contextual information framework is composed of information objects that represent the constituents that comprise events. People, organizations, concepts, ideas, places, natural phenomena, events themselves, cultural artifacts (including records, books, works of art), and digital cultural heritage resources could all be defined as entities and play the role of constituent. The mapping of relationships between these entities creates a network of nodes and arcs that mimics actuality. The selective use of entity and relationship types can convert otherwise impossibly complex sociotechnical environments into information architectures or networks with remarkable and useful properties. As

Nowotny suggests, we invent means of coping with uncertainty and ways of reducing complexity. The Internet could be used as the basis for establishing such a network and making it widely accessible.

An electronic information network of this type has the properties of a scale-free complex network where many nodes (information objects created to meet international standards) are interlinked via arcs (using defined relationship types) (Barabási 2002). They therefore can be referenced or cited and new nodes can be added at any time, upgraded, or copied to new locations. The system would be both recursive and reflexive in structure, with nodes grouped in clusters reflecting groupings that occur in actuality, and the nodes themselves composed internally of similarly interlinked or networked objects.

One of the interesting aspects of a contextual information framework is that the same entity may appear in many clusters or contexts, where it is likely to have different roles or functions. The establishment and implementation of an identity relationship type to systematically link these multiple representations of the same agent creates a higher level network structure that greatly assists in the navigation and usability of the system. In fact, it is a conscious implementation of the small-world effect which enables the human-scale navigation of vast and complex information networks. A critical issue that has been identified in the library and museum worlds, in particular, is the difficulty they have in matching entities and establishing that they are indeed the same entity (Rahm and Do 2000). The matching or identity-mapping of authority entities from information silos has been tackled at various levels, but as a rule it has been within the context of creating a union catalog of unique entries and not a network of entities that are linked to each other by a defined relationship that establishes that the two entities are indeed referring to the same thing.

One of the ancillary benefits of this type of system is that core information relating to key constituents becomes duplicated in a variety of separate clusters or nodes depending on local contextual informational requirements. This redundancy is viewed as one of the key tools for preservation of information in the long term, just as the Rosetta Stone provides keys for decoding linguistic, semantic, and ontological disjunctions.

The World Wide Web is an example of an "open network" that is evolving through time. However, the current use of hypertext markup language (HTML) as the standard Web markup language lacks the semantic elements that would allow the development of a fully empowered contextual information network. The increasing use of extensible markup language (XML) is seen as a move towards a mechanism that will add significant usability and functionality to contextual information frameworks. A network of this type could be implemented on any scale (from local to international), although it is recognized that in certain settings stricter management processes might be desirable in order to control the quality of the information sources.

## Open Information Systems and Security

It is accepted that there may be concerns in the community about the use of an open network to reference potentially sensitive information about people and organizations. Open access to information may infringe either personal privacy or cultural practices. The security aspects of this would need to be carefully considered. However, it is not being suggested that all data and information be placed on the open network but that references can be made to it. The actual location of these digital resources and, indeed, the precise location of the materials need not be placed in the public domain for this concept to be effective.

Implementation of a knowledge network to cover the whole of the cultural heritage community worldwide would be a major achievement and would take many years to develop. However, studies and small-scale projects already implemented have shown that the basic framework can be established very quickly, sometimes in a matter of days. By way of example, the key sources utilized in compiling a contextual information framework focused around a Hungarian low-level radioactive waste facility comprised:

- Local records: data related to raw, conditioned and packaged waste, implicit and explicit information on the sources of the waste, references to contextual information (for example, specifications, local rules, safety cases), information on record creation (source of record, storage location, validity period, responsibilities).
- Organizational information: structure, mission statement, goals and objectives, timescales, programs, key milestones, regulatory requirements, history.
- State information: organization entities, roles and responsibilities, regulation, international cooperation, reporting requirements, principal skills and disciplines.
- Community information: State profiles, key organizations, roles and responsibilities, guidance and regulation, legislation, international cooperation programs, agreements, and protocols.

There are a number of benefits in using a contextual information framework approach, which include:

- Making knowledge transfer easier and encouraging the sharing of experiences within and between organizations.
- Enabling existing knowledge sources to reside within a structured and visible system.
- Increased visibility of knowledge sources will promote their preservation and value.
- Introduction of a nonintrusive technique that complements existing business practice.
- Supporting the decision-making process by making available a wide range of information giving a view of "the bigger picture."
- Improving transparency within the cultural heritage community and for external observers (transparency is fundamental to building trust).
- Referring to sensitive information within the system without it being reproduced.
- Vastly improved discovery, accessibility, and comprehensibility of resources.

From an implementation point of view, a significant aspect of the contextual information framework concept is that it is neither necessary to identify all the information nodes at the start, nor immediately populate those that have been identified. Further knowledge sources can be created and populated at any time and linked to other sources as the knowledge base grows, in much the same way as the World Wide Web developed. A contextual information framework evolves in parallel with the changes in the society it maps.

As with any information system, the information sources that form the network nodes must be reliable and relevant. It would be the responsibility of the separate organizations and the standard frameworks governing the various elements of the cultural heritage community to ensure that the contextual information is properly prepared and managed and that a focus on quality is of signal importance.

One of the characteristics of the World Wide Web is that there is no one body in overall control or with the responsibility for overseeing the placement of information in the system. It is envisioned that contextual information frameworks would adopt a similar strategy and thus evolve the robustness that comes with dispersed but shared responsibility. For peace of mind for the cultural heritage community it could be possible for leading national cultural heritage bodies to establish themselves as the principal nodes in the network, providing both major resources for their respective communities and a management function, particularly in the area of standards. The recent endorsement by the National Library of Australia of a project to establish a "People Australia" portal based on the principles of the contextual information framework is a case in point.[6]

## The Need to Maintain Knowledge of Digital Cultural Heritage Resources

Managing and preserving digital cultural heritage resources for the long term may be considered as a potentially costly and possibly a low-priority activity that uses valuable human resources while contributing little or nothing to current views of business viability. Indeed, experience suggests that records and information management is often viewed as an overhead that can be relegated to the "back end" of a major project. A common outcome of such an approach is that the contextual information vital for understanding the outcomes of the project, or indeed the technology of the resource itself, particularly in the long term is either not captured or not linked to the resultant resources. One reason for this is that people continually accumulate their own implicit knowledge, or contextual information, enabling them to understand the work they undertake and comprehend the records associated with their work. This implicit knowledge may be highly contextualized to their specific job or organization. When work places or communities accumulate implicit knowledge through shared experience it becomes common knowledge and it is generally not appreciated how important the documentation of that shared context is for comprehension and understanding by outsiders.

## Insiders and Outsiders

For those without that implicit knowledge, the task of comprehending information will increase in difficulty and complexity the further they are removed either in time, culturally, or in a sociotechnical sense, from the originating context. For them to understand the information they will be required to undertake additional research, investigation, and consultation. This division between the insider and outsider is the essential problem facing intergenerational information transfer and the viability of digital resources. As can be seen, the problem is not just confined to the transfer of information between generations but applies to the transfer of information within generations when access to contextual information is required by an outsider.

Unfortunately, it has not been common practice to systematically document and manage contextual information. This is a task that has often been left to archivists and historical researchers whose primary endeavor is to make information about the past accessible but who are often saddled with the burden of making comprehensible the inadequately documented records left behind by others.

### It Starts with Records

The information transfer process starts with the creation of records documenting daily activities involved in the creation and eventual curation of resources. Relegating records management to the backend of a project is a high-risk strategy. When a major project reaches a conclusion, it is not uncommon for the staff responsible for creating the records to move on, taking their implicit knowledge with them and leaving others to make well-meaning judgments regarding priorities for long-term preservation. Given the emphasis that should be placed on the long-term accessibility and usability of digital resources, the associated contextual information or knowledge management must be considered a high priority activity that cannot be postponed to a later date—it must be an integral part of the project being properly planned and resourced with clearly defined outputs and objectives, just as with any other product-critical activity.

### Resources, Society, and Time

Whatever the view on any particular digital resource or any particular digital technology, digital cultural heritage resources have been created. They exist now and will undeniably impact on both present and future societies. The questions surrounding the viability of these resources means that the present generation, with its knowledge of the resources, has a clear obligation to preserve that knowledge and pass it on to future curators so informed decisions on future management can be made.

There are numerous sources from which knowledge about resources could be accumulated. Our present knowledge includes an implicit understanding of its significance, provenance, and the societal value of the resource and the technology used to build it. It may seem inconceivable to us today that a future society would not posses a similar understanding, particularly when dealing with what we regard as valuable cultural materials, but history has shown that societal values do change and both explicit and implicit knowledge can be lost. The drivers influencing changes in societal values are not the subject of this paper, but we cannot shrink from our responsibility to take reasonable measures to empower the next generation with the knowledge currently in our possession.

A number of very credible reasons for preserving digital cultural heritage resources could be cited; however, it is suggested that the fundamental and most important reason is to give future societies the opportunity to make decisions based on the best information available.

## Conclusion

The creation and curation of digital cultural heritage resources is moving into a new era, following the emergence of the Internet as a major cultural force enabling the linking and sharing of information. The costs to individual organizations of these activities are diminishing on a relative scale, and through the use of open-source technologies it is possible for small institutions and individuals with the relevant technical and creative skills to actively participate. Justifiable and informed decisions are being made now and will continue to be made on the basis of comprehensive knowledge about resource characteristics, their source, enabling technologies, and potential impact on society. Important decisions have yet to be made and, due to the hoped-for longevity of some of the resources, can only be made many years hence. It is therefore incumbent on this generation to develop and maintain a knowledge base and to implement systems that ensure the information remains accessible to those having to make critical decisions.

The purpose of this chapter has been to extend current thinking by conceptualizing how the elements of our cultural heritage knowledge base might be managed. The long-term preservation of the resources containing data and explicit information is only a part of the challenge and we must recognize that informed decisions are also influenced by our implicit knowledge of the resources and the contextual framework within which they are created and curated. We live in a world of perpetual change and there is a very real threat that the accumulated knowledge that guides the digital cultural heritage resources today will be lost to the next generation.

This generation is unable to accurately predict either the technical capabilities or societal values of the future and therefore the standards and institutional environment in which decisions will be made. As a consequence, we are unable to predict accurately what information will be required for a future generation to make informed judgments. However, a first step in empowering the future custodians of our legacy is to recognize the information sources and to develop bold strategies for ensuring that they remain accessible.

A strategy based on a contextual information framework concept has been described. This type of network linking information nodes at organizational, national, and international levels would have a number of benefits, including clear recognition of the location of important knowledge sources and duplication of information, thus reducing the likelihood of catastrophic loss. Much information exists in the public domain to which links could be made to other standalone sources.

There is still clearly more work to be undertaken in developing, first, an acceptable case for contextual information frameworks and, second, a strategy that is acceptable to all potential contributors in the cultural heritage community. However, the volatility of the knowledge base that has been created and its potential value to future generations is so great that concerted action at all levels must be taken to preserve it and ensure we avoid a potentially disastrous epistemic failure.

## Notes

1. World Commission on Environment and Development 1987, 46.

2. For background on this issue, see *UKOLN Digital Preservation—Publications* 2004.

3. See *The CIDOC Conceptual Reference Model* 2004.

4. See *Functional Requirements for Bibliographic Records* 1998.

5. See *Austehc Web* 1999.

6. See *The Proposed People Portal Project* 2005.

## References

*Austehc Web*, Australian Science and Technology Heritage Centre, University of Melbourne. 1999. Available at ⟨http://www.austehc.unimelb.edu.au/austehcweb.html⟩ (accessed 16 April 2005).

Barabási, A.-L. 2002. *Linked: The New Science of Networks*. Cambridge, Mass.: Perseus.

*The CIDOC Conceptual Reference Model*, International Council of Museums, July 2004. Available at ⟨http://cidoc.ics.forth.gr/⟩ (accessed 16 April 2005).

Crofts, N., M. Doerr, and T. Gill. 2003. "The CIDOC Conceptual Reference Model: A Standard for Communicating Cultural Contents." *Cultivate Interactive* 9, February.

Doerr, M. 2003. "The CIDOC CRM—An Ontological Approach to Semantic Interoperability of Metadata." *AI Magazine* 4, no. 1.

Doerr, M., K. Schaller, and M. Theodoridou. 2004. "Integration of Complementary Archaeological Sources." Paper presented at *Computer Applications and Quantitative Methods in Archaeology Conference, CAA2004*, Prato, Italy, 13–17 April. Available at ⟨http://cidoc.ics.forth.gr⟩ (accessed 16 April 2005).

*Functional Requirements for Bibliographic Records: Final Report*. 1998. International Federation of Library Associations and Institutions, UBCIM Publications New Series, volume 19 (March). Available at ⟨http://www.ifla.org/VII/s13/frbr/frbr.pdf⟩ (accessed 16 April 2005).

Harris, H. 1997. "Remembering 10,000 Years of History: The Origins of Migrations of the Gitksan." In *At a Crossroads: Archaeology and First Peoples in Canada*, ed. G. P. Nicholas and T. D. Andrews. Burnaby, British Columbia: Simon Fraser University: Archaeology Press.

*ISAAR(CPF)—International Standard Archival Authority Records for Corporate Bodies, Persons and Families*, 2d ed. 2004. Vienna: International Council on Archives. Available at ⟨http://www.ica .org/biblio.php?pdocid=144⟩ (accessed 16 April 2005).

*ISAD (G)—General International Standard Archival Description*, 2d ed. 1999. Madrid: International Council on Archives (endorsed 2000). Available at ⟨http://www.ica.org/biblio.php?pdocid =1⟩ (accessed 16 April 2005).

Lagoze, C. 2000. "Business Unusual: How "Event-Awareness" May Breathe Life into the Catalog?" *Bicentennial Conference on Bibliographic Control for the New Millennium: Confronting the Challenges of Networked Resources and the Web*, Library of Congress Cataloguing Directorate, November. Available at ⟨http://www.loc.gov/catdir/bibcontrol/lagoze.html⟩ (accessed 16 April 2005).

Luhmann, N. 2000. *Die Politik der Gesellschaft*. hg.v. André Kieserling. Frankfurt: Suhrkamp.

McCarthy, G. 2000. "The Structuring of Context: New Possibilities in an XML enabled World Wide Web." *Journal of the Association for History and Computing* 3, no. 1 (April). Available at ⟨http://mcel.pacificu.edu/JAHC/JAHCIII1/ARTICLES/McCarthy/index.html⟩.

Nowotny, H. 1999. "The Need for Socially Robust Knowledge." *TA-Datenbank-Nachrichten*, no. 3/4, 8. Jahrgang-Dezember, S.12–16. Available at ⟨http://www.norfa.no/_img/ nowo99a.htm⟩ (accessed 16 April 2005).

Nowotny, H. 2003. "Coping with Complexity: On Emergent Interfaces Between the Natural Sciences, Humanities, and Social Sciences." In *The Founding of International University Bremen*. Bremen: International University Bremen, 66–78. Also available at ⟨http://www.nowotny .ethz.ch/publikationen_en.html⟩ (accessed 16 April 2005).

*The Proposed People Portal Project, February 2005*. Internal report, National Library of Australia, February 2005.

Rahm, E., and H. H. Do. 2000. "Data Cleaning: Problems and Current Approaches." *Bulletin of the IEEE Computer Society Technical Committee on Data Engineering* 23, no. 4, December.

Scott, P. 1966. "The Record Group Concept: A Case for Abandonment." *American Archivist*.

Sheth, A., S. Thacker, and S. Patel. 2002. "Complex Relationship and Knowledge Discovery Support in the Info Quilt System." *VLDB Journal* 12, no. 1: 2–27.

Sugiyama, K., J. Ohuchi, H. Takao, and T. Tsuboya. 2003. *Record Preservation Study on Geological Disposal—Significance and Technical Feasibility*. Radioactive Waste Management Funding and Research Centre, RWMC-TRE-03001, March.

*UKOLN Digital Preservation—Publications*. 2004. University of Bath: UKOLN. Available at ⟨http://www.ukoln.ac.uk/preservation/publications/⟩ (accessed 16 April 2005).

World Commission on Environment and Development. 1987. *Our Common Future: Brundtland Report*. Oxford: Oxford University Press.

# 13 Engaged Dialogism in Virtual Space: An Exploration of Research Strategies for Virtual Museums

Suhas Deshpande, Kati Geber, and Corey Timpson

This chapter examines within a theoretical framework an audience-centered strategy of researching the optimal performance of virtual museums.[1] This draws on two distinct theories: Appraisal theory—a modern model focused on the assessment of discourse and its audience—and classical rhetoric, focused on the persuasive use of language, and its impact upon an audience.

The benefit of developing and applying this double theoretical framework is that it allows for a deeper insight into the interaction between audiences and cultural knowledge in the virtual realm. Knowledge of how a varied audience uses or interprets the material contained in a virtual museum is crucial to the designers of such cultural programs. It allows them to better plan, create, and monitor the performance of exhibits and institutions in the virtual space.

An understanding of the key features of virtual landscapes is central to the development of any analytical tool that can be fruitfully applied to the design and evaluation of virtual museums. Some of the major characteristics of this contemporary environment, applicable to virtual museums in particular, are identified and discussed in this study.

Our present report makes reference to a research study conducted at the Canadian Heritage Information Network (CHIN) on the future of virtual museums. Although the CHIN research study, entitled *Virtual Museums: The Next Generation*,[2] only made passing reference to rhetoric and did not consider Appraisal Theory, it offered insightful analyses of audience behavior and the direction of future development for virtual museums. The examples given below to illustrate our theoretical tenets and the realities of the virtual environment are selected from CHIN's experience with the Virtual Museum of Canada (VMC) (2001), and the Smithsonian National Museum of American History's virtual exhibit entitled "September 11: Bearing Witness to History."[3]

Finally, we discuss how a research strategy that makes use of the extrapolation of Appraisal Theory concepts and of classical rhetoric could be developed further to create a useful tool for the design and evaluation of next generations of virtual museums. Additional issues for future research are identified.

## Appraisal Theory and Classical Rhetoric

The Appraisal theoretical framework is a relatively recent development of the last fifteen years and starts from Systemic Functional Linguistics mentored by M. A. K. Halliday. Its research leader is Professor James Martin from the University of Sydney. Appraisal Theory deals with the linguistic resources through which a speaker "expresses, negotiates and naturalizes particular intersubjective and ideological positions."[4] It represents a particular approach to exploring, describing, and explaining the way language is used to evaluate, to adopt stances, to construct textual personas, and to manage interpersonal positioning and relationships. Its two core concerns are:

1. How speakers/writers adopt and indicate positive or negative attitudes in their discourse.

2. How they negotiate these attitudinal and other types of positioning with actual and potential dialogic partners (White and Eldon 2004).

Therefore, for the purpose of our research, the Appraisal Theory concepts may be extrapolated to explore the way in which writers/creators of a virtual site relay their system of values and their ideology to their users; these attitudes, judgments, and affective responses are implicitly or explicitly present in the virtual discourse and they are conveyed in such a rhetorical manner to engage similar attitudes, judgments, and affective response from the dialogic partners, i.e., the audience/users.

In the context of the virtual environment, where senders' discourse and users' reaction overlap, the audience has the most important role of evaluating the shared discourse site. Whether the message is conveyed via image, text, video, or in an interactive format, the audience receives *and* evaluates the message. The effectiveness of both the message and its modality of transmission depends on audience engagement. We propose to extrapolate the Appraisal Theory concept of engagement to measure the users' arousal of interest, the spurring of their curiosity, and the satisfaction of their informative needs.

The multitiered approach of Appraisal Theory, as discussed in detail below, contains a simple yet comprehensive view of audiences. Classical rhetoric, which maintains a straightforward yet comprehensive view of audiences, provides an excellent parallel foundation for our analysis. That is why the present study adopts this classical view, and in particular the Aristotelian lore,[5] and blends it with the view of audiences from Appraisal Theory.

The audience was an important consideration in Ancient Greece, as reflected in the Assembly of Athens, which was a direct democracy.[6] Aristotle's treaty provides a systematic analysis of his public addressees, their psychology, as well as an outline of how best to impart them. The *Rhetoric* is Aristotle's successful effort to delineate a systematic account of the art of persuasion; in it, he defined rhetoric as "the faculty of observing in any given case the available means of persuasion."[7] The audience is a crucial element in Aristotle's work, as proven not only by the attention given explicitly to the target of discourse within the analysis of the process of communication, but also by his attempt to find the most effective strategy of audience persuasion.

According to the Aristotelian theory, the importance attached by the skillful rhetorician to the understanding of his audience will ultimately dictate his choice of persuasive devices. The same process applies to the virtual space, as the skill of the authors to understand their users *before* they generate a virtual landscape will affect the users' evaluation of it as satisfactory for their intellectual and emotional needs.

> A successful rhetorician structures his speech to elicit emotions that are connected with stable motivational structures. That is, after all, why he needs to understand the character of his audience. To bind their convictions, he must appeal to their strongly entrenched, as well as to their immediate, interests and desires. (Rorty 1996, 21)

Aristotle defined the three key elements that are the foundation of a persuasive argument in relation to the audience response.[8] These elements are:

1. The ability of the rhetorician to appear credible and trustworthy, so as to be persuasive—*ethos* (Crowley 1994, 84);

2. The ability of the rhetorician to evoke emotion in his audience through the way he communicates and consequently energizes them into action—*pathos* (Corbett 1971, 319); and

3. The ability of the rhetorician to persuade by "systems of reasoning" or mostly by means of logic—*logos* (Covino and Jolliffe 1995, 20).

Aristotle devised these criteria relying on "rough empirical generalizations about the psychology of various types of audiences" (Rorty 1996, 8). His theoretical framework could be applied to the virtual landscape and its communication, as we observe audience's reactions to various elements of a virtual museum. The credibility of the institution, organization, or individuals who created the virtual museum is crucial to the credibility of the site as a whole. This is the element considered by classical rhetoric to be *situated ethos*, the ethos that comes with the vested authority of the sender of a communication (Crowley 1994, 126–127). The *situated ethos* impacts directly on the *invented ethos*. In other words, the initial trustworthiness of a Web site and of its creators generates the audience's acceptance of the content; even more than that, the trustworthiness of the creators of a virtual museum site generates the credibility of all the multiple voices that contribute to the respective virtual museum.

Emotion (*pathos*) plays a very important, although not always obvious role in the persuasiveness of a virtual museum site. The emotions generated in an audience may vary between: visual pleasure, surprise, contentment, interest, enhanced or satisfied intellectual curiosity, and others of a similar kind. These types of predictable audience responses are helpful in evaluating the feedback of the audience responding to virtual museums.

It may be easy to see how the creators of virtual spaces transmit their discourse, a logically ordered and organized content (*logos*), as well as emotional messages that will in their turn elicit emotional reactions (*pathos*), in addition to rational and informative feedback (*logos*), due to their empowerment with credibility and authority (*ethos*). However, for the first time in the history of communication, the receivers of communication, the users of the virtual landscape, are empowered with such an authority, discernment, and credibility (*ethos*) based on their competence, prior knowledge/preknowledge, and choices (*logos*) that are absolutely key factors for the very existence of the virtual space communication. Through their unprecedented *ethos* they also influence the later development of the virtual landscape; their positioning fluctuates between users/receivers and creators/senders. Any set of their choices may lead to the discovery of new patterns or arrangements.

Aristotle's most important statement about the central role he granted the audience was intricately linked to the importance he attached to the *enthymeme* in the act of persuasion. He considered the *enthymeme* to be the "strongest of the proofs," the "body of persuasion," as the audience is most easily persuaded when they think that a certain matter has been demonstrated (Rapp 2002). Aristotle called *enthymeme* a "rhetorical syllogism" (Covino 1995, 21). An *enthymeme* leaves out the second premise that is already accepted by the audience, due to commonsense knowledge or a shared ideology and set of values. For example, the following is a syllogism: All beings are mortal. Socrates is a human being. Socrates is mortal. As an *enthymeme*, this would become:

"All human beings are mortal, and Socrates is mortal"; the audience implicitly fills in the missing premise.

Aristotle saw the *enthymeme* as the strongest of all arguments because it involves both audience and speaker in a joint enterprise: "Because they are jointly produced by the audience, *enthymemes* intuitively unite speaker and audience and provide the strongest possible proof. . . . The audience itself helps construct the proof by which it is persuaded" (Bitzer 1959, 409).

What makes the *enthymeme* the strongest of all arguments in Aristotle's theory is not to be found in the elements of speech, but in the fact that the senders of communication and their audience are involved in the same thought process on a certain subject. By sharing the premise of an argument—walking through it together as it were—the speaker/senders and their audience share the conclusion.

The *enthymeme* may be translated in numerous ways in the realm of virtual space and its virtual museums. The general idea remains the same: the creators of a virtual museum and their audience reach a conclusion together—a conclusion that is already known by the authors, who have given their audience the necessary means to reach it. An excellent example of an *enthymeme* is apparent after a virtual visit to the *September 11: Bearing Witness to History* exhibit created by the Smithsonian Institute to document the terrorist attacks of September 11, 2001, in the United States.

In the continuum of the virtual environment, however, there are numerous possibilities and paths available to the audience, enabling multiple *enthymemes*. As researchers have pointed out, a well-executed *enthymeme* means that the content and the number of its premises are adjusted to the intellectual capacities of the public audience (Rapp 2002). Hence, a well-designed virtual environment will accommodate many levels of intellectual capacity for its variety of users.

The nature of communication in the virtual landscape makes Appraisal Theory concepts, when extrapolated, a useful tool for further analyzing virtual museums. Appraisal Theory has three main categories: *attitude*, *engagement*, and *graduation*.

*Attitude* covers the positive or negative assessments present in the discourse; for the purpose of this research document, *attitude* is present in the virtual landscape and its users' response. These assessments rely on the discourse creators' emotions and/or culturally determined systems of values and ideas. The *attitudinal positioning* is concerned with what might be thought of as "praising" and "blaming," with meanings by which creators/writers indicate either a positive or negative assessment of people, places, things, happenings, and states of affairs as presented by the discourse/virtual landscape. The *dialogistic positioning* that "anticipates or acknowledges likely responses, reactions, and objections from actual or potential dialogic partners," and the *intertextual positioning* that "takes into account or responds to prior utterances" are varieties of the larger category of *attitudinal positioning* (White and Eldon 2004). It is clear how these varieties of *attitudinal positioning* apply to the situation of "virtual space" discourse, because *dialogic positioning* towards the various needs of users creates virtual museums, because the authors anticipate those needs or acknowledge their visitors' responses. Virtual museums are also created by *intertextual positioning*, because they take into account previous linguistic feedback of their users, and often refer to other documents and sites.

*Attitude*, in its turn is divided into three secondary concepts:

1. *Affect*    The characterization of phenomena by reference to emotion.

2. *Judgment*    The evaluation of human behavior with respect to social norms.

3. *Appreciation*    The evaluation of objects and products (rather than human behavior) by reference to aesthetic principles and other systems of social value.

The concept known as *affect* roughly corresponds to *pathos* in classical rhetoric, as it is concerned with the emotions expressed in the discourse. *Judgment* in Appraisal Theory roughly corresponds to *logos and ethos* of classical rhetoric. It is concerned with the assessment of human behavior and language as positive or negative on the background of organized systems of ethical values, or accepted social norms and beliefs (*ethos*). *Judgment* is more aware of the context of communication than *logos* within classical rhetoric, as it recognizes that the audience will evaluate the *truth* of information presented to it based upon its own system of norms, beliefs, and values.[9] Acceptance or rejection of the content is not simply an objective evaluation of information (*logos*); rather, evaluation occurs within the cultural framework of shared norms and values (*ethos*). Appraisal Theory recognizes this and further research may demonstrate how its concepts could be applied to the evaluation of virtual museums' performance.

The third and final concept of the *attitude* category of Appraisal Theory is called *appreciation*, which roughly resembles the *logos and ethos* of classical rhetoric. *Appreciation* is concerned with the positive or negative evaluation of mostly objects, processes, material circumstances, and states of affairs. It assesses values that fall under the general heading of aesthetics, as well as a non aesthetic category of "social valuation" (*ethos*). Translated into the virtual landscape, these evaluations may deal, for instance, with the quality of transmitted images or messages (*logos*), measured against a set of established values and norms (*ethos*). Users evaluate virtual spaces in the same manner as in the Appraisal Theory *appreciation*: virtual museums are viewed as entities, and examined for the quality of their products. *Appreciation* is useful in our evaluation of virtual museum performance because it has three specific, relevant elements:

1. *Reaction*     The audience evaluates a virtual museum or exhibit based upon its quality or the impact it makes.
2. *Composition*     A virtual museum or exhibit is evaluated according to whether it conforms to various sets of rules and norms.
3. *Social Virtual Value*     The audience evaluates a virtual exhibit or museum according to various social conventions that are shared, but may also be specific to each user.

*Engagement* in Appraisal Theory deals with the authors' positioning of their voice with respect to the various information and emotions conveyed by their discourse; their position either acknowledges or ignores the diversity of points of view taken by their audience in relation to their discourse, and they negotiate an interpersonal space for their own positions within that diversity. This is applicable to the virtual discourse of virtual museums, as users may have a variety of reactions to the information and emotional content of the site. However, the authors *must* acknowledge the position taken by their addressees and they are compelled to negotiate the subsequent generations of their virtual museums in direct response to these evaluations. After the analysis of the specific process of *engagement* of *both* senders and receivers in the virtual space communication, *engagement* is seen as an essential factor to be taken into account for the future development of virtual museums. *Engagement* could be beneficial for both the design and the enhanced performance of virtual museums, as our own experience at VMC has proven when we have used some simple and reliable techniques.[10]

*Graduation*, the last main category of Appraisal Theory deals with elements which provide *grading* or *scaling*, related to "the interpersonal force which the speaker attaches to an utterance (*force*) or in terms of the preciseness or sharpness of focus with which discourse expresses an idea (*focus*)" (White and Eldon 2004). As before, the analysis of virtual museums will borrow and slightly change these concepts borrowed from the Appraisal Theory to prioritize the audience's reaction to the *force* of the site content, or to the sharper or duller *focus* of the ideas and images presented. These reactions may be measured in a variety of ways, as illustrated below.

Taken together, classical rhetoric and Appraisal Theory lend our audience-centered research a conceptual framework and a theoretical vocabulary that could lead to the implementation of an efficient tool for the design and evaluation of virtual museums. This framework allows for the recognition of the creators' message from both an interpersonal and interactive evaluation perspective. Because both theories take into account the significant role of the audience, we may consider how users engage and position themselves in the virtual realm within the context of the "virtual" situation of a "virtual" culture. The users and the creators of the virtual site coauthor, up to a certain point, the virtual landscape of virtual museums, whose key characteristics will be discussed next.

### The Virtual Landscape and its Competent Users

The virtual landscape is generated through an extremely dynamic and essentially dia- logic discourse. In her essay "Virtual Bodies and Flickering Signifiers," Katherine Hay- les analyzes the role of producers and users of virtual messages. She remarks that there is a shift of attention in virtual communication from presence and absence to pattern, a software interface, randomness, and a hypertext information space. This shift is explained in relation to the specific features of this medium, where coding technolo- gies change the conditions of communication. Producer and user communication in the virtual museum's cyberspace can be seen as a "complex dance" between pattern and randomness, or predictability and unpredictability (Hayles 1993).

In spite of all the changes undergone, this discursive environment is still amenable to the dialectic that gave rise to both classical rhetoric and Appraisal Theory. In marked difference with the physical world, where space and time boundaries are more or less stable, the virtual landscape, created by a dynamic discourse, has fluid boundaries. There is no previously set path or route for the audience/users to follow. Rather, each of their interactions with the virtual environment is a unique experience. Individuals construct their own landscape and meaning by selecting destinations in their own chosen order; they visit Web sites and interact in different ways with each one of them, with varying intensity, depending upon their changing individual interest.

This is particularly true in the case of virtual museum sites, which are uniquely ex- periential. The design of a virtual museum, viewed as content, structure, functionality, interaction, and visual design, aims to create, for the time being, a highly patterned, user-constructed experience. This may change in the future. For virtual museums, at present, a predictable spectrum of choices represents an essential goal and a measure of success. Randomness is narrowed mostly to search/selection (browse) functions, within data prearranged into arrays or tables.

The Virtual Museum of Canada (VMC), for example, was created with this view of the virtual landscape in mind. A visual metaphor of the museum is employed in the case of the home page navigation, which is endowed with museum arches leading to open spaces. Visitors are thus invited into the virtual museum, where they are able to visit artifacts and stories from institutions across Canada. Once they enter the museum through the national heritage gateway, they can interact in two ways with the exhibits, using standard hyperlinked navigation, or a powerful search engine. The resulting discourse occurs through a clear and predictable navigation, and a set of information-retrieval mechanisms, governed by our visitors' individual points of view. The power of individual choice is highlighted by the different manners in which various people travel within the virtual landscape. Distances are measured not in miles or yards, but in personal interest. Visitors are always moving in the direction of their choice. The VMC builds upon the essential quality of the virtual landscape leading to a dialogic engagement between its environment and its audience; the journey is wholly defined through this dialogic interaction. The Internet is blurring space and time boundaries for virtual museums and creates a synchronic communication continuum. Space and time have become a function of the *preknowledge* and *competence* of the visitor.

Another set of key characteristics of the virtual experience is its speed and rhythm. "Internet time" is "different in pace and different in control. Transactions and communication on the Web unhindered by matter and its inertia go faster" (Weinberger 2002, 44).

The "September 11: Bearing Witness to History" virtual exhibit makes excellent use of the speed and rhythm of virtual space to connect with its audience. The audience is able to move quickly between images, and indeed between sections of the Web site. In this way, they build firsthand their own experience of September 11. This is something that can only be achieved in a limited way—at best—in the physical landscape. Aside from limitations of space for displaying artifacts, physical exhibits are subject to the laws of physics, as they can only be configured in one way at any point in time. For example, the "September 11" exhibit uses this characteristic of the virtual landscape through a feature named "Tell Your Story." The audience is invited to tell their own experiences of September 11, 2001.

The instantaneous nature of the virtual environment allows for the transmission of massive amounts of data in a relatively short period of time. This is important for the VMC, which was initially created to engage "audiences of all ages in Canada's diverse heritage through a dynamic Internet service freely available to the public" (CHIN 2001). It means that rich data can be displayed to its audience. The initial goal of placing a diverse range of content in front of our visitors was due to the fact that the VMC was created to serve a mass audience. The fact that another set of boundaries is redefined, the one between producers and users, affects mass and interpersonal communication modes, generating a need to reevaluate production and usage concepts, as well as the roles of producers versus users, their presence and absence during the communication process.

The present study suggests that in order to be truly effective in reaching an audience, content that is *personally* meaningful should be presented. Researchers are increasingly aware of the importance attached to a continuous process of learning. Learning is not confined to "brick and mortar"[11] and is frequently based either on a continuum or combination of a traditional and a virtual learning space environment. Virtual museums are a clear example of the latter case. This leads to the concept of *personalization*. Rather than delivering a standard interface to all visitors, customized Web sites present content to suit different users based upon carefully researched profiles. This is the case of the Personal Museum at VMC, in which visitors are able to store their selected information for later retrieval and/or development (figure 13.1). The "Virtual Museum" research study at CHIN suggested that audience-based interfaces be utilized using a combination of techniques that include:

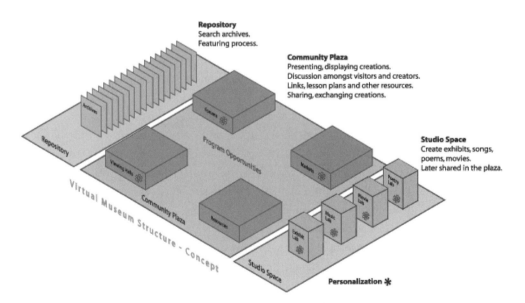

Figure 13.1   "Based on 'My Personal Museum Concept' presented by eMaterial Studios Inc./Nov. 3, 2003." © Corey Timpson.

mapping the specialized vocabulary of curators into vernacular through the use of thesauri and other vocabulary tools; creating layered architectures with middleware mapping between the object description and the user interface; and/or developing mark-up
standards for describing objects in different ways. (Dietz et al. 2004)

The key aspect of building an audience/user profile is to identify actionable characteristics (Dietz et al. 2004). A problem that was identified in the "Next Generation"
study at CHIN is that simply determining demographics (such as age, country, marital
status, education) is not as significant as identifying interests (such as history, or
science, or art, or jazz), preferences (such as number of results to display at one time
on a page, or whether to include images, or preferred language), or learning styles
(Dietz et al. 2004). It is possible for two teachers, for example, to have very different
preferences. Classical rhetoric tells us that audience/users' preferences are more important for affecting persuasion than demographics, yet virtual museums and exhibits
are often identified as simply being directed to one audience or another. The discrepancy between virtual museum visitors' demographic profiles and their actual-use patterns leads one to hypothesize that the best way to meet the audience's needs is to
allow for the tailoring of content based on what they do and/or want, rather than
who they are.

The challenge remains exactly how to determine audience preference. Significant
research is currently being conducted by Falk and Dierking who identify the set of factors that affect audience's participation in the process of learning within a traditional
and a new environment, outlined in their "contextual model of learning."[12] We believe that while some valuable demographic information can be acquired explicitly
from the users via questionnaires, membership registration, etc., we may attempt to
infer additional significant information from usage patterns via statistics and through
the concepts. There are, however, some difficulties that arise in this process. Web
logs, which record how users traverse a Web site, do not capture demographics that
can be linked to user behavior. That is, one cannot distinguish between the manner in
which a teacher and a student interact with an educational Web site only from viewing
and analyzing Web logs. Analyzing interfaces and functionalities based on the concepts
presented above potentially provides a better match to our users' actual needs. This
statistical information reflects a relatively accurate image of the number of visitors, return visits, entry points, exit points, and the time spent in specific portions of the site.

Appraisal Theory concepts extrapolated for our research could teach us some interesting facts about the audience. The main category of *attitude* could be seen to roughly

correspond to the audience's usage patterns, the manner in which they interface with the virtual landscape of the virtual museum, establishing a *dialogic positioning*. The concept of *affect*, part of *attitude*, points out that audience behavior is impacted as it reacts through emotion to the emotional and information content of the exhibit, whereas *appreciation* teaches us that users evaluate the quality of the virtual product, and not just the content-related. Both of these aspects of the Appraisal Theory, that fall under the main concept of *attitude*, clearly impact on navigational behavior. For example, a visitor who initially came to a Web site out of interest in a particular subject area, may be moved by a specific aspect of that topic, or by the way a topic is presented. Their increased focus on the content in that particular area represents their changed behavior as a result of an emotional response. Alternatively, they may deem that the quality of an exhibit is poor and leave, or limit their interaction to what is essential, rather than explore other aspects of the site. A site deemed by its users to be of very high quality, on the other hand, may result in a longer engagement. Whereas *affect* and *appreciation* serve to highlight important, but fairly obvious qualities of the virtual realm, it is the concept of *judgment* that is most revealing to our current study.

The concept of *judgment* recognizes that when presented with a virtual museum or exhibit, the users will evaluate the validity of its content based upon their own pre-existing systems of values. In applying the *judgment* to our study we should once again extrapolate it to cover not only a system of values, but especially our users' required *e-literacy*[13] and *competence*,[14] in other words the *preknowledge* that results in their intention and controls their usage patterns (browsing, searching, etc.) on entering our virtual museum site. The existence of *preknowledge* manifests itself in a variety of ways. Identifying and qualifying the existence of this *preknowledge* is challenging for the creators of virtual museums, but we believe there are ways to do it. For example, using a predefined search indicates a different kind of user from those who enter our virtual landscape and have a specific concept or idea in mind when searching. On the other hand, depending upon the nature of the predefined searches, users may be just browsing (when searches are very general) or they may show extensive *preknowledge* (when specialist language is used). Similarly, the speed with which users browse the site and select links may indicate familiarity with the site, and knowledge of what they are looking for (as distinct from browsers). The specificity of the search spans a wide scale from a random one to a search that follows a precise algorithm and denotes an advanced *competence*.

Determining the *preknowledge* of the audience implies the possibility of employing it. In establishing user profiles, one may target interfaces and content to these identified profiles, by dynamic content generation. A different, maybe more promising, approach could be to design tools that are specifically targeted to different audiences. Appraisal Theory tells us that users bring with them *preknowledge* and a certain ideological and values bias; classical rhetoric recommends us to direct persuasive elements at our targeted audience. Therefore, creating targeted tools recognizes both theoretical frameworks.

The creation of targeted tools was the approach taken by CHIN in the design of a section of the VMC entitled "Community Memories." The objective of Community Memories is fourfold:

- To create a national online portrait of Canada's history by connecting individual local histories;
- To engage Canadians in sharing their personal heritage with others;
- To stimulate community/museum partnerships in the development of online local history exhibits; and
- To strengthen the capacity of smaller museums to create digital content for use on the World Wide Web and in local programming (Dietz et al. 2004).

Museums from across Canada create exhibits for the Community Memories program, which are hosted by CHIN and accessed through the Virtual Museum of Canada. Recognizing that this resource is aimed at two distinct user profiles, namely, "dedicated researchers" such as genealogists, local historians, and reference librarians, and "casual browsers" such as the general public, tools were provided for each group of users. Each contribution is a standalone exhibit. Users can access each exhibit individually and browse by accessing images and text. Each exhibit tells its own story, and users can view each story from start to finish. For users focused on professional research, and for more sophisticated browsing, there is a digital assets library tool that allows users to search across exhibits. This allows for more thematic and subject-oriented searching. The design of distinct tools for distinct audiences recognizes *preknowledge* and the existing motivation that users bring with them to a virtual museum.

It is ultimately their values and *preknowledge* that users draw upon to evaluate, in their turn, the information with which they are presented in a virtual museum or exhibit. Appraisal Theory gives us an increased awareness of this behavior, which brings to light the importance of audience's evaluation and makes us deeply aware of the *preknowledge* and *competence* that users bring with them as they first interact with and then evaluate Web sites and virtual exhibits. Designing virtual museums for different profiles allows the message of the site to reach its intended audience.

Using this approach, one can establish a dialogic and dynamic communication experience, the site of interactive rhetorical persuasion, as we understand it to be in the virtual landscape. Our virtual communication does not use only the spoken or written word, as has often been the case in the application of classical rhetoric. Rather, the language of the virtual environment is multimedia, which does not mean only the association of text, sound, still, and moving images, but also the richness of the experience enabled by new language-related activities in the virtual landscape.[15]

## Conclusion

The theoretical framework presented in this paper draws upon Aristotelian classical rhetoric, Appraisal Theory, and a recognition of certain key characteristics of the virtual landscape and its users' varied competence. It is a loosely constructed theory, as it is intended to be used anywhere that insight and analysis are required to better understand the ways audiences are engaged while visiting a virtual museum. Falk and Dierking have also addressed the users' varied competence and the essential role played by the individual factors shaping the audience, when exploring the efficiency of the process of learning within the contextual model of learning.

Classical rhetoric identifies several key characteristics of audience behavior. Creators of virtual museums can draw upon the simplicity and comprehensive nature of the Aristotelian key concepts of *ethos*, *pathos*, and *logos*. Appraisal Theory is a currently evolving field that recognizes the importance of social context, and provides insight into the role of social values in audience decision-making. Both theories are important in the creation of virtual museums. Although essential, the identification and understanding of the target audience is too often overlooked in the creation of virtual museums and exhibits. Great care must be taken to identify and understand one's target audience, as well as to identify the reaction desired from this audience.

The current study paid particular attention to two key aspects of the virtual landscape: the centrality of its interactive discourse that leads to a specific fluidity of its time and space boundaries, and the speed and multipatterned manner with which virtual discourse occurs. We view the nature of the virtual landscape as particularly suited to an analysis conducted with tools inspired by the concepts of classical rhetoric and Appraisal Theory. Creators of virtual museums would do well to recognize the unique nature of virtual landscape. Navigation should be clear and allow for a nonlinear interaction of the site with the audience. The design of virtual museums displays an architectural and pattern-based intention. In the case of the virtual museums analyzed, predictability represents an essential goal and a measure of success. Randomness is limited to browsing functions, such as search functions within memory-objects, prearranged in databases. The real challenge is to create a new form, a more fluid architecture that will encompass both predictability and randomness factors, creating continuity rather than two opposing aspects.

This chapter proposes a theoretical framework and represents a starting point of research on audience-centered rhetoric in the development of virtual museums. Future studies conducted within this framework should pay particular attention to Appraisal Theory and explore the psychology of the audience.

## Notes

1. The authors gratefully acknowledge the assistance of Dana Geber for her comments on many sections of this paper. The authors are responsible for any errors or omissions in the paper.

2. Dietz et al. 2004.

3. The National Museum of American History 2002.

4. For most of our theoretical concepts related to Appraisal Theory, we have quoted or paraphrased P. R. R. White and Eldon 2004.

5. Aristotle's *Rhetoric* describes a system of persuasion that includes ethical and political considerations. Knowledge and consideration of the audience is one of these sets of skills.

6. Cole 1991, 23. Cole remarks that in ancient Athens there existed democratic political regimes "in which eloquence might well count for more in the pursuit of fame and power than wealth and family." The ability to persuade an audience was therefore crucial, and it is no surprise that a great deal of attention was given to the subject by Sophists and philosophers alike.

7. Aristotle, *Rhetoric*, 24; 1355 b.

8. Ibid.

9. The truth information is a complex term, but here by *truth* we mean correct information, related to the audience in a certain context of virtual situation and culture.

10. For our views on *engagement* please see some of the ideas selected from our research in Gauvin, Geber, and Timpson 2003.

    We consider the engagement elements to be the content, the architecture, and the visual design, all contributing to the overall experience and building the relationships between a Web site and its visitors. The Engagement Factor is evaluated with three simple and basic metrics:

- Number of visits;
- Number of unique visitors; and
- Duration of visits.

There are other measurable elements, which could indicate engagement, such as heavy visits (those involving very high page-view counts, which is an indication of the volume of content viewed by visitors; if you are targeting your visitors properly this metric should increase over time) and committed visits (visits similar in nature to the heavy visits described above, only using time-based measurements in addition to page views). If you want visitors of a virtual museum to stay on your site for a long time, you want the second metric to move higher, but if you want them to perform an action (select an option, find information, etc.), you may want to see this metric go down. Virtual exhibit creators likely want to see the second metric move up over time. No movement would mean the need for increased marketing or featuring. A constant decrease would show that you are probably not attracting the right kind of visitors.

Since the purpose of the Engagement Factor concept is to provide easy-to use metrics, accessible to all participants and contributors, and to provide a means of comparing various types of products, it was decided to include only the three metrics mentioned above, in the calculation of the Engagement Factor.

The formula for the Engagement Factor, as developed by Lyn Elliot Sherwood, is as follows: Visits/Visitors X Duration. This formula incorporates the average length of visits (duration) and the average number of return visits (number of visits divided by number of visitors) to determine the depth of interest visitors have in a particular product.

Each of the three basic metrics of the Engagement Factor provides important data regarding the quality of the users' relationship with the product. However, each one on its own tells only part of the story. It is for this reason that the Engagement Factor, which is based on the correlation of all three, has proven, thus far, to be effective in determining the users' overall engagement with a product.

Any increase in the number of visits, number of unique visitors, or visit duration triggers changes in the value of the Engagement Factor. An increase in the number of visits or the length of duration produces an equal increase in the Engagement Factor, while an increase in visitors results in a decrease in the Engagement Factor (due to the decrease in repeat visits). As has been previously demonstrated, by using these three measures in conjunction with one another a more comprehensive perspective is afforded regarding visitor engagement than would be possible by looking at each measure individually. In essence, by incorporating three modes of a user session, the Engagement Factor produces an easy-to-use and actionable metric.

While the Engagement Factor, as an absolute value, allows for cross-product comparison and can provide an indicator of the quality of online visits, its strength as an actionable metric rests in its movement and evolution over a given period. It provides actionable clues for those involved in conception, production, distribution, publishing, evaluation, and funding of online cultural heritage digital products.

11. "Brick and mortar"—a phrase that has been used in current research on interactive learning as a formula that describes traditional classroom environment.

12. Falk and Dierking (2000) developed a contextual model of learning that grew out of a framework that they developed back in the early 1990s (see Falk and Dierking 1992). They did it after spending a lot of time in museums, surveying thousands of people throughout their visits, and observing them in specific exhibitions, as well as conducting innumerable interviews. Their model is based on sociocultural theory, and deals with three overlapping contexts—the Personal Context, the Sociocultural Context, and the Physical Context—that all contribute to and influence the interactions and experiences that people have when engaging in free-choice learning activities such as visiting museums. Thus, the experience, and any free-choice learning that results, is influenced by the interactions between these three contexts. For further details, see ⟨http://www.ilinet.org/contextualmodel.htm⟩ (accessed 17 April 2005).

13. We use *e-literacy* in the same way as the American Library Association that defines information literacy as a set of abilities requiring individuals to "recognize when information is needed and have the ability to locate, evaluate, and use effectively the needed information" and states that "information literacy is a survival skill in the Information Age. . . . Information literacy forms the basis for lifelong learning. It is common to all disciplines, to all learning environments, and to all levels of education. It enables learners to master content and extend their investigations, become more self-directed, and assume greater control over their own learning." For more information, see ⟨lib1.bmcc.cuny.edu/lib/help/glossary.html⟩ (accessed 17 April 2005).

14. We use *competence* in a slightly different way than Noam Chomsky, as *competence* in relation to accessing virtual space communication, rather than just the linguistic *competence* he discussed and defined in Chomsky 1965. Linguistic theory is concerned primarily with an ideal speaker-listener, in a completely homogeneous speech-communication, who knows its (the speech community's) language perfectly and is unaffected by such grammatically irrelevant conditions as memory limitations, distractions, shifts of attention and interest, and errors (random or characteristic) in applying his knowledge of this language in actual performance. Chomsky made a clear difference between *competence*, an idealized capacity, and *performance*, the production of actual utterances. His definition of linguistic *competence* has come to be associated exclusively with grammatical competence.

15. See Heylighen, Joslyn, and Turchin 2000. The four types of linguistic activities, as described in this article, are: (1) the association of concrete and unformalized languages: the art; (2) the association of abstract and unformalized languages: philosophy; (3) abstract and formalized: mathematics, theoretical sciences; and (4) concrete and formalized: applied sciences. They write,

> Art is characterized by unformalized and concrete language. Words and language elements of other types are important only as symbols which evoke definite complexes of

mental images and emotions. Philosophy is characterized by abstract informal thinking. The combination of high-level abstract constructs used in philosophy with a low degree of formalization requires great effort by the intuition and makes philosophical language the most difficult type of the four. Philosophy borders with art when it uses artistic images to stimulate the intuition. It borders with theoretical science when it develops conceptual frameworks to be used in construction of formal scientific theories. The language of descriptive science must be concrete and precise; formalization of syntax by itself does not play a large part, but rather acts as a criterion of the precision of semantics.

By *new language-related activities* we mean creative associations of concrete, abstract, formalized, and unformalized language activities possible only in the virtual realm.

## *References*

Aristotle. 1972. *Rhetoric*. London: Penguin Classics.

Bitzer, Lloyd. 1959. "Aristotle's Enthymeme Revisited." *Quarterly Journal of Speech* 45.

Canadian Heritage Information Network (CHIN). 2001. Department of Canadian Heritage. Available at ⟨http://www.virtualmuseum.ca⟩ (accessed February 2005).

Chomsky, Noam. 1965. *Aspects of the Theory of Syntax*. Cambridge, Mass.: MIT Press.

Cole, Thomas. 1991. *The Origins of Rhetoric in Ancient Greece*. Baltimore: Johns Hopkins University Press.

Corbett, Edward P. J. 1971. *Classical Rhetoric for Modern Students*. New York: Oxford University Press.

Covino, William A., and David A. Jolliffe. 1995. *Rhetoric: Concepts, Definitions, Boundaries*. Boston: Allyn and Bacon.

Crowley, Sharon. 1994. *Ancient Rhetoric for Contemporary Students*. New York: Macmillan.

Dietz, Steve, Howard Besser, Ann Borda, and Kati Geber with Pierre Lévy. 2004. Canadian Heritage Information Network. "Virtual Museum (of Canada): The Next Generation," Available at ⟨http://www.chin.gc.ca/English/Members/NextGeneration/index.html⟩ (accessed 16 April 2005).

Falk, John H., and Lynn D. Dierking. 1992. *The Museum Experience*. Washington, D.C.: Whalesback Books.

Falk, John H., and Lynn D. Dierking. 2000. *Contextual Model of Learning: Learning from Museums: Visitor Experiences and the Making of Meaning*. Oxford: AltaMira Press.

Gauvin, Kim, Kati Geber, and Corey Timpson. 2003. "Use of the Engagement Factor in the Analysis of Visitor Engagement in the Virtual Museum of Canada (VMC)." Canadian Heritage Information Network (CHIN).

Hayles, Katherine. 1993. "Virtual Bodies and Flickering Signifiers." *October Magazine and MIT* 66 (fall). Available at ⟨http://www.english.ucla.edu/faculty/hayles/Flick.html⟩ (accessed December 2004).

Heylighen, F., C. Joslyn, and V. Turchin. 1999. "Four Types of Linguistic Activities." Principia Cybernetica Web (Prinicipia Cybernetica, Brussels). August. Available at ⟨http://pespmc1 .vub.ac.be/LINGACT.html⟩ (accessed January 2005).

National Museum of American History, Smithsonian Institution. 2002. "September 11, Bearing Witness to History." Available at ⟨http://americanhistory.si.edu/september11/⟩ (accessed February 2005).

Rapp, Christof. 2002. "Aristotle's Rhetoric." In *The Stanford Encyclopaedia of Philosophy* Summer 2002 edition. Available at ⟨http://plato.stanford.edu/archives/sum2002/entries/ aristotle-rhetoric/⟩ (accessed January 2005).

Rorty, Amelie Oksenberg. 1996. "Structuring Rhetoric." In *Aristotle's Rhetoric*, ed. A. Oksenberg Rorty. Berkeley: University of California Press.

Weinberger, David. 2004. *Small Pieces Loosely Joined: A Unified Theory of the Web*. Cambridge, Mass: Perseus Books Group, American Library Association. Available at ⟨lib1.bmcc.cuny .edu/lib/help/glossary.html⟩ (accessed 17 April 2005).

White, P. R. R., and Eldon. 2004. "The Appraisal Website Homepage." Verbosity Enterprises. October. Available at ⟨http://www.grammatics.com/appraisal⟩ (accessed January 2005).

# 14 Localized, Personalized, and Constructivist: A Space for Online Museum Learning

Ross Parry and Nadia Arbach

This paper examines the confluence of user-driven software, learner-centered education, and visitor-led museum provision—and the relevance of each of these to the challenges of online museum learning. What emerges is a paradigm of increased *personalization*, *localization*, and *constructivism* characterized by a greater awareness of and responsiveness to the experiences, preferences, and contexts of the distant museum learner. This model recognizes that the visitor-central online museum learning environment relies not only upon institutional vision (embedding these resources within a wider institutional strategy) and technological agents (from intelligent agents to creative information architectures), but also upon sound pedagogical principles regarding flexibility, sociability, and activity. Examining both theory and practice, we observe that the aspiration for more personalized, localized, and constructivist online experiences is impacting upon the status, value, and role of new media within organizations; the readiness and capacity of museums to use technological innovation; and the ways museums conceive and articulate their involvement with Web-based learning.

## Localization: The Online Learner as Distance Learner

Environment is an integral aspect of learning.[1] Influential models of museum education maintain that spatial context (the physical environment) has an important effect on how learning happens (Falk and Dierking 2000). Consequently, the role, design, and value of physical learning spaces in museums today command significant attention.[2]

Traditionally, the space of the museum building itself (sometimes rarefied, sometimes liminal, frequently multivalent, invariably social) is the place the learner is helped to confront and negotiate.[3] This spatial element of the museum (which carries with it, like any other medium, a set of personal and social consequences and meanings) contributes to making museum learning unique. And yet, despite museum educators continuing to find creative ways to reimagine and recast this gallery space, the dynamic remains constant: "the museum" endures as a discrete site and "other" venue to be visited (Hooper-Greenhill 1994). However, with the continued adoption of online media tools and their capacity for interactive multimedia communication at a distance, parts of this dynamic become susceptible to remodeling. Instead of coming to the space of the museum, the online provision of the museum enters the space of the user.

With some mass media, such as television, we are familiar with the notion of broadcasters "entering the homes" of a viewing public. With the Web, however, the metaphor of "visiting" a site (and effectively leaving one's own space) remains more enduring. For the museum's sector, this may be attributed in part to the ways it continues to translate and impose many of the conventions of the onsite museum onto its online provision. And yet, the reality of using these two media (especially in an age of multichannel digital interactivity through the TV) perhaps does not warrant such an opposition. To some extent, it is a misapprehension that a user who downloads pages from a museum's server has somehow "entered" the museum—perhaps the more significant threshold that has been crossed is that of the museum into the space (whether personal, professional, or public) of the individual user. Online museums enter learners' own physical and social spaces. By contrast, the spaces that the learner "enters" are conceptual and figurative. This distinction is perhaps too frequently underestimated or misunderstood in the development of online learning resources.

Museum educators have been effective at recognizing the diversity of experiences, learning styles, and methods of meaning-making that learners bring to the museum, and the subsequent impact that this has on the formulation of education provision.[4] However, learning with the museum at a distance brings with it an extra layer of complexity—not least the diversity of spatial contexts in which the scattered learners are potentially located. This diversity is captured in the UK government's vision of its "Learning Society," where museums and other learning institutions are envisioned as being linked "to homes, the workplace, hospitals, the high street and the street corner in the same way that public utilities like the telephone are now universally available."[5] With the museum effectively coming to the everyday contexts of the user, the singular space of the museum is replaced by a variety of localized *learning spaces*.

Whereas, usually the onsite museum can control the *physical* context of learning (but, of course, not the reaction of the learner to this physicality), at a distance (online) the museum cannot affect the learner's surroundings. Therefore, when museums are supporting learning in this way, recognizing and valuing the specificity of place of the learner becomes very important. In their work on collection classifications and the "local production of order" across different institutions, Dave Randall and his colleagues prompt us to think even further about specificity of place in terms of the localization of technology, encouraging us to understand new media in "socially situated terms" and to reflect upon how its effective deployment can depend on contexts of use including (as they put it), "the ordinary, practical orientations and competencies of those who are to work with it" (1997, 114–115).

In terms of these practicalities and competencies that locally surround a technology, museums are perhaps more attuned to *formal* learning contexts, such as the needs of classroom teachers. In the *E-box* project, for instance, the University of Leicester worked with the local museum's service and a company (VEd) specializing in developing 360-degree virtual learning environments to produce a virtual handling box for secondary schools in Leicester (Parry and Hopwood 2004, 69–78). The focus group, convened by the museum service's education officer, made very clear the needs of the city's schoolteachers: as well as being flexible, differentiated, and centered upon the curriculum (characteristics all largely anticipated by the producers), one of the most important qualities was reliability. With limited time to prepare for classes and tightly scheduled time with the schools' computer facilities, *E-box* had to be dependable and fully functional in the small timetable slots when the teachers needed it to work.[6] For this specific situation with these particular teachers, schoolchildren, schools, and resources, the physical and social context in which the resource was to be used turned out to be a critical (if not the prime) factor. In other words, *E-box* came to be built with the characteristics of the distant and localized environment of these particular learners in mind.

Perhaps the greater challenge for museums is the variety of *informal* localized learning contexts that online learners might potentially occupy. Outside of museums, educators involved in distance education continue to give significant thought to building learning resources that are sympathetic to the specific physical context of the distance learner. Traditionally taking the loss of proximity as a starting point,[7] writers on distance learning have emphasized how the support provided must take into account local conditions (Melton 2002, 14), and have also stressed the advantages of building flexibility into programs to accommodate the personal circumstances of each individual learner.[8] Taking its cue from these areas of "open learning," "distance learning," and "e-learning," museum practice can benefit from developing an informed and robust critical apparatus from which to think about online and distance learning.

Concepts such as "situated learning," for instance, can potentially help museum educators to give shape, direction, and focus to their work online (Lave and Wenger 1991). As an example, the City Heritage Guides Project, managed by the 24 Hour Museum and funded by the UK government's Culture Online initiative, creates a series of accessible and inspiring cultural city Web guides for an international, national, and also local audience. The online guides offer up-to-date information about museums, galleries, and heritage attractions. As well as using contributions from journalists and museum professionals, the guides use real-life experiences, opinions, and creative content written by a range of communities and individuals within each city—each encouraged to use the "Storymaker" tool to build text-based features or trails with photographs. Just as the 24 Hour Museum is the UK's virtual national museum, so the *City Guides* aim to serve as virtual museums for each of the chosen localities. Locality and specificity of place are celebrated in each *City Guides* site, with attention given to the individual resident's everyday life.

These transactions are, in a sense, a form of situated learning. The *City Guides*' online learners engage with aspects of their locality while physically remaining in that context and while the museum generates an online experience that is consistent with (or sensitive to) that same local environment. Though the outcome is not to develop a particular expert skill set or to create "full practitioners" through apprenticeship, the functions and values of the guides are nevertheless remarkably consistent with the value the situated cognitive perspective gives to context, participation, and activity (Lave and Wenger 1991, 122). For the *City Guides* (as with more and more online museum learning experiences), learning is becoming grounded in the locality and everyday life of the learner—a result of a social process, made up of actions and activities.

A resulting question is how museums will choose to use "situated activity" as part of their provision. It is likely that, within the broader context of increased localization, museums recognizing these kinds of online learning spaces as opportunities for situated learning will become empowered with an array of pedagogical tools and rationales that may help strengthen and expand their institution's educational provision.

The physical context of the learner helps shape the learning process; museums are grasping the significance of this in relation to localization and the Web. Likewise, they are beginning to conceptualize their online learners as "distance learners"—with all the assumptions and strategies that established term brings with it.[9] And yet, as they do this, another lever is being worked in relation to the nature of learning that museums aspire to support. This is the constructivist approach.

## Constructivism: The Online Learner as Producer

Following the advocacy of writers such as George Hein, today more museum educators tend to support a theory of communication where *knowledge* is constructed internally by the learner rather than existing as an absolute independent entity, and where *learning* is the individual's own construction of meaning as opposed to an incremental aggregation and assimilation of external information (1995, 21–23). Therefore, a challenge for online (and distant) museum learning—and an area of debate for many commentators—has been the extent to which this constructivist approach can operate productively within the museum's online provision.[10]

An important part of this debate has focused upon how online learning can acknowledge and use the learner's prior knowledge. Constructivist pedagogy centers upon the previous experience and interests of the learner and asserts that these experiences shape the process of learning. The challenge, therefore, for a museum attempting to support constructivist learning online is how to identify and use what Peter Clarke calls the learner's "starting point." But whereas Clarke advocates rigorous evaluation as a means of establishing this starting point,[11] David Schaller and Steven Allison-Bunnell deploy Kieran Egan's notion of developmental "kinds of knowing" as a way of negotiating the constructivist's obligation to account for the prior knowledge of online learners (2003). Whatever the solution may be—Clarke's pragmatism or Schaller and Allison-Bunnell's theoretical circumspection—both point toward the priority afforded to sound pedagogical principles within the design of online resources. This practice demonstrates the value now attributed to—and the intellectual investment in—the delivery of credible constructivist online museum learning resources.

Another important issue in the development of genuinely constructivist Web-based experiences has been the challenge of providing a motivating, reciprocal, and flexible environment. Lynne Teather and Kelly Wilhelm's vision of a constructivist Web site is based on differentiated content creation, choice, and open information architectures: "The constructivist Web site . . . would invite visitors to construct their own knowledge. It would employ a wide range of active learning approaches, present a wide range of points of view, and provide many entry points, with no specific path, no identified beginning and end. Such a site would enable visitors to connect with objects and ideas through a range of activities and experiences that employ their life experiences" (Teather and Wilhelm 1999). While this open approach can certainly support a free-choice and active learning environment consistent with the constructivist approach, it may be less successful in offering a constant and sensitive facilitator for the learner's experience. As Clarke (2001) reminds us, the role of this facilitator is "to orchestrate a meaningful and memorable series of experiences and motivate the learner to actively engage with them." Essentially, the facilitator engages the learner, responding thoughtfully (and perhaps spontaneously) to the learner's experiences, needs, and preferences. Consequently, in the constructivist model, it is through a *dialogue* that the online learner comes to an understanding of a concept (Weller 2002, 65).

This model of a dialogue has provided the inspiration for the work of the Research Centre for Museums and Galleries (at the University of Leicester) and Mackenzie Ward Research in a joint research project.[12] This collaboration has contributed to the development of the Making the Modern World (MMW) Web site commissioned by the National Museum of Science and Industry, UK. At first glance the site clearly aims to engage general users, and the "Stories" section covers an extensive spread of content encompassing 250 years of history. In the "Learning Modules" section, this online experience incorporates an example of a technologically sophisticated and pedagogically well-informed learning resource that aspires to a dialogic exchange between a learner and a museum. To support the dialogic exchange, MMW has made use of an online learning system and methodology called Understand, created by MacKenzie Ward Research.

The Understand system uses interactive activities—interspersed throughout a narrative that leads the learner through a chosen pathway—to simulate a dialogue between the system and the learner. The learner is able to enter free-text, which is then analyzed by an intelligent agent (an expert system) from which customized answers are generated. Consequently, dialogues are targeted at each individual learner as a result of their individual actions, and aim to encourage the learner to revise or enhance their answers as they learn about a subject. If authored to do so, a system such as Understand can value all contributions entered by the learner, provide positive interventions, and allow any exchange to influence the content of the next step on the learning trajectory.

Though writing several years before the Web had transformed learning and museum provision, George MacDonald and Stephen Alsford (1991, 309) described vividly the importance and nature of such a supportive dialogue: "It is easy to introduce interactivity at relatively superficial levels, but challenging to provide meaningful participation. Nevertheless, a conversational model is what museums need to achieve: a transactional learning situation that is not simply a response to stimulus, but a response that acts on the environment in a way that gives rise to further stimuli. This conversation may take place in the museum, or it could equally take place from a re-mote location, along the electronic highways . . ." Though still perhaps behaviorist in some aspects of its current iteration (the system still has model answers in mind), Making the Modern World and the free-text dialogues that it can support offer a glimpse of the sort of responsive experiences MacDonald and Alsford had hoped for.

The dialogic exchanges of Making the Modern World also signal the importance of participation and production within constructivist learning. Constructivism's emphasis on personally constructed or socially-constructed knowledge apportions significant value to activity and interactivity—with the active "production" of new ideas key to the learning process. By actively engaging in an experience, the constructivist learner is able to test, apply, and integrate existing and newly acquired knowledge (Roussou 2004, 4). Consequently, in online learning experiences informed by constructivism the learner is frequently in a position to actively engage with and produce content.

The Moving Here Web site, as an example, is a collaboration between thirty English museums, libraries, and archives. It allows visitors to submit their own stories about migration to England. Users are encouraged to add stories of up to 1,000 words for others to read, and are able to illustrate them with digitized images from the collaborators' collections. Crucially, in this case (and many others like it)[13] there is a shift in the use of the technology. Digital media moves from being simply a *delivery medium* for the institution, to being a *communications tool* for creation and expression by the visitor. Concurrently, there is a change in the role of the online learner from that of *consumer* to that of *consumer and producer*.

With such a site, we witness a shift in the control of media production, as anticipated by David Anderson (1991, 21), that results in a shift in the control of the learning process "from transmitters to receivers, teachers to learners." With it comes a more intimate role for the learner in the process of interpretation (MacDonald and Alsford 1991, 309). This might mean resisting the appearance of Peter Walsh's "unassailable voice" of the museum (1997, 77–85) and getting used to the transparency of what others have called the "curtainless stage" (Anderson 1997, 30). And yet, such a role for the learner—who is empowered to contribute in a substantive, conspicuous, and valued way—recognizes the complex nature of knowledge production and consumption online, and the fluid boundaries between the two.[14]

Just as online museum learners are moving from consumption and production, so they are also going from learning alone to learning with others. In the emergent context of constructivist online museum learning, it is notable that museums are increasing the provision for socializing activities. Collaboration is another important aspect of constructivist learning (Pereira 2000), and the parlance of "distance learning" and "e-learning" tends to divide it into either synchronous or asynchronous activities (Salmon 2002). Typical of the creative and technologically ambitious ways in which museums have fostered real-time (synchronous) collaborations is the use of shared virtual worlds and 360-degree virtual environments, in which learners appear as digital avatars. Early work into the use of these Collaborative Virtual Environments (CVEs) in cultural heritage contexts has highlighted the need for more research into understanding how learners communicate (and might communicate more effectively) with each other in such experiences.[15] And yet, there is evidence to suggest that CVEs can be engaging, motivating, and intuitive learning environments—particularly for younger users familiar with the culture and conventions of gaming (Parry and Hopwood 2004).

By contrast, Tate Online's agoraXchange is an example of asynchronous collaboration. One of the Net Art publications on Tate Online, the agoraXchange site, by artists Natalie Bookchin and Jacqueline Stevens, is a forum for the exchange of ideas that encourages online visitors to work together. Contributors collaborate to develop the rules, design, and code for an online game involving politics, economics, and sociology, and work with a committee of artists, activists, and political theorists to decide upon a final prototype of the game. Forums such as this allow participants to explore their own views and to share them with others. The debates that have evolved from this process bear witness to how collaboration not only inspires self-discovery but also brings about a heightened awareness of others' ideas—outcomes that are entirely consistent with Martin Weller's rationale for using collaborative approaches (2002, 69).

In-gallery, there is some evidence to suggest that digital interactives can "'impoverish' the social interaction with and around exhibits" (Heath and von Lehn 2002). By contrast, the work of Tate Online (and others) is indicative of online media providing effective ways of supporting synchronous and asynchronous collaborative learning activities between learners and across different groups. Museums can now enable social interaction and learning through practical situations by encouraging the online museum learner to act as *collaborative producer*.

### Personalized: The Online Learner as User

Behind the direct contribution and involvement (independent and collaborative) of users hangs a meaningful backdrop relating to a deeper political and societal move towards more personalization in public service. This paradigm is described with great clarity by Charles Leadbeater. His discussion piece is specifically written within the context of UK governance and policy, but is clearly aware of its potential for much wider influence. It makes a powerful case (based on individual self-actualization, self-realization, and self-enhancement) for the role of personalization in the reshaping of public services (Leadbeater 2004, 84). The work presents evidence of moves within society to more customer-friendly interfaces; learners having more control over the pace and style at which they learn; and users becoming codesigners and coproducers of content. The personalization that Ledbeater identifies and advocates is not of "top down" provision to passive and dependent users, but a more pervasive and powerful "bottom up" action by active participants in the process (2004, 16–18).

In a manner echoing the reciprocal way that technology and society play off each other, developments in digital media serve as both cause and effect of this personalization of public provision. Digital media seems to be set within a developmental paradigm that perceives a move towards greater connectivity (the use of networks), greater mobility (portable, wireless media), and greater individualization (media driven by intelligent agents that are responsive to the user's specific needs). Instead of producing computer terminals with fixed or generic content and immovable hardware, the emergent technological landscape is witnessing the evolution of more ambient, bespoke, and user-driven digital media. The role of personalization (of content, software, and hardware) appears to be driving innovation and strategic development.[16]

Together, the emphasis on user-driven digital media and the push for greater personalization in public services are acting as a third key-driver on the shape of museums' online learning space. The discourse of personalization is by no means new to museums aware of the largely self-directed nature of visitors' experiences, as well as their diverse abilities, ages, cultural backgrounds, and preferred learning styles. What is significant, however, is that digital media is increasingly identified as an effective tool for generating content (sometimes on the fly) that responds to individual needs.[17]

Fantoni unpacks an important distinction between personalizing technology that is driven by the user (*adaptable* media) and that which contains its own intelligent agents (*adaptive* media) (2003, 1–16). The two types of media also differ from each other in the frequency and success of their use. On the one hand, the use of *adaptable* media (where users control content or configure their own user profile) is well established in many online museum resources. The Virtual Museum of Canada (VMC), for instance, has allowed users to register and log-in to create a personalized site through collecting, interpreting, and exhibiting up to thirty of the thousands of images on the VMC site. Similarly, users of the Metropolitan Museum of Art can register to access several personalized features, including a gallery of their favorite Met pieces and a customized events calendar.

On the other hand, the application of *adaptive* media (where the system responds automatically to the individual needs of the user) is still not as widespread and proves somewhat more problematic. Even though research has highlighted users' preferences for *adaptive* solutions (Fantoni 2003, 1–16), such systems still have persistent limitations: they are usually developed on an individually-tailored basis; they require programing skills to build, and are time-consuming to develop; they may frustrate learners if they do not respond to their queries; and they may miss or misinterpret important information (Weller 2002, 137–138).

Nevertheless, technical capability, audience expectation, and professional and institutional priorities are becoming increasingly receptive to the personalization of online museum provision. The importance of personalization in making information relevant and meaningful for users on the Web is already evident (Frost 2002, 86, 88). For the museum aiming to support a constructivist learning experience for localized (distant) learning, the part to be played by personalization and by adaptive media that can proactively and reactively meet the learner's individual needs seems worthy of much greater attention.

### Embedded, Reasoned, and Circumspect: Delivering the Space for Online Museum Learning

Together, user-driven software, learner-centered education, and visitor-led provision are leveraging the creation of more *localized*, *personalized*, and *constructivist* online museum learning experiences. The new space for online museum learning is taking shape at the convergence of these three discourses, where museums consider their online learners as *distance learners*, as *producer-authors*, and as individually *profiled users* (figure 14.1). Here, the new toolkit of online museum learning (virtual dialogues, collaborative environments, and adaptive technology) is being assembled and deployed.

The new space for online museum learning represents a shift in the conception and design of Web-based education provision for museums (as summarized in table 14.1), carrying with it a greater emphasis on specificity, reciprocity, and activity. And yet with these values come challenges. The new space for online learning must address a multiplicity of localities, compared to the singularity of the museum gallery. It must allow the online learner to contribute and collaborate, with all the technical and curatorial issues associated with shifting the axes of curation and creation away from the

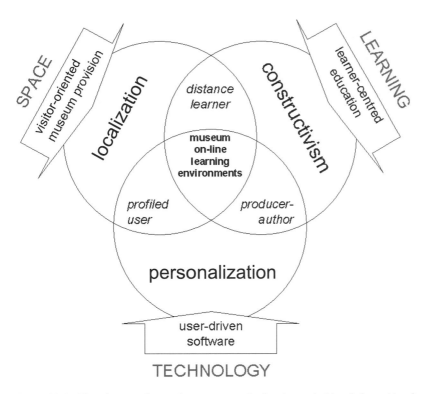

SPACE — visitor-oriented museum provision

localization

distance learner

constructivism

LEARNING — learner-centred education

museum on-line learning environments

profiled user

producer-author

personalization

user-driven software

TECHNOLOGY

Figure 14.1    User-driven software, learner-centered education and visitor-led provision leveraging the creation of museum online learning environments. © 2005 Parry and Arbach.

institution. To power its intelligent adaptive technology, it must also find ways of modeling online museum learning itself.

To successfully deliver these learning environments the museum must overcome each of these barriers. However, the success of visitor-centered learning environments is not dependent solely upon pedagogical theory and technical capability. Of equal importance is the extent to which the new environment is embedded in the rest of the museum's provision and the rationale that is used for its construction.

Today an individual's experience of a museum can exist onsite and offsite, and in both cases online and offline. Web-based museum learning represents part of the offsite/online provision (figure 14.2). The museum must have a clear sense of how this provision relates to (and is distinct from) the other aspects of the matrix—especially as the "blend" that museums offer and individuals pursue between online and offline, onsite and offsite learning can be complex. For instance, the Interactive

**Table 14.1**
The Emergent Space of Online Museum Learning

| Spaces for online museum learning | |
|---|---|
| Existing | Emergent |
| *Localization* | |
| The learner entering the *museum's space* | The museum entering the *learner's space* |
| The online museum celebrates its "*worldwide*" quality | Celebrating *locality* |
| *Constructivism* | |
| *Didactic, discovery*, and *behaviorist* models of learning | *Constructivist* learning |
| Online user as *consumer-operator* | As *producer-author* |
| *Personalization* | |
| *Targeted* at audience segments | Also providing capacity for *individualization* |
| Information architectures built on *static* hyperlinks | A more *dynamic* information architecture powered by intelligent agents |
| The *independent* learner (or groups of learners) | More synchronous and asynchronous *collaborative* activities between learners and across different groups |

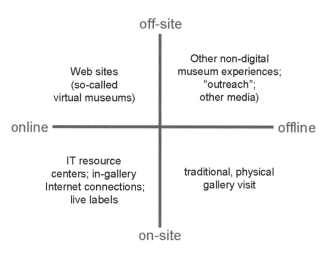

Figure 14.2   The blend of onsite/offsite and online/offline museum experiences. © 2005 Parry and Arbach.

Media Group (IMG) at the Minneapolis Institute of Arts (MIA) tested the potential of learning possibilities for onsite and online learning with a dynamic exhibition that blended these two approaches. The MIA's exhibit "Restoring a Masterwork" allowed the public to watch online in real-time over a six-week period as conservators restored a 300-year-old masterpiece. Visitors could also access the restoration of the painting onsite, with multiple viewing areas that promoted maximum interaction between the public and the conservators. Although the online version had been created to cater to those who could not visit the museum, MIA found that onsite visitors used the Web version to continue watching the progress of the restoration, while originally online visitors developed a strong interest and came to the museum (Sayre 2000). In the new space for online learning, museums must understand the subtle dynamics and potential of "blended learning," and formulate a clear rationale for online learning.

Just as a museum must ask itself why it wants or needs a Web site (Bowen 2000, 4), so it must also ask why it wants to build a space for online learning.[18] The Tate's conscious decision to put digital learning on an equal footing with onsite museum learning—the organization considers Tate Online as its fifth gallery—acknowledges that online experiences are essential and can match onsite resources in terms of conception, reach, scope, and intent. The Tate has adopted a reasoned approach to the development of digital programs, especially focusing on the relationship between online and offline, and how Tate Online can act as a gallery in its own right while complementing onsite activities. Tate Online, according to its Head of Digital Programmes, provides "the unique and now very familiar characteristics of the internet ('always open,' 'highly flexible,' provides 'limitless space'), [giving it] a central role in the ever-evolving relationship between visitors, the collection and the offline galleries" (Rellie 2004). To deliver effective spaces for online learning, museums need to understand their online provision with similar confidence and clarity.

Such rationales must always be on the move—responding to changes in technology as well as to levels and cultures of Web usage. As it appears that we are still at the bottom of the S-curve, the characteristics of the Web medium (functionality, usage, and culture) are unlikely to remain constant (Brown 2000, 12–13). Consequently, the construction of online museum learning resources remains in a fluid and variable state. However, by drawing upon digital theory (of technological development), sociological theory (of personalization), and communication theory (of constructivism), a model of the new space for online learning has come into sharper focus. Within these new practices, the knowledge and application of not just learning theory but *e*-learning theory will continue to prove crucial to museum learning online.

*Notes*

1. Consider, for instance, the importance that the concept of situated learning gives to "place," "setting," "environment," and "learning in context" within the creation of meaning. See Lave and Wenger 1991.

2. See, for example, the Clore Duffield Foundation's *Space for Learning: A Handbook for Education Spaces in Museums, Heritage Sites and Discovery Centres*, at ⟨http://www.art-works.org .uk/⟩. The main objective of the Space for Learning research project was to facilitate the creation of high-quality, flexible, and sustainable learning spaces for the future. Its research was based upon responses from ninety-one nonnational sites and twenty-one national sites, as well as eleven in-depth case studies of learning spaces at a range of small- and large-scale sites across the United Kingdom.

3. See Duncan 1995 and Parry 2002, 67–70.

4. Durbin 1996. For a reflection on the impact of these styles on the design of digital interactives, see Schaller and Allison-Bunnell 2003.

5. See *Connecting the Learning Society* 1997, 4.

6. In fact, such was the anxiety of trying to access sites that were down, or reliant upon the schools' slow or unreliable connections, that the group asked for a version of the resource to be produced on CD-ROM.

7. See Peters 2000, 17, and chapter 2.

8. See McLachlan-Smith and Gunn 2001, 45–50, and Collis 1996, 66–68.

9. Anderson has already alluded to the conjunction between developments in new media technologies, distance learning, and the future of museum learning. See Anderson 1999, 31.

10. For a paper that surveys some of the early reflections on this "visitor-centered philosophy of constructivist education for the museum on the Web," see Teather and Wilhelm 1999.

11. Though written mainly for a British audience (with one eye on the UK government's policies and strategies on lifelong learning and social inclusion), Peter Clarke's 2001 report considers how online learning projects might preserve the unique characteristics of learning from museum collections—free choice, active, oriented toward real objects, memorable, enjoyable, and full of wonder.

12. Making the Modern World Online was partly funded by £1.4 million from the UK government's "Invest to Save Budget." The association between the University of Leicester and the project was funded by a £97K grant from the Department of Trade and Industry through the Knowledge Transfer Partnership program.

13. Also see Bearman and Trant 1999, 22. The authors cite the Museum of the Person, São Paolo, Brazil, where "the WWW provides the space in which the visitors create the very museum by recording their own lives."

14. For a lyrical description of the Web helping to establish "a culture that honors the fluid boundaries between the production and consumption of knowledge," see Brown 2000, 20.

15. For two evaluations of projects that have used Collaborative Virtual Environments and have identified issues of communication between collaborators, see the reflections on the Shrine Educational Experience, a cooperative project developed jointly by Shrine of the Book at the Israel Museum, Jerusalem, and the Politecnico di Milano, in Di Blas, Paolini, and Hazan 2003 and Mitchell and Economou 2000, 149–154.

16. For a museological review of the European Commission's approach to innovation in the area of digital heritage, see Knell 2003, 132–146.

17. See Anderson 1999, 26. Jennifer Trant (1998, 123) also stresses the importance of information that "appears appropriate and relevant to an individual's experience and circumstances."

18. The answer, Clarke suggests, should be in the museum's education policy and the institution's understanding of the purpose of education.

## References

Anderson, David. 1999. *A Common Wealth: Museums in the Learning Age*. London: Department of Culture, Media and Sport/TSO.

Anderson, Maxwell. 1997. "Introduction." In *The Wired Museum: Emerging Technology and Changing Paradigms*, ed. K. Jones-Garmil, 11–32. Washington, D.C.: American Association of Museums.

Bearman, David, and Jennifer Trant. 1999. "Interactivity Comes of Age: Museums and the World Wide Web." *Museum International* 51, no. 4: 20–24.

Bowen, Jonathan. 2000. "The Virtual Museum." *Museum International* 52, no. 1: 4–7.

Brown, John Seely. 2000. "Growing Up Digital: How the Web Changes Work, Education and the Ways People Learn." *Change* 32, no. 2: 11–20.

Clarke, Peter. 2001. *Museum Learning on Line*. London: Resource: The Council for Museums Archives and Libraries.

Collis, Betty. 1996. *Tele-Learning in a Digital World: The Future of Distance Learning*. London: Thomson Computer Press.

*Connecting the Learning Society: The National Grid for Learning.* 1997. UK Government Consultation Paper. London: Department for Education and Employment.

Di Blas, Nicoletta, Paolo Paolini, and Susan Hazan. 2003. EE Experience: Edutainment in 3D Virtual Worlds." Paper presented to *Museums and the Web 2003.* Available at ⟨archimuse.com/mw2003/papers/diblas/diblas.html⟩ (accessed 17 April 2005).

Dierking, Lynn D., and John H. Falk. 1998. "Audience and Accessibility." In *The Virtual and the Real: Media in the Museum,* ed. Selma Thomas and Anne Mintz, 57–70. Washington, D.C.: American Association of Museums.

Duncan, Carol. 1995. *Civilising Rituals: Inside Public Art Museums.* London: Routledge.

Durbin, Gail. 1996. *Developing Museum Exhibitions for Lifelong Learning.* London: Stationery Office.

Falk, John H., and Lynn D. Dierking. 2000. *Learning from Museums: Visitor Experiences and the Making of Meaning.* Walnut Creek, Calif.: AltaMira Press.

Fantoni, S. F. 2003. "Personalization through IT in Museums, Does it Really Work? The Case of the Marble Museum Web Site." In *Cultural Institutions and Digital Technology,* ICHIM Paris, 8–12 September 2003, 1–16. Paris: Archives and Museum Informatics Europe.

Frost, C. Olivia. 2002. "Where the Object Is Digital: Properties of Digital Surrogate Objects and Implications for Learning." In *Perspectives on Object-Centered Learning in Museums,* ed. S. G. Paris, 79–94. Mahwah, N.J.: Erlbaum.

Heath, Christian, and Dirk vom Lehn. 2002. "Misconstruing Interaction." In *The Proceedings of Interactive Learning in Museums of Art and Design,* ed. M. Hinton. London: Victoria and Albert Museum.

Hein, George. 1995. "The Constructivist Museum." *Journal for Education in Museums* 16: 21–23.

Hooper-Greenhill, Eilean. 1994. *Museums and Their Visitors.* London: Routledge.

Knell, Simon. 2003. "The Shape of Things to Come: Museums in the Technological Landscape." *Museum and Society* 1, no. 3: 132–146.

Lave, Jean, and Etienne Wenger. 1991. *Situated Learning: Legitimate Peripheral Participation.* New York: Cambridge University Press.

Leadbeater, Charles. 2004. *Personalisation through Participation: A New Script for Public Services.* London: Demos.

MacDonald, George F., and Stephen Alsford. 1991. "The Museum as Information Utility." *Museum Management and Curatorship* 10: 305–311.

McLachlan-Smith, Claire, and Cathy Gunn. 2001. "Promoting Innovation and Change in a 'Traditional' University Setting." In *Innovation in Open and Distance Learning: Successful Develop-*

*ment of Online and Web-Based Learning*, ed. F. Lockwood and A. Gooley, 38–50. London: Kogan Page.

Melton, Reginald F. 2002. *Planning and Developing Open and Distance Learning: A Qualitative Approach.* London: Routledge/Falmer.

Mitchell, William L., and Daphne Economou. 2000. "The Internet and Virtual Reality in Heritage Education: More than Just a Technical Problem." In *Virtual Reality in Archaeology*, British Archaeological Reports International Series S843, ed. J. A. Barceló, M. Forte, and D. Sanders, 149–154. Oxford: Archaeopress.

Parry, Ross. 2001. "Including Technology." In *Including Museums: Perspectives on Museums, Galleries and Social Inclusion*, Ed. J. Dodd and R. Sandell, 110–114. Leicester: RCMG.

Parry, Ross. 2002. "Virtuality, Liminality and the Space of the Museum." In *Clicks and Mortar: Building Cultural Spaces for the 21st Century*, 67–70. Cambridge, Mass.: mda.

Parry, Ross, and John Hopwood. 2004. "Enabling the Soft Museum: Virtual Reality and the Community." *Journal of Museum Ethnography* 16: 69–78.

Pereira, Márcia A., Rachel A. Harris, Duncan Davidson, and Jennifer Niven. 2000. "Building a Virtual Learning Space for C&IT Staff Development." Paper presented to *The European Conference on Educational Research*. Edinburgh, 20–23 September.

Peters, Otto. 2000. *Learning and Teaching in Distance Education: Pedagogical Analysis and Interpretations in an International Perspective.* London: Kogan Page.

Randall, Dave, Dave Francis, Liz Marr, Colin Divall, and Gaby Porter. 1997. "Situated Knowledge and the Virtual Science and Industry Museum: Problems in the Social-Technical Interface." In *Museums and the Web 97: Selected Papers*, ed. D. Bearman and J. Trant, 111–125. Pittsburgh: Archives and Museum Informatics.

Rellie, Jemima. 2004. "One Site Fits All: Balancing Priorities at Tate Online." In proceedings of *Museums and the Web 2004.* Archives and Museum Informatics. Available at ⟨http://www.archimuse.com/mw2004/papers/rellie/rellie.html⟩ (accessed 17 April 2005).

Roussou, M. 2004. "Learning by Doing and Learning through Play: An Exploration of Interactivity in Virtual Environments for Children." *Computers in Entertainment* 2, no. 1: 10.

Salmon, Gilly. 2002. *E-tivities: The Key to Active Online Learning.* London: Kegan Paul.

Sayre, Scott. 2000. "Sharing the Experience: The Building of a Successful Online/On-Site Exhibition." In proceedings of *Museums and the Web 2000.* Archives and Museum Informatics. Available at ⟨http://www.archimuse.com/mw2000/papers/sayre/sayre.html⟩ (accessed 17 April 2005).

Schaller, David T., and Steven Allison-Bunnell. 2003. "Practicing What We Teach: How Learning Theory Can Guide Development of Online Educational Activities." In proceedings of *Museums and the Web 2003.* Pittsburgh: Archives and Museum Informatics.

*Space for Learning: A Handbook for Education Spaces in Museums, Heritage Sites and Discovery Centres.* 2004. Clore Duffield Foundation.

Teather, L., and K. Wilhelm. 1999. "'Web Musing': Evaluating Museums on the Web from Learning Theory to Methodology." In proceedings of *Museums and the Web 1999.* Pittsburgh: Archives and Museum Informatics.

Trant, Jennifer. 1998. "When All You've Got Is 'The Real Thing': Museums and Authenticity in the Networked World." *Archives and Museum Informatics* 12: 107–125.

Walsh, Peter. 1997. "The Web and the Unassailable Voice." *Archives and Museum Informatics* 11, no. 2: 77–85.

Weller, Martin. 2002. *Delivering Learning on the Net: The Why, What and How of Online Education.* London: Kogan Page.

# III Cultural Heritage and Virtual Systems

# 15 Speaking in Rama: Panoramic Vision in Cultural Heritage Visualization

Sarah Kenderdine

*The pictorial expression of a "symbolic form," of a specifically modern, bourgeois view of nature and the world . . . serves both as an instrument for liberating human vision and for limiting and imprisoning it anew . . . it collects all details . . . and puts an end to the uncertainty of relationships by claiming them all in one fell swoop.*
—Stephen Oettermann, *Panorama*

Virtual and augmented reality in cultural heritage research[1] (hereafter referred to as virtual heritage) asserts to act at the interface between advanced scientific visualization and the rich cultural datasets, texts, and extant remains of the discipline. In contemporary virtual heritage practices, panoramic vision systems are often employed for comprehensive data capture (omnidirectional photography, video, and scanning technologies). In addition, interfaces that use a panoramic scheme to "enclose" the visitor are used for display and communication purposes. The (re-) emergence of the panoramic scheme as "the new image vogue"[2] is based on the desire to design virtual spaces and places that can be inhabited by the viewer—to maximize a sense of immersion and ultimately "presence" in either past environments (digital recreations), or remote real-world locations (panoramic enclosures for archaeological site visualization; for instance, see figure 15.1). This trend is often in tension with the projected requirement for the pedagogical outcomes that are the basis for heritage interpretation.

*Speaking in Rama* extrapolates the phenomenon of panoramas in the rich history of creating illusion spaces—to investigate past, current, and future forms of panoramic immersion. As Oliver Grau, new-media art historian states, "images have always been subject to media technologies of spatial illusion, immersion and display" and "every epoch uses whatever means available to create maximum illusion" (2003, 5). The intention below is to acknowledge the embedded sociocultural characteristics of the panorama and to examine the strategies of spatiality and navigation it supports—while

Figure 15.1    Stereographic projection cylinder (AVIE), currently under construction at Centre for Interactive Cinema Research (iCinema), UNSW, Australia, showing Banteay Srei from Sacred Angkor: stereographic panoramas of the temple complex at Melbourne Museum 2004–2005. Visualization courtesy of iCinema. Angkor image. © APSARA-Authority, Cambodia, photography Peter Murphy, 2004.

also examining ways to extend its historic horizon. A future for the use of the panoramic scheme must point to the ways in which the panoramic imagination can be extended for the purposes of virtual heritage mandates,[3] mindful of the historical and cultural discourse of place and space, and its representation, that precedes us.

Grau suggests that incorporating an examination of the historical does not marginalize the principles of virtual reality (its technical developments, modalities of navigation, interaction, and interface) but seeks to place them in a historical context:

> In the historical context, new can be revitalized, adequately described, critiqued in terms of phenomenology, aesthetics and origin. . . . New media do not render old ones obsolete but rather assign them new places in the system. (2003, 8)

The panorama reveals itself as a navigable space, persistent throughout media history, which is charged with sociocultural implications. Considering the reemergence of the panoramic scheme in contemporary virtual reality reinforces the notion that breakthroughs in technology often emerge from the preexisting fabric of cultural discourse.

Extrapolating from the historical record, we can observe how panoramic "immersive" visualization strategies (where message and medium become merged) act specifically against distanced and critical reflection considered necessary for scientific rational inquiry (Grau 2003, 339). In these spaces the technologies of production and display must remain invisible to the viewer to allow for maximum intensity of the message to be perceived. Creators of virtual heritage are required to understand such issues to validate the use of these visual strategies as they reinforce particular dynamics of contemporary culture—the regime of visual controls and reification of spectacle characteristic of the modern era, embedded in consumer culture and global media.[4] The "Disneyfication of culture" is a common criticism leveled against virtual heritage, which is often seen more akin to theme park amusements than scientific augmentation. As Alison Griffiths theorizes in her analysis of museum spaces:

> The discursive oppositions between science and spectacle, information and entertainment, and passive and interactive spectators first articulated in relation to visual technologies one hundred years ago have repeatedly resurfaced in contemporary debates over multimedia exhibits in public museums. (2003, 376–377)

### Panoramas in a Historic Context

The rich technical and sociocultural record of the panoramic scheme (in its broadest sense) embraces the history of perspective and representation from as early as cave paintings, through the painterly illusions of the Baroque (Klein 2004) and Renaissance, on to the machinery of the Great Exhibitions in the nineteenth and twentieth centuries. A historical trajectory could include everything from rock and cave painting, scroll painting, interior frescoes, and church interiors through magic lanterns, *mondo nuovo*, various phantasmagorias, all manner of optical devices, *cabinets des curieux*, *wunderkammern*, the Great Exhibitions, glasshouses, and winter gardens.[5] This elaborate exhibitionary apparatus was also complemented by portable optical toys which brought vision systems into drawing rooms and parlors in the nineteenth century (Bruno 2002), for personalized examples of virtual travel. The optical devices of the latter period were dedicated to further expanding the "optical unconscious"[6] and took a multiplicity of forms, including an assortment of "—scopes" and "—oramas."

The mass public screen entertainment of the panorama made its debut in the late 1700s and is the subject of a number of extensive analytical histories.[7] It has been hailed as "the first true mass medium" (Oettermann 1997, 7). In its most basic form, the panorama of the nineteenth century could be described as a long circular set that surrounded the spectators and often included props inserted between the viewer and the plane of the image, complete with dynamic (and natural) lighting effects. The technology necessitated the development of architectures suitable to convey the illusions, and panoramas were initially conceived of in purpose-built designs. For example, Robert Barker's[8] patent in 1787 for a panorama relied on the creation of a suitable rotunda for its visual impact, and for the "performance." His desire was to make the viewer "feel as if really on the spot" (Druckrey 2003, 62). The circular building ensured that the viewer was encapsulated by the scene.

The specific venues that housed painted panoramas relied on both visual spectacle and optical scale for their appeal. Due to the enormous resources that these projects demanded, and the limited turnover of images shown (one per year), most were short-lived. Multiactivity complexes (panoramas and roller skating for instance) allowed some large-format pictures to continue to be produced, but they remained limited in supply, novelties. Panoramas did find a viable market in the latter part of the nineteenth century when the 360° screen became the standard at the Universal Exposition in Paris in 1889 (which housed a total of seven different panoramas) and the Universal Exposition of 1900, which housed the Mareorama, Pleorama, Stereorama, Cinéorama, and the Lumiere Brothers Photodrama.

In the Mareorama the spectator traveled landscapes between Marseille and Yokohama, including Naples, Singapore, and China. The platform for spectators was disguised as a transatlantic ship seventy meters long, for the berth of over 700 passengers. This ensemble lay on a suspension system that allowed the structure to undulate with the action of the "waves." Actors were employed to simulate shipboard life, and a ventilation system diffused marine scents. Lightning followed a diurnal transit mapping the voyage (Oettermann 1997, 77–81).

The Cinéorama was a circular building of 100 meters circumference. Ten projectors were used to create a continuous image across the screen, and in the center of the room was a balloon with all its accoutrement of anchors, ropes, ladders, and counterweights. The basket was lifted and lowered while a series of images simulated take-offs, landings, and the trips themselves (1997, 85).

A noteworthy development was the miniaturization of the panorama in the form of the Myriorama (Greek word for "ten thousand"). Described as a landscape kaleidoscope, it was fashioned from fanciful or exotic landscapes cut into sections of equal width, and on each one the horizon line matched its adjacent image. The scope for arranging new scenes from the cards was almost unlimited. The miniaturization of the panorama made it both portable and accessible as a souvenir and, as a novelty for children (Oettermann 1997, 67).

Panoramas, and their associated forms (to name a few more: Diorama, Georama, Giorama, Cyclorama, Betaniorama, Cosmorama, Kalorama, Kineorama, Europerama, Typorama, Neorama, Uranorama, Octorama, Poecilorama, Physiorama, Nausorama, Udorama), were part of the nineteenth century vogue of "speaking in rama" as chronicled by the French novelist, Balzac (1917, 234), in *Père Goriot*. These forms were collectively known as "*machines en rama*" or "rama contraptions." William Uricchio (1998), theoretician in media studies, notes that panoramic vision was considered more an act of "perception" than a "parameter of representation," thus the term has found its way into nonvisual media to describe a "comprehensive coverage," as titles of books, theses, and movies suggest (e.g., *Political Panorama* (1801); *Panorama of Youth* (1806)).

The panoramic form is also closely aligned to that of the panopticon. Jeremy Bentham's design for a new form of prison in 1787 incorporated a central tower that made the prisoners visible to the guards while they themselves were impervious to being seen:

> As "schools of vision," the panorama and the panopticon are at the same time identical and antithetical: in the panorama the observer is schooled in the same way of seeing that is taught to prisoners. (Oettermann 1997, 41)

### Photographic and Cinemagraphic Panoramas

> [The] visual media of the moving image has embraced the prospect of vision as unlimited travel. (Cubitt 1998, 79)

The history of the panorama is closely interwoven with that of photography (both the use of panoramic views by photographers and the use of photos by panoramic scene painters), and each played a role in the other's development. Damoizeau in 1891 and later the Lumière Brothers in 1888 developed the first rotational equipment for creating photographic panoramas. Eadweard Muybridge, the celebrated photographer of motion, was also a great panoramic photographer, and one of his works included an eleven-segment panorama of Guatemala City.[9] Photographic panoramas[10] can be interpreted as images that embodied movement before the advent of cinema. They provided surrogate reality for those who were unable to travel, or a framework of experience for those planning to visit the location depicted. What can be observed is a reliance on overwhelming sensory experience rather than narrative structure—and on photographic realism combined with instruments of spectacle.

Of all the films copyrighted in the United States between 1896 and 1912, the largest single number of entries use "panorama" in the title. Panoramic film, however, broke with the traditional form of horizontal horizon of painted panoramas, to further explore the space vertically, and with the use of tracking shots. William Uricchio describes how these films also "charted the texture of movement itself" (1998, 125). By comparing characteristics of the painted panorama with cinema, he observes that the two forms share similarities, and notes "the panorama's cycles of popularity; the shifting class and age of its audiences, its placement within a pedagogical framework, its pricing, its accompaniment with music and lectures; its variation" (1998, 127).

In this way the panorama functioned as a photo-real cinema of attractions, and "hinted at the dream of complete spectacle, of 'total cinema'."[11] It laid the basis for future large-screen formats such as the Cinemascope, Cine 180, IMAX, and OMNI-MAX. Indeed, the archaeology of the panorama is also found in the history of cinema. But as cinemagraphic studies suggest: "The viewer's mastery of a visual domain is subverted by the marks of the mediator," that is, the film maker controls the subjectification of the view, and does not embody "movement" in the same way as a panorama (Uricchio 1998, 127).

## Limits of the Illusion

Records contemporary to the heyday provide useful observations on the limits of panoramic illusion. We find the concerns raised are precisely the problems addressed in contemporary research in virtual reality and virtual heritage.[12] The nineteenth century panorama visitor was treated to increasingly sophisticated scenes and elaborate rotundas and structures in which to observe them—requiring less and less from the spectator's imagination to fill the gaps. The effect of the increasing realism was the demand for a totalizing experience which mixed all the genres of sensation (acoustic, atmospheric, olfactory) to be part of the viewer's experience. Johann Eberhard, for example, spoke of the unease he had felt when confronted with a painted panoramic view of

London, a large bustling city but weighed down as he described by "melancholic silence."

The panorama format allowed the spectator to move, and thus transformed the disincarnated single eye "gaze" to the "glance"[13] of two mobile eyes. Lack of movement in the scene was also often cited with disapproval—where were the scudding clouds, the atmospheric variation? To complete this list of faults these static landscapes failed to provide temporal, or narrative flow.

This limit of movement in the scene was especially observable with regard to human figures painted into the scenes. Ingenious attempts to set the panorama in motion were made, but spatial and temporal verisimilitude was impossible to achieve because of the problems of parallax. Indeed, as one onlooker experienced, the combination of effects, both physiological and psychological, could be quite nauseating. His diagnosis:

> I am swaying between reality and unreality...between truth and pretence. My thoughts, my whole being are given a movement which has the same effect as spinning or the rocking of a boat. Thus I explain the dizziness and sickness which overcomes the concentrated onlooker in the Panorama. (Eberhard 1805)

## Decline of the Panorama

The panorama fell from the popular embrace in the early twentieth century. Stephen Oettermann, author of one of the most significant histories on the panoramas across the world, maps the demise as it changed from standing as a symbol of the Enlightenment to a "menacing presentiment of the society of surveillance" (Jay 1998). Oettermann even suggests that is was already obsolescent even at the time it appeared and was most reified. It was only by producing countless new up-to-date scenes and configurations of the form that it was able to stave off its obsolescence as a novelty. Lieven de Cauter suggests that, in fact, the twentieth century soon began to favor "vertigo machines" (e.g., roller coasters), instead of educational pavilions. He contrasts the thrill of motion with the contemplative experiences of panoramas. Jonathan Crary, theoretician on nineteenth-century art, perceives that the embodied eye of the spectator demanded its own "ocularcentric privilege" in contrast to the "all seeing eye of God."[14] Soon the democratized view was sought. As Martin Jay (1998) notes, the heyday of the panorama was one that:

still had time for moments of arrested motion and visual contemplation, a time before the eye was thrust back into a libidinally charged body continuous with networks of information and stimulation, where all terrains seem equally false.

These introductory historical notes on the panoramic form above provide a background for deeper analysis into the cultural and perceptual aspects of the media form—explored below.

## Virtual Travel

The proliferation of the panorama was driven by three primary sociocultural dynamics (Comment 2000). The first of these was a desire to make sense of oneself in the world and to regain control of an increasingly complex urban space fuelled by the Industrial Revolution. The peculiar reaction to this anxiety was the propagation of enormous painted canvases that detailed the same city in which they were shown. The second element in the history of panorama was its use in the propaganda of the war machine, with an all-encompassing depiction of military strength and war heroes. The third theme that emerged was that of travel to distant lands, historic cities, and imposing landscapes associated with the colonial and imperialist policies. The panorama as a device was central for expounding the theory that travel was good for developing one's mind. The geographical form of panorama is perhaps best illustrated by the Giant Globe and the Géorama (Oettermann 1997). The latter's patent of 1822 reads:

> With the aid of this machine, one could embrace in one single glance the whole surface of the earth: it consists of a sphere of 40 feet in diameter at the centre of which the spectator is positioned on a platform of 10 feet in diameter, from which is discovered all parts of the globe. (Oettermann 1997)

As "a substitute for travel and a supplement for newspapers,"[15] social and media historian Richard Altick points out, "fidelity to the fact was a prime consideration."[16] Mobility, visibility, and information were linked to an "ultra realism" that combined to produce a static temporality in the spectator. "Simulated travel" was integral to many forms of optical device.[17] By 1870, for instance, the stereograph was firmly associated with exotic locales, becoming a necessary "adjunct to the telescope and microscope, showing us the true form and configuration of the distant world" (Earle 1979). The stereoscope can be called the first "media machine" addressed specifically for the home audience (Huhtamo 1995). As Antoine Claudet stated of the stereoscope in 1860:

> The general panorama of the world. It introduces to us scenes known only from the imperfect relations of travellers. By our fireside we have the advantage of examining them, without being exposed to the fatigue, privation and risks of the daring and enterprising artists, who, for our gratification and instruction, have traversed lands and seas, crossed rivers and valleys, ascended rocks and mountains with their heavy and cumbrous photographic baggage.[18]

### The Giddy Horizon

Stephen Oettermann states that the notion of "horizon" was, most notably, defined by the panorama. He writes:

> The discovery of the horizon, the liberation of the eye and at the same time the era's diffuse sense of imprisonment all have a perfect counterpart in the panorama: while seeming to offer an unconfined view of a genuine landscape, it in fact surrounds observers completely and hems them in far more than all previous attempts to reproduce landscapes. At the same time that the panorama celebrates the bourgeoisie's ability to "see things from a new angle" it is also a complete prison for the eye. The eye cannot range beyond the frame because there is no frame.[19]

The increasing popularity in the nineteenth century of experiences related to the bodily sensation of giddiness, views from elevated positions, and the hot air balloon were all reflected in the appeal of the panorama. The "horizon" also meant that there was "something" beyond it, a hint of promise. The "unlimited horizon"[20] of the panorama projected a "universal gaze suggesting the image was inexhaustible, and in this respect panoramas anticipated the database" (Druckrey 2003, 102).

### Panoramic Vision

Maurice Merleau-Ponty, who wrote extensively on the philosophies of vision, has argued that seeing is a kind of possession. One of the recurring features of the nineteenth century relationship with the screen is the desire for interactivity. Optical devices such as the stereoscope subverted the conventional rules of perspective, and in so doing they disturbed the viewers' comfortable relationship with the world:

> The thrill of many optical toys derives from the unstable subjectivities they produced: where the spectator, nevertheless remained in control of the image. . . . Large scale screen formats intensified the spectating process provided by small optical toys. Like the kaleidoscope, panoramas offered a disorientating optical experience. In contrast to the singular spectating position assumed by conventional painting, the panorama offered an unending simultaneity of viewpoints. (Plunkett 2002)

Panoramas themselves represent an epistemological breakthrough in the previous scopic regime of images "as important as the emergence of pictorial abstraction" (Parente 1999). The proliferation of optical inventions using the supernatural illusion of natural magic was superseded by a new technical visuality.

Oliver Grau succinctly describes one of the key parameters of the panoramic device as being to draw "immense crowds into closed environments, with the world laid out in spectacle" into which "one could project oneself imaginatively" (2003, 8). This has been described by media researcher André Parente as the "aesthetic of transparency" (1999). This new type of visibility relied on the observer's appearance on stage. How this user-centered aspect of panoramic vision is activated in virtual heritage spaces is demonstrated in the examples found later in this chapter.

### The Electronic Baroque

Recently, several theorists[21] have drawn a trajectory from the illusionistic, spatial, and political qualities of the Baroque—in art, and architectural and theatrical constructions dating from the mid-fifteenth century, to new media forms—and these theories have relevance to the discussion below. The authors extend parallels to the histories of pre-cinema, cinema, and special effects, and into the labyrinthine scripted narratives of our shopping malls, and virtual reality environments of today. Norman Klein, writing on mass culture, media, and urban studies, employs the delightful term—the "Electronic Baroque" (2004, 403–405). Essentially, the Electronic Baroque is formed by a merger between the top-down ceiling baroque and the all-absorbing 360-degree panorama. Again, inherent in the function of the Electronic Baroque is a desire to evoke states of transcendence that amplify the viewers' experience of illusion. The underlying aesthetic of astonishment is achieved through the ambivalent relationships generated in the spectacle's construction of a spatial perception that emphasizes rational and scientific principles, while evoking a state of amazement in the audience that has little to do with rationality:

The techniques of the electronic baroque are not concerned with similitude and reproduction of a parallel reality: that is, they do not seek to look into a world that is like our own. Existing reality is devalued in favour of a space that postpones the "traditional" narrative and instead evokes heightened emotion. (Klein 2004)

This tension, between the scientific and illusionism is prevalent in all optical technologies and media devices. As writer and cultural critic Marina Warner (2004, 16) observes, "the media of visual trickery, deception and illusion move away from evermore intense techniques of scientific scrutiny towards entertainment, as instruments of knowledge become part of show business."

## Politics of Illusion

As noted above, the discussion of new and not-so-new optical tools cannot be considered without reference to the sociocultural frameworks through which they are constructed. Jonathan Crary (2001) studies the relationship of ocular devices and modernity and has suggested some of the elements that made some artificial "ways of seeing" more successful than others. Rather than accepting the dominant history of an evolutive narrative culminating in cinema, he shows a history of politics of the conformation of the body. For him, the optical devices that survived were the ones that primarily combined two factors: 1) they were sufficiently "phantasmagoric," meaning not only their capacity for creating illusion, but also (in his quotation of Adorno), their "concealment of the process of production," and 2) they had the ability to create a visual experience that presupposes the body as immobile and passive.

Thomas Edison, inventor of the motion camera, is used by many theorists to illustrate the transformation of perceptual awareness through his contributions to the creation of optical devices (namely the kinetoscope). With Edison we can see a move away from earlier forms of preindustrial practice and early nineteenth century forms of display, exhibition, and consumption—to those of "quantification and abstraction." Crary, concerned primarily with the cultural economic implications of Edison's work states:

Edison saw the marketplace in terms of how images, sounds and energy, or information could be reshaped into measurable and distributable commodities and how a social field of individual subjects could be arranged into increasingly separate and specialised units of consumption. The logic that supported the kinetoscope and the photograph—that is, the structuring of perceptual experience in terms of a solitary rather than collective subject—is replayed today in the increasing centrality of the computer screen as the primary vehicle for the distribution and consumption of electronic entertainment commodities. (2001, 31)

Edison recognized the integration of software and hardware that foreshadowed the late twentieth century: "the indistinction between information and visual images, and the making of quantifiable and abstract flow into the object of attentive consumption" (Crary 2001, 33). Edison was aware that the products were, in fact, inseparable from the creation of new needs in consumers, thereby continually altering the network of relations. Crary extrapolates the need for "continual repatterning of the ways in which a sensory world can be consumed" with regard to the methods employed by the information technology manufacturers of today. He articulates the prevalent consumer paradigm of our times:

> Throughout the changing modes of production, attention has continued to be a disciplinary immobilization as well as an accommodation of the subject to change and novelty—as long as the consumption of novelty is subsumed within repetitive forms. (Crary 2001)

### Virtual Reality Panoramas

Since the demise of the nineteenth century panorama for mass entertainment, experiments and research in the use of the panoramic scheme (other than in the development of widescreen cinemas) can be found observed in the electronic arts.[22] Some of these works reenact the cinematographic device, combining the immersive architecture of the panoramas, and the interactive language of the new digital interfaces with the movement of the cinema image. These works, indeed, have laid the foundation for how we think of the future possibilities of an immersive and interactive cinema, and as we will observe, virtual heritage.

Artists such as Jeffrey Shaw and Michael Naimark have been working within the oeuvre of extended narratives and augmented devices for panoramic images since the mid-1980s, and offer useful examples in the context of this chapter. As Lev Manovich (2001, 282) describes, Jeffrey Shaw "evokes the navigation methods of panorama, cinema, video, and virtual reality. He 'layers' them side by side." Installations such as *Place* (1995) and *Place Ruhr 2000* surround the visitor (who stands on a rotating platform) with a 360-degree panoramic screen. The works reframe the traditional panorama within the new one of virtual reality. The interface allows the visitor to navigate between the various locations, each of which are depicted in panoramic cylinders that have been distributed throughout the landscape map. Once inside the individual panoramic cylinder the user confronts a scene augmented by small animated effects. These works act both as "representation and documentation of social and economic histories of the places depicted" (Grau 2003, 240–242).

Michael Naimark, for *Be Now Here* (1995), traveled to heritage sites around the world to take his panoramic views. Using a 35 mm three-dimensional (3D) stereographic camera mounted on a motor-driven tripod, he was able to capture 360-degree motion scenes at places such as Angkor in Cambodia, Dubrovnik in Croatia, Timbuktu in Mali, and Jerusalem. His immersive display consisted of 3D video projected onto a 360-degree screen, combined with a spatial soundscape—in an anthropological approach to both virtual travel and site documentation (Grau 2003).

In the late 1980s panoramic and spherical image maps were popularized by Apple with their QuickTime VR (QTVR) software. Since that time, there has been a profusion of panorama softwares for generating Internet applications, including Photovista (Live Picture), IPIX (Internet Pictures), SmoothMovie (Infinite Picture), and Videobrush (Live Picture). Indeed, the largest-ever photographic panorama, taken of the city of Delft, was heralded in November, 2004 and all 2.5 billion pixels are available online.[23]

A recent New York Times article (Mirapaul 2003) highlighted the powerful effect of Web-based panoramic content. As noted by Fred Ritchin, Associate Professor of Photography and Communications at New York University, "the full potential of [online] panoramas has yet to be realized." Ritchin said that viewers—and photographers—are still accustomed to the conventional photograph's single viewpoint. He said he would like to see the panoramic image present multiple viewpoints to better effect. Because panoramic images can contain links to other panoramic images, Ritchin also imagines their use in elaborate nonlinear narratives.

Interesting developments for the navigation and interpretation of panoramas include the enclosure of avatars (real-world video footage or virtual characters) into the scenes, and special engines for panorama browsing online.[24] These experiments are welcomed by Erik Goetze, who maintains a blog about panorama technology. "One of the things about panoramas is they're fairly static," he writes. "They typically just show a place; they don't tell a story."[25]

### The Ocular Gastronome of Distributed VR Panoramic Image Bubbles

The touristic gaze has been invigorated by the advent of mass digital photography and the panoramic "photo bubble." To paraphrase Susan Sontag, the photograph has been used to: appropriate and transcribe reality; aestheticize and neutralize experience;

miniaturize the real; democratize everything to be the same; give shape to travel; and complete the cycle of proof that one has actually traveled.[26] John Urry, noted for his analysis of global tourism dynamics, demonstrates how landscapes are transformed into objects and transferred from person-to-person in gestures of "mastery" over the environment. They become at once subjective ("I was really here") and objective ("and this is 'the real thing'").[27]

The appeal of "navigation" as a construct for cyberspace has been the subject of much new media analysis. Analysis often expounds upon Charles Baudelaire's description of the anonymous observer ranging the streets of Paris—the *flâneur*,[28] an "ocular gastronomer," who samples at will from the exotic atmosphere of the Parisian arcades,[29] alone while in a crowd. Walter Benjamin suggests that the *flâneur* transforms the space of the city with her perceptions and imagination. In this way, navigation creates a virtual (and subjective) space.

Anne Frieberg, pioneer in the field of visual culture studies, presents an "archaeology of perception" that characterizes the modern cinematic, televisual, and cybercultures. In this, the "mobilized virtual gaze" is "a received perception mediated through representation" and travel "in an imaginary flânerie through an imaginary elsewhere, and imaginary elsewhen."[30]

### Space, Navigation, and Presence

> In virtual reality, the panoramic view is joined by sensorimotor exploration of an image space that gives the impression of a "living" environment. (Grau 2003, 7)

The navigation of virtual space is the scopic regime for an evolution of the panoramic image. The use of the panorama requires a lexicon for navigable space that is "not only a topology, geometry, and logic of static space" but is transformed by "new ways in which space can function in computer culture," that is a "trajectory" rather than a place. To extrapolate, place is the product of cultural producers while space itself (or "non-places" as Lev Manovich puts it) are produced by the activities of the user within the system (2001, 280).

Analysis of virtual reality space by Manovich maintains that screen tradition presupposes the body as linked to the machine, (as in all representational forms, for example: paintings, mosaics, cinema), but also provides the "unprecedented new condition of requiring the viewer to move" (2001, 112). The interface for the spectator is one single coherent space, made from both a physical space and the space of representation which is described as a space of simulation. Manovich notes that the nineteenth century panorama can be considered as the transitional form between classical simulation (wall paintings, human size sculpture, dioramas), and virtual reality, where the viewer could move around a central viewing area, and where the real space is subordinated to the virtual space. When the physical space can be totally disregarded, conditions for virtual reality space are created, and these spaces are, by their nature, navigable.

Ethical considerations are obvious if one considers that virtual reality tends to encourage users to suspend belief and render the interface "invisible." Terry Harpold (2001), researcher on new media culture, warns of "direct manipulation" (possible when there is the illusion of user immersion) pronouncing it "among the most privileged methods of psychic and political coercion of the post-Enlightenment period." He goes on to suggest, however, that we may—through understanding the history of contested spatiality of science, art, and politics away from the computer—"provide strategies of design that break the epistemic confines of direct manipulation" through the computer (Harpold 2001, 18).

## Matters of Perspective

Both Martin Jay (2001) and James Elkins have argued that linear perspective and the cult of "Cartesian perspectivism," used in Western science and art, is just one among a variety of optical regimes. However, the panorama, situated as it is within nineteenth century forms of exhibition installations, architectural spaces, and optical devices, utilizes a lexicon of these specific visual cues from Cartesian and Euclidean world views (projection, vanishing points, lengthening shadows, and so forth).

The last decades of ubiquitous use of immersive image media across the world have helped reinforce Euclidean perspective and representation of space as the dominant perspective form. This is occurring throughout the nonwestern world resulting in a homogeneity of representation obscuring different forms of spatial understanding and representation.

The way that perspective seemingly replicates visual perception has helped to reinforce its use, and has tended to inhibit culturally mediated alternatives. Perspective as a global vision system has been the subject of scrutiny since the 1920s. This research interrogates the idea of perspective as either a learned convention or a physiological given (Wyeld 2004, 2). Indeed, it is argued that perspectival devices have become so integral to our visual understanding that they affect our seeing. "Perspective is a universalizing method," that ultimately eliminates "uncertainties of intuition from all technical operations related to the construction of the physical world."[31] As Harpold notes:

> Our responses to these visual conventions are always—if not always consciously—adaptable. We take them to be markers of a reliable representation of the realms of the eye. Yet we also understand implicitly that they belong to a domesticated, geometrically sanitized version of those realms. In this way, the "visibility" of a GUI's spatial forms is a function of both a tacit acceptance of visual conventions, and a pragmatic willingness to suspend some of them, if circumstances require it. (2001, 11)

The mathematical translation of perspective onto the computer screen does not allow for the distortions that are necessary for an aesthetically pleasing result. In other words, "pictorial perspective does not coincide with the math of its geometry" (Wyeld 2004, 16). The peripheral distortion inherent in computer-generated models can only be resolved through projection onto curved surfaces, and not solved through mathematical compensation (without introducing its own distortions).

Erwin Panofsky's *Perspective as Symbolic Form* has dominated all art historical and philosophical discussions on the topic of perspective in this century. He describes "perspective as metaphor"—symbolic of the rationality of the Renaissance that relied on "adherence to the norms of the apparatus that created it" (Wyeld 2004, 20). Following, it is documented that non-Westernized peoples do not read perspectival images (including photographs) in the way that Westerners do, and they use a variety of representations and conventions. India has an isometric tradition which exists within a regional consciousness for its activation; China adopts the scroll form which imparts narratives rather necessarily depicting real scenes, and the Japanese adapted this form to a folding screen with architectural and metaphorical depth. Australian Aborigines use dot paintings to retain the memories of both virtual and real place and landscapes. As Theodor Wyeld (2004) notes:

> How much these aesthetic and symbolic meanings survive in translation to contemporary media remains in contest, and it can generally be thought that the translatability of an aesthetic art form relies on the aesthetic education (exposure) of the form to the viewer.

## Virtual Heritage and Panoramic Vision

Projects that use panoramic technologies for virtual heritage are numerous. A burgeoning number of Web sites now promote visits and views to heritage and museum sites through a panoramic eye. Two of the seminal works to extend the earliest Web-based applications included The Hermitage Museum Web site,[32] produced by IBM in 1998, which used a large number of "zoomable" panoramic images in the mode of a virtual tour. *1000 years of the Greek Olympic Games: Treasures from Ancient Greece*,[33] was an award-winning and seminal work, produced by The Powerhouse Museum in conjunction with Intel, in celebration of the Olympics in Sydney in 2000. The work used a combination of real-world spherical panoramas of the archaeological site mapped to a 3D virtual reconstruction (figure 15.2). Panoramas were also augmented with zoomable object movies of archaeological artifacts such as the pedimental sculptures found in the museum at Olympia. The work existed as a CD-ROM, a Web-based work, and as a passive stereo (linear polarized) installation with a "fly-through" of a virtual model (in 3D) augmented with real-world site panoramas (in 2D).[34]

Claims that the virtual reality panorama is a tool for "preserving" heritage are prevalent. Most notable to be found on the Internet to date is the work of Tito Dupret

Figure 15.2   Inside the Temple of Zeus reconstruction panorama from *1000 years of the Olympic Games: Treasures from Ancient Greece*. Project included a digital reconstruction of Olympia polarized 3D installation, Web site, and CD-ROM, launched 2000. © Powerhouse Museum, Australia, 2000.

who has initiated World Heritage Tours.[35] The WHTour is a private non-profit organization dedicated to creating a documentary image bank of panoramic pictures and virtual reality movies for all sites registered as World Heritage by UNESCO. Mr Dupret embarked on an international mission to photograph the 754 sites, from the Statue of Liberty to the Taj Mahal, using panoramic images for the Internet. He works under the auspices of The World Monument Fund. Similarly, Web sites such as that of the Danish photographer Hans Nyberg[36] (which contains over 30,000 panoramas and features a new full-screen example every week) host a large number of these heritage-related sites. One can observe how precisely this proliferation fuels the theories of the cyber *flâneur*, and the mobile global (virtual) tourist gaze of which John Urry (among others) speaks (2001). In education too, the prevalent use of QTVR as the new form of visual education content is exemplified by the Visual Media Centre's History of Architecture Web site at Columbia University. This Web site of over 600 panoramas to date, encompasses dozens of buildings from temples in Greece to great churches in Europe, shrines in the Yemen and Iran, to Frank Lloyd Wright's houses. The developers of the project believe that the QTVR format is "revolutionizing the teaching of architecture," and panoramic "nodes" are rendered in both low resolution for the Internet and high resolution for classroom disciples of "animated architecture."[37]

Panoramic devices in installation-based works for the interpretation of cultural heritage (either onsite or in galleries), allow for more control over additional sensory inputs, and multimodal views and locations. Examples include the Brazilian-authored Visorama,[38] a telescope interface which, using panoramas and image-based rendering techniques, becomes a center of hypertextual commuting. It contains images and sounds that allow the observer to navigate through space and time in any given actual landscape, as if he were making use of a system of dynamic cartography.

Other projects include the kiosk installations at Ename (Belgium), as well as Tervuren (Belgium) and Wieringen (Netherlands). Based on the 360-degree virtual panorama, the key approach of the systems is "readability," as they help the visitor to "read" the landscape elements and features in the village or town when exploring it (Pletinckx et al. 2004).

The following two projects draw on the use of stereoscopy to powerfully enhance the immersive qualities of the panoramas. A recent application of augmented stereographic panoramas of the temple complex at Angkor, Cambodia (figure 15.3) was developed for display in The Virtual Room at Melbourne Museum (2004/2005).[39] The high-resolution stereographic panoramas capture the potent sacred space of Angkorean temple architecture and relief sculpture, and allow users to travel through a

Figure 15.3    Panorama of Angkor Wat at dawn, from Sacred Angkor: Stereographic Panoramas of the Temple Complex project installed at Museum Victoria, displayed in The Virtual Room 2004–2005. Angkor image. © APSARA-Authority, Cambodia, photography Peter Murphy, 2004.

landscape of celestial palaces, rich with Khmer iconography illustrating the narratives of Hindu and Buddhist mythologies. The work uses a combination of technologies to bring panoramic scenes and viewers into a new degree of intimacy. Real-world 3D photographic landscapes, together with spatial soundscapes, audio spotlights, animations, and real-world stereo video bring new life to each scene (Doornbusch and Kenderdine 2004). The panoramas are displayed at life-sized scale, and point to the way in which sophisticated "augmentation" of panoramic scenes can generate rich narratives of experience. The Virtual Room is an eight-sided octant stereo display which allows users free circumambulation around the panoramic scene (figure 15.4). Advances in computer graphics hardware have meant that the augmented animated features are not distinguishable from the real-world photography. In addition, the stereographic nature of the work allows for the geometry of the scene to be derived so 3D models can be made, providing for scientific image-based modeling requirements.

*Conversations* is a distributed multiuser virtual environment comprising three discrete virtual reality stations produced by the iCinema Centre for Research in Interactive Cinema.[40] Each station consists of a head-mounted display, a head tracker, a navigation device, headphone, and microphone, providing three users access to a dynamic virtual world. The story simulates the escape of two prisoners from Pentridge Prison in Melbourne in 1965. The three virtual reality stations, connected by a high-bandwidth

Figure 15.4   *The Virtual Room*, Melbourne Museum. © VROOM Inc. 2004.

Figure 15.5  *Conversations*, distributed multiuser environment equipped with head-mounted displays and camera tracking. Incorporating stereographic panoramas and spatial sound for the interpretation of the famous escape of prisoners from Pentridge Prison, Melbourne in 1965. © Centre for Interactive Cinema Research (iCinema), UNSW, Australia, 2004.

network, enable the three users to be simultaneously immersed in and navigate through a digitally generated 3D environment comprised of computer graphic, photographic, and videographic components. This environment is based on the contemporary real-world stereographic panorama of a prison setting in which avatars (of the users) and video characters interact with the historical accounts of the escape and subsequent trial (figure 15.5).

To further articulate the emphasis on real-world graphics in the aforementioned projects two current research projects, funded out of the European Union IST frameworks, are noteworthy. Benogo (2005) and Create[41] explore the use of real-world photoimaging for virtual worlds and cultural heritage. Both examples seek to investigate the contribution of real-world places and objects to the experience of "being there" using a range of projection technologies from head-mounted displays to CAVE-type environments,[42] and represent some of the research into evaluation of these virtual environments that is currently being conducted for heritage.[43] Significantly, these researchers (Roussou and Drettakis 2003) make the distinction in virtual heritage applications about the need for "realness." Through aligned research in the project Archaeos, which uses nonphoto realistic (NPR) representation, they concluded:

> Photorealism may be less important with the NPR techniques that resemble traditional methods of depiction used by specialists. . . . [These] may suffice, or even be preferable. In the case of the general public, VR representation is developed to provide a window into the past which incorporates the interpretations made by specialists into engaging presentations. In this case, we believe that photorealism is a necessary aspect for educational and recreational value of representation. (Roussou and Drettakis 2003)

### A Note on Panoramic Capture Technologies

As a computer science, the creation of panoramic vision systems in the form of omnidirectional cameras has largely mimicked the natural world (the reflecting eye of Gigantocypris, and human vision, foveal and peripheral sense systems). The systems developed include the catadioptric (i.e., using mirrors) for wide field-of-view single-image capture without parallax; a variety of hardware-based approaches to stereographic panoramic capture with wide angle of visualization with parallax and depth recovery; and various software approaches for creating both equirectangular or spherical and cylindrical mappings (which have a limited vertical angle). Applications in panoramic vision research include computer vision, robotics, image/video processing, 3D environment modeling, tracking (surveillance), and video representation.

Computer science is directing research away from the algorithmic solutions for the problems of panorama capture and display (problems inherent to off-the-shelf cameras) to follow the path of explicitly designing equipment that will alleviate many of the "difficulties" associated with processing images taken with traditional camera systems. Developments in technologies for the capture of very wide angles of view prompt reexamination of stereographic problems and the representation of 3D.[44] Panoramic capture technologies form the basis not only of capturing real-world scenes but also allow for model generation.[45]

## Beyond the Horizon

> Visible landscapes are like icebergs: only a small proportion of their real substance lies above the surface. (Roberts 1987, 83)

This focus on the panorama and its virtual derivatives has demonstrated how closely virtual reality engages with the historical scheme that precedes it. The purpose of virtual cultural heritage research is to develop ways in which cultural heritage assets can be accessed and engaged with—and panoramic virtual worlds are a persuasive interface paradigm for these purposes. Key issues in the current renegotiation of virtual spaces are the metaphors for navigation and "bodily presence"—both of which come with attendant sociocultural conditions and responses, examined here through an historical lens. Virtual heritage is charged with the task of providing rich learning experiences, and in the examples above we see projects simultaneously negotiating the traditional panoramic strategies used in creating spaces of immersion and illusion—in conjunction with its pedagogical functions. The provocative tension that exists for virtual heritage as a tool for scientific and cultural visualizations resonates between the scientific requirement to reproduce rational material reality, and those "sensations" of the "Electronic Baroque" that intrigue visitors to the space. Examining the methods that successfully mediate these apparent opposites is the inspiration for this chapter. Virtual heritage developers can observe the current *and* historical critique to promote design that supports the mandate of the emerging discipline. For instance, a deeper understanding of how virtual environments, (created within predominantly Cartesian-based perspectival systems), both appropriate and distort other culturally derived views of the world, allows us ample opportunity for experimentation, evaluation, and further research.

The examples of virtual heritage described above range from engaging the traditional virtual tourist or cyber *flâneur*—to those that demonstrate, with increasingly sophisticated techniques (and experiments), aspects of immersion and user interaction. As such, these projects foreground the trajectory of future applications. Some of the issues introduced in this chapter posit ways in which virtual heritage spaces can negotiate the criteria for "presence" and even "cultural presence" (Champion 2005) so as to also engage with narrative structures and interactions between virtual bodies—and foster cultural heritage awareness and education. The complexities of such tasks are the subjects explored in the following two chapters of this book—by authors Erik

Champion and Bharat Dave in "Dialing Up The Past" (chapter 16), and Bernadette Flynn in "The Morphology of Space in Virtual Heritage" (chapter 17).

*Notes*

1. Virtual and augmented heritage refers primarily to the creation of 3D and 4D (including temporal frameworks for navigation) digital spaces, and/or overlays of virtual space on real-world views (augmentation). The discussion concentrates on visual information architectures created in digital cultural heritage.

2. André Perente, proposal for a panorama exhibition, unpublished, 2005 (supplied in personal communication with the author).

3. Refer to theses on virtual heritage (as pertaining to archaeological virtual reality) such as Barceló, Forte, and Sanders 2000; Frischer et al. 2000; and Sideris and Roussou 2002.

4. For example, as discussed in Debord 1995. Also see Crary 2001, as further explored in this chapter.

5. For a comprehensive list, see the Dead Media Project, available at ⟨http://www .deadmedia.org⟩ homepage (accessed 5 January 2005).

6. By 1931 Walter Benjamin was using the phrase "optical unconscious" in his "Small History of Photography" (see Benjamin 1979). Also see this idea later in Benjamin 1968, 217–251.

7. For examples, see Oettermann 1997; Hyde 1988; Comment 2000; Altick 1978; and Avery 1995. Annotated references to panorama projects can be found at the Australian Centre for the Moving Image Web site, at ⟨http://www.acmi.net.au/AIC/PANORAMA .html⟩ (accessed 5 January 2005).

8. Robert Barker defined his invention in the patent of 19 June 1787: "An entire new contrivance or apparatus, which I call *La Nature a Coup d'Oeil*, for the purpose of displaying views of Nature at large by Oil Painting, Fresco, Water Colours, Crayons, or any other mode of painting or drawing," from "The Repertory of Arts and Manufactures: Consisting of Original Communications, Specifications of Patent Inventions, and Selections of Useful Practical Papers," in vol. 4, no. 14 of *The Transactions of the Philosophical Societies of All Nations* (London: T. Heptinstall, G. G. and J. Robinson, Paternoster-row; P. Elmsly, Strand; W. Richardson, Cornhill; J. Debrett, Piccadilly; and J. Bell, 1796).

9. Refer to Harris and Sandweiss 1993.

10. For an excellent online bibliography refer to American Memory from the Library of Congress, "A Brief History of Panoramic Photography," available at ⟨http://memory.loc .gov/ammem/collections/panoramic_photo/pnbib.html⟩ (accessed 15 February 2005).

11. For an extensive discussion of panoramic spaces as intermediary forms to immersive VR, refer to Grau 2003, Manovich 2001, Shaw and Weibel 2003, among others.

12. For a comprehensive summary of projection research issues and solutions for stereographic panoramas see Paul Bourke's Web site, *Centre for Astrophysics*, Swinburne University, Melbourne, Australia, at ⟨http://astronomy.swin.edu.au/~pbourke/projection/⟩ (accessed 15 February 2005). For technologies associated with panoramic vision systems, see Benosman and Kang 2001.

13. Norman Bryson's "glance," as quoted in Jay 1998.

14. Jonathan Crary, as referenced in Jay 1998.

15. See Scott Wilcox, in his introduction to Hyde 1998; see also Altick 1978 and Avery 1995.

16. See Altick, as quoted in Druckrey 2003, 62.

17. For examples of the Italian *mondo nuovo* refer to Bruno 2002.

18. Antoine Claudet, 1860, quoted in Huhtamo 1995, 1.

19. Quoted in Jay 1998.

20. Druckrey, referring to Wolfgang Schivelbusch's (1995) text in *Disenchanted Night*.

21. For examples, see Ndalianis 2004 and Klein 2004.

22. For example: *Moving Movie* (1977), *Displacement* (1984), *Be Now Here* (1995) by Michael Naimark; *Place—A User's Manual* (1995), *Place D'Uhr* (2000), *EVE* (1993–2004), and *Panoramic Navigator* (1997) by Jeffrey Shaw; *Landscape One* (1997) and *Panoscope* (2001) by Luc Courchesne, Passagen (1997); *Wall of Death* (1999) by Graham Ellard and Stephen Johnstone, as listed in Parente 1999.

23. Panorama of Delft, "How to make a gigapixel picture," at ⟨http://news.bbc.co.uk/2/hi/technology/4022149.stm⟩ (accessed 16 March 2005).

24. For example, see "SPi-V engine," FieldOfView Web site, at ⟨www.fieldofview.nl/spv.php⟩ (accessed 15 January 2005).

25. Erik Goetze, "VRlog," at ⟨www.vrlog.com⟩ (accessed 17 January 2005).

26. As summarized in Urry 2001, 125–127. Urry's synthesis follows the works by Sontag 1979; Berger 1972; Barthes 1981; Albers and James 1988; and Osborne 2000.

27. Urry 2001, 125–127.

28. Charles Baudelaire, *Les Fleurs du mal* (1861), as referred to in Olalquiaga 2002, 23.

29. Refer also to Benjamin 1968, 217–251.

30. Anne Frieberg, quoted in Manovich 2001, 273.

31. Wyeld 2004, quoting Perez-Gomez and Pelletier 1997, 110.

32. Technologies implemented by Intel, see ⟨http://www.hermitagemuseum.org/html_En/00/hm0_8.html⟩ (accessed 5 January 2005).

33. *1000 years of the Greek Olympic Games: Treasures from Ancient Greece*, Powerhouse Museum Web site, at ⟨http://projects.powerhousemuseum.com/ancient_greek_olympics/⟩ (accessed 5 January 2005).

34. See Kenderdine et al. 2003; Kenderdine and Ogleby 2001; Kenderdine 2001a; Kenderdine 2001b; and Kenderdine, Ogleby, and Ristevski 2000.

35. World Heritage Tours, at ⟨http://www.world-heritage-tour.org/⟩ (accessed 5 January 2005).

36. Danish photographer Hans Nyberg, who maintains a Web site with links to 400 sites containing 30,000 panoramas, features a new full-screen example every week, at ⟨http://www.panoramas.dk⟩ (accessed 5 January 2005).

37. Visual Media Center's History of Architecture Web site, at ⟨http://www.mcah.columiba.edu/ha⟩, as reported in Wishna 2005.

38. The Visorama is a multimedia and virtual reality system developed by the Image Technology Centre (N-Imagem) from the School of Communication University of Rio de Janeiro (ECO-UFRJ), in partnership with the Visgraf Project from the Pure and Applied Mathematics Institute (IMPA).

39. The Virtual Room, at ⟨http://www.vroom.org.au⟩. For the Sacred Angkor project refer to Sarah Kenderdine's online publications at ⟨http://www.vroom.org.au/research.asp⟩ (accessed 30 March 2005), versions of which were also published in *Virtual Systems and Multimedia, Gifu, Japan* (2004) and presented at the International Cultural Heritage Informatics Meeting (ICHIM), Berlin, 2004.

40. iCinema: Centre for Interactive Cinema Research, at ⟨http://www.icinema.unsw.edu.au/⟩ (accessed 16 February 2005).

41. *Constructivist Mixed Reality for Design, Education and Cultural Heritage*, Create, at ⟨http://www.cs.ucl.ac.uk/research/vr/Projects/Create/⟩ (accessed 5 January 2005).

42. The CAVE (CAVE Automatic Virtual Environment) was developed by the Electronic Visualization Laboratory at the University of Illinois in the early 1990s as a real-time stereoscopic multiple screen projection (3-, 4-, 5-sided) immersive environment. The system uses head and hand tracking and shutter glasses to simulate user interaction in the virtual space.

43. See Roussou 2004a and 2004b. See the author's Web site "MakeBelieve," at ⟨http://www.makebelieve.gr/mr/www/mr_publications.html⟩ (accessed 16 February 2005).

44. For an in-depth, current account of all technologies associated with panoramic vision systems refer to Benosman and Kang 2001.

    An increasingly ambitious range of capture technologies range from the ladybug for video capture, to the ASTRO Sensor Series, the series of omnidirectional, stereo cameras by ViewPLUS. This video cluster provides not only the color images from all directions but also the stereo information (3D) can be obtained. The video feeds are stitched together on the fly (with variable success), at ⟨http://www.viewplus.co.jp/products/sos/astro-e.html⟩ (accessed 18 January 2005). Other capture technologies include the Ladybug, at ⟨http://www.ptgrey.com/products/ladybug/⟩ (accessed 16 February 2005). See also Kimber and Foote 2001.

45. For 3D Street Model Generation from route panoramas using video footage and motion blurring to provide depth, see Shi, Zheng, and Kato 2004.

    Similar is the 240 × 360-degree video camera in production at the Centre for Interactive Cinema Research (iCinema), Australia, progressing from earlier developments at ZKM, Germany, at ⟨http://www.icinema.unsw.edu.au/projects/prj_panocam.html⟩ (accessed 17 February 2005). These sorts of technologies confirm that the appetite for panoramic vision systems as a basis for immersion environment design is again on the increase.

## References

Altick, Richard. 1978. *The Shows of London*. London: Cambridge University Press.

Avery, Kevin J. 1995. "The Panorama and Its Manifestation in American Landscape Painting, 1795–1870." Ph.D. diss., Department of Art History and Archeology, Columbia University.

Balzac, Honoré de. *Old Goriot*, vol. 13, part 1. New York: P. F. Collier & Son.

Barceló, Juan; Maurizio Forte, and Donald Sanders, eds. 2000. *Virtual Reality in Archaeology, British Archaeological Reports*, International Series, no. 843. Oxford: Archaeopress.

Barker, Robert. 1796. "The Repertory of Arts and Manufactures: Consisting of Original Communications, Specifications of Patent Inventions, and Selections of Useful Practical Papers." *The Transactions of the Philosophical Societies of All Nations* 4, no. 14. London.

Bearman, David, and Franca Garzotto, eds. 2001. *International Cultural Heritage Informatics Meeting, September, Milan*. Pittsburgh: Archives and Informatics.

Bearman, David, and Jennifer Trant, eds. 2004. *International Cultural Heritage Informatics Meeting, Berlin*. Pittsburgh: Archives and Informatics.

*Being There Without Going*, Benogo ⟨http://www.benogo.dk/⟩ homepage (5 January 2005).

Benjamin, Walter. 1968. "The Work of Art in the Age of Mechanical Reproduction." In *Illuminations*, ed. H. Arendt, trans. H. Zohn, 217–251. London: Jonathan Cape.

Benjamin, Walter. 1979. "Small History of Photography." In *One-Way Street and Other Writings*. Trans. E. Jephcott and K. Shorter. London: New Left Books.

Benosman, Ryad, and Sing Bing Kang, eds. 2001. *Panoramic Vision: Sensors, Theory, and Applications*. New York: Springer-Verlag.

Bourke, Paul. "Stereographic 3D Panoramic Images" homepage. Available at ⟨http://astronomy.swin.edu.au/~pbourke/stereographics/stereopanoramic⟩ (accessed June 13 2004).

Bruno, Giuliana. 2002. *Atlas of Emotion: Journeys in Art, Architecture, and Film*. New York: Verso.

Champion, Erik. 2005. "Cultural Presence." In *Encyclopedia of Virtual Communities and Technologies*, ed. S. Dasgupta. Washington, D.C.: George Washington University.

Comment, Bernard. 2000. *The Painted Panorama*. New York: Harry N. Abrams.

*Constructivist Mixed Reality for Design, Education and Cultural Heritage*, Create. Available at ⟨http://www.cs.ucl.ac.uk/research/vr/Projects/Create/⟩ homepage (5 January 2005).

Crary, Jonathan. 2001. *Suspension of Perception: Attention, Spectacle, and Modern Culture*. Cambridge, Mass.: MIT Press.

Cubitt, Sean. 1998. *Digital Aesthetics*. London: Sage Press.

Debord, Guy. 1995. *The Society of the Spectacle*. Trans. Donald Nicholson-Smith. New York: Zone Books.

Doornbusch, Paul, and Sarah Kenderdine. 2004. "Presence and Sound: Identifying Sonic Means to 'Be There.'" In *Qi Connectivity and Complexity*, ed. R. Ascott, 67–70. Plymouth, UK: Planetary Collegium.

Druckrey, Timothy. 2003. "Fugitive Realities, Situated Realities, 'Situational Realities', or Future Cinema." In *Future Cinema: The Cinematic Imaginary After Film*, ed. J. Shaw & P. Weibel, 60–66. Karlsruhe: ZKM; Cambridge, Mass.: MIT Press.

Earle, Edward W. 1979. "The Stereograph in America: Pictorial Antecedents and Cultural Perspectives," *Points of View: The Stereograph in America—A Cultural History*. Rochester: The Visual Studies Workshop Press.

Eberhard, Johann Augustus. 1805. *Handbuch der Ästhetik*, Part 1, Letter 28, Halle, 175.

Frischer, Bernard, Franco Niccolucci, Nick Ryan, and Juan Barceló. 2000. "From CVR to CVRO: The Past, Present, and Future of Cultural Virtual Reality." In *Proceedings of VAST 2000* (British Archaeological Reports 834), ed. F. Niccolucci, 7–18. Oxford: Archaeopress.

Grau, Oliver. 2003. *Virtual Art: From Illusion to Immersion*. Cambridge, Mass.: MIT Press.

Griffith, Alison. 2003. "Media Technology and Museum Display: A Century of Accommodation and Conflict." In *Rethinking Media Change: The Aesthetics of Transition*, ed. D. Thornburn and H. Jenkins, 376–377. Cambridge, Mass.: MIT Press.

Harpold, Terry. 2001. "Thick & Thin: 'Direct Manipulation' & the Spatial Regimes of Human-Computer Interaction." 1.2.1. Available at ⟨http://www.siggraph.org/artdesign/gallery/S01/essays/0386.pdf⟩ (accessed 13 June 2003).

Harris, David, and Eric Sandweiss. 1993. *Eadweard Muybridge and the Photographic Panorama of San Francisco, 1850–1880*. Cambridge, Mass.: MIT Press.

Huhtamo, Erkki. 1995. "Armchair Traveller on the Ford of Jordan: The Home, the Stereoscope and the Virtual Voyager." *Mediamatic* 8, no. 2, 3. Available at ⟨http://www.mediamatic.net/article-200.5952.html&q_keyword=200.277⟩ (accessed 15 January 2005).

Hyde, Ralph. 1988. *Panoramania!: The Art and Entertainment of the "All-Embracing" View*. London: Trefoil Publications.

Jay, Martin. 1998. "The Panorama: History of a Mass Medium." *ArtForum International Magazine*, March 1, LookSmart's Find Articles Web site, 2000. Available at ⟨http://www.findarticles.com/p/articles/mi_m0268/is_n7-v36/ai_20572943/print⟩ (accessed 10 February 2005). Book reviews.

Jay, Martin. 2001. "Scopic Regimes of Modernity." In *Vision and Visuality*, ed. H. Foster. Seattle: Bay Press.

Kenderdine, Sarah. 2001a. "1000 Years of the Olympic Games: Treasures from Ancient Greece: Digital Reconstruction at the Home of the Gods." *10th International World Wide Web Conference*, Hong Kong. CD-ROM. American Computer Manufacturers.

Kenderdine, Sarah. 2001b. "A Guide for Multimedia Museum Exhibits: 1000 Years of the Olympic Games: Treasures of Ancient Greece." *Museums International* 53, no. 3.

Kenderdine, Sarah. 2004. "Stereographic Panoramas of Angkor, Cambodia." In *10$^{th}$ Conference on Virtual Systems and Multimedia*, Gifu, Japan, 612–621. Los Alamitos, Calif.: IEEE.

Kenderdine, Sarah, Kate de Costa, Cliff Ogleby, and John Ristevski. 2003. "VROOM (Virtual Reconstruction of Olympia Model): The Creation of a Virtual Tour from a Digital Model." In *The Sydney 2000 Olympics Book: Sport and Festival in the Ancient Greek World*, ed. D. Phillips and D. P. Pritchard. Wales: Classical Press of Wales.

Kenderdine, Sarah, and Cliff Ogleby. 2001. "Anicent Olympia as a Three-Dimensional Experience." In *International Cultural Heritage Informatics Meeting, September, Milan*, vol. 2, ed. D. Bearman and F. Garzotto, 333–341. Pittsburgh: Archives and Informatics.

Kenderdine, Sarah, Cliff Ogleby, and John Ristevski. 2000. "1000 Years of the Olympic Games: Treasures from Ancient Greece—The Digital Reconstruction of Olympia, 3D Zeus and Website." 6th International Conference on Virtual Systems and Multimedia, presented Gifu, Japan, October 2000. In *International Society of Virtual Systems and Multimedia*, 104–116. New York: ACM.

Kimber, Don, and Jonathan Foote. 2001. "Flyabout: Spatially Indexed Video." In *Proceedings of ACM Multimedia 2001*, Ottawa, Canada. New York: ACM. Available at ⟨http://www.fxpal.com/people/foote/papers/flyabout/flyabout.htm⟩ (accessed 16 February 2005).

Klein, Norman. 2004. *The Vatican to Vegas: The History of Special Effects*. New York: The New Press.

Manovich, Lev. 2001. *The Language of New Media*. Cambridge, Mass.: MIT Press.

Mirapaul, Matthew. 2003. "The Sweeping View From Inside a Digital Bubble." New York Times on the Web, September 25. Available at ⟨http://www.nytimes.com/2003/09/25/technology/circuits/25virt.html?ex=1065510061&ei=1&en=440bad650dac2a9d⟩ (accessed 14 June 2004).

Ndalianis, Angela. 2004. *Neo-Baroque Aesthetics and Contemporary Entertainment*. Cambridge, Mass.: MIT Press.

Oettermann, Stephen. 1997. *Panorama: History of a Mass Medium*. New York: Zone Books.

Olalquiaga, Celeste. 2002. *The Artificial Kingdom: On the Kitsch Experience*. Minnesota: University of Minnesota Press.

Parente, André. 1999. "Project Panoramic Vision." Lecture delivered at Goldsmith's College, London.

Perez-Gomez, Alberto, and Louise Pelletier. 1997. *Architectural Representation and the Perspective Hinge*. Cambridge, Mass.: MIT Press.

Pletinckx, Daniel, Lars De Jaegher, Truus Helsen, Iris Langen, Neil Silberman, Marie-Claire Van der Donckt, and Jan Stobbe. 2004. "Telling the Local Story: An Interactive Cultural Presentation System for Community and Regional Settings." In *The 5th International Symposium on Virtual Reality, Archaeology and Cultural Heritage*, ed. D. W. Fellner and S. N. Spencer, 233–239. Brussels: The Eurographics Association.

Plunkett, John. 2002. "Screen Practice Before Film." Exeter University, Centre for British film and Television Studies, Bill Douglas Centre for the History of Film and Popular Culture, 4 November. Available at ⟨http://www.bftv.ac.uk/projects/exeter.htm⟩ (accessed 13 June 2004).

Roberts, Robert. 1987. "Landscape Archaeology." In *Landscape and Culture: Geographical and Archaeological Perspectives*, ed. R. Wagstaff. Oxford: Blackwell.

Roussou, Maria. 2004a. "Examining Young Learners' Activity within Interactive Virtual Environments." In *Proceedings of the 3rd International Conference for Interaction Design & Children*, 167–168. New York: ACM Press.

Roussou, Maria. 2004b. "Learning by Doing and Learning through Play: An Exploration of Interactivity in Virtual Environments for Children." *ACM Journal of Computers in Entertainment* 1, no. 2.

Roussou, Maria, and G. Drettakis. 2003. "Photorealism and Non-Photorealism in Virtual Heritage Representation." In *First Eurographics Workshop on Graphics and Cultural Heritage*, ed. A. Chalmers, D. Arnold, and F. Niccolucci. Brighton, U.K: The Eurographics Association. Available at ⟨http://www.makebelieve.gr/mr/research/papers/VAST/vast_03/submission/1024_mroussou_vast03_cameraReady.pdf⟩ (accessed 30 July 2004).

Schivelbusch, Wolfgang. 1995. *Disenchanted Night: The Industrialization of Light in the Nineteenth Century*. Berkeley: University of California Press.

Shaw, Jeffrey, and Peter Weibel, eds. 2003. *Future Cinema: The Cinematic Imaginary After Film*. Karlsrulhe: ZKM, Cambridge, Mass.: MIT Press.

Shi, M., J. Y. Zheng, M. Kato. 2004. "3D Street Model Generation from Route Panoramas for Culture Heritage." In *10th International Conference on Virtual Systems and Multimedia*, Gifu, Japan, 644–653. Los Alamitos, Calif.: IEEE.

Sideris, Athanasios, and Maria Roussou. 2002. "Making a New World out of an Old One: In Search of a Common Language for Archaeological Immersive VR Representation." *Proceedings of 8th Int. Conference on Virtual Systems & Multimedia*, 31–42. Seoul: VSMM and Kiwisoft Co. Ltd.

Uricchio, William. 1998. "Panoramic Visions: Stasis, Movement, and the Redefinition of the Panorama." *La nascita dei generi cinematografici*, Atti del V Convegno Internazionale di Studi sul Cinema, Udine, 26, 28 March.

Urry, John. 2001. *The Tourist Gaze*. 2d ed. London: Sage Publications.

The Virtual Room ⟨http://www.vroom.org.au⟩ homepage (20 January 2005).

Warner, Marina. 2004. "*Camera Lucida*." In *Eyes, Lies, and Illusions*, ed. L. Mannoni, W. Nekes, and M. Warner. London: Hayward Gallery in association with Lund Humphries.

Wishna, Victor. 2005. "A New Look at Old Buildings." *Humanities* (march/April). Available at ⟨http://www.neh.fed.us/news/humanities/2005-03/anewlook.html⟩ (accessed 30 March 2005).

Wyeld, Theodor. 2004. "Perspective as Learned Convention or Physiological Fact." Unpublished manuscript.

# 16 Dialing Up the Past

Erik Champion and Bharat Dave

## Introduction

The early 1980s witnessed the first application of computer graphics to historical reconstruction in a project on the Roman Baths (Woodwark 1991, 18–20). Since then, digital media have been used to create increasingly complex, rich, and interactively-driven projects focusing on historical sites and contexts around the world (Forte and Siliotti 1996). These visually and aurally stunning digital representations of objects and events raise problematic issues for the creation of meaningful virtual heritage environments. One of the central objectives in many of these environments is to communicate experiential dimensions of another place from another time. If a sense of *place-centeredness* is an important goal in virtual heritage projects, how should they support something more than, to borrow Naimark's terms (1998), either *moving around* or *looking around*? Can virtual environments help us question or modify people's understanding of the cultural significance of heritage sites? What kinds of activities may foster a richer sense of place?

Despite the centrality of *place* as a structuring concept in virtual heritage environments, what this specific sense of place might be remains an elusive question. Instead of grappling with a potentially circular argument such as *here is a place because it was designed to be a place*, this essay develops a more graduated approach to understanding features of different kinds of virtual places and the cultural and social functions they facilitate. It is hoped that such an approach will enable us to both analyze and design various virtual environments. According to this approach, most current virtual heritage environments can be categorized—based on designer intention—into spatially visualized, activity-based, or hermeneutic environments.

The essay is organized as follows. Section 16.1 provides a brief overview of the passage from the early days of digital imagery to the current interactive multimedia representations, and how the notion of place has come to occupy a central position in interactive digital environments. In the context of virtual heritage applications, we discuss place and "placeness" and suggest that it is more than just visual realism of the highest possible order. We draw upon concepts from architecture and cultural geography to suggest that place is much more than just a locator of objects.

Section 16.2 extends the preceding general discussion into specific elements of "placemaking." Using selected perspectives from literature on virtual environments, the discussion identifies a number of specific features of placemaking. It is suggested that virtual heritage environments need to support hermeneutic features of place.

Section 16.3 introduces a graduated matrix for correlating types of virtual environments, features of place, and the types of relations they engender. Section 16.4 describes a research project that explores some of the issues introduced in this essay in a specific virtual heritage project, followed by concluding remarks.

## 16.1    Finding Place

Once Ivan Sutherland likened computer graphics to "a looking glass into a mathematical wonderland," it was only a matter of time before other wonders and lands became visible through this looking glass (1996, 506–508). The developments in digital media that have followed have led to production of many seductive and mesmerizing wonderlands based on real and imagined environments.

Aspen Moviemap, for example, was one of the earliest projects to allow interactive exploration of the city of Aspen (Lippman 1980, 32–42). While Aspen Moviemap was an experiment in re-presenting real-world using digital media, more recent projects have been developed with a view to digitally rendering not just existing but also past environments. The projects in virtual archaeology range in scope from single monuments to entire landscapes, and represent a singular moment in time or encompass historic development of place over time (Forte and Siliotti 1996). Even if many of these projects are driven by what the technology can do, there is increasing awareness that such projects are interpretive exercises[1] and a place may be only partially reconstructed through digital representations.

Some interesting responses to the nature of place are explored in multiuser and role-playing environments (especially in games and entertainment). These projects revolve around the notion of users doing something besides moving around or looking around. Further, the awareness of actions of others interacting with the information at the same time is an essential design element. Thus besides a three-dimensional setting

Figure 16.1    Parallel worlds (Palenque project).

we now see the introduction of multiple layers, e.g., awareness of others, causality, moveable artifacts, etc., being introduced in interactive digital environments.

Although sense of place did not dominate Sutherland's mathematical wonderland, we now regard place as one of the most recurring and binding themes in the current projects, especially in virtual environments. For example, historical reconstructions need a grounded anchor whether it is a landscape or a map. Even fictitious Active Worlds communities revolve around territorial demarcations in an imaginary space (figures 16.1 and 16.2).

This raises another nagging question. Is the notion of place in interactive games somehow identical to the notion of place in virtual heritage projects? If not, in what ways are they different? Is there any "there" *there*?

As suggested earlier, it is not enough to claim that these digital environments constitute places simply because they are designed as places. Something more is needed to be able to distinguish one place from another and also to tell if it succeeds in its version of "placeness."

Since many virtual heritage projects invest significant effort in achieving a high degree of visual fidelity, it is tempting to use a degree of realism as one of the yard sticks to distinguish different kinds of places. For we have many technologies at our disposal

Figure 16.2    Crocodile mountain (Palenque project).

to increase the realism of virtual places, 3D laser scanning, dynamic level of detail, and procedurally generated texture maps for dynamic and fluid objects.

However, saturating a virtual environment with detail does not necessarily create a strong sense of "place." In fact, if viewers of such information only remain as spectators to overwhelmingly precise details, we would not have advanced any further than being passive wanderers in such environments.

Photorealism certainly has value but not as the most essential and differentiating feature to foster a sense of place. For example, fog and London appear as almost inseparable in literature, paintings, and most people's memories of the city. This suggests that we often remember a place through its atmosphere rather than through accurate recall of details of objects that populate a place.

Our knowledge of a place is also deepened not by being passive receivers but by our activities as shaped by that place, and our identification with or against that place. Instead of virtual environments simply being locators of objects, we need some way to communicate the cultural dimensions of such environments. Instead of evoking "cyberspace," such environments need to support interactive *construction* of a shared notion of "place" around some cultural beliefs and encounters of human transactions in real space.[2]

If the above suggests that photorealism may be seductive but not enough, and that interactive *construction* of something exceeding that which is visible is needed to convey a sense of place, where can we begin?

Although interactive digital media encountered placemaking rather recently, there are other richer and older disciplines that have grappled with these issues. For example, architecture, literature, and film revolve around designing place as a locus of possibilities through suggestion rather than through attention to exhaustive details that exclude any other possibilities.

Architectural design, for example, is more than just an assemblage of building elements or physical objects. Truly successful design of spaces uses composition of physical objects, spaces, and contextual conditions to evoke associations that may be metaphorical, allegoric, or thematic. In this process, the interactions between buildings and spaces, the environmental context, and people and their beliefs and values all come into play to create unique places.

As Christian Norberg-Schulz (2000) notes, place is where there is "a dynamic unity of architectural elements, inhabitants and interactions between/among them." Architectural experience is thus interactive. Memorable places, accordingly, are referential and suggestive of possible associations waiting to be made—individually and collectively—only when someone actually experiences the architecture.

For cultural geographers, culture has a setting and this setting is enabled through a perceived sense of place. As culture requires a setting, it must be "embedded in real-life situations, in temporally and spatially-specific ways" (Crang 1998).

While unique, place is further an "integration of elements of nature and culture . . . interconnected . . . part of a framework of circulation."[3] The interactions between these objects and their setting may be quite complex (Canter 1974).

Thus culture is a feedback loop. A visitor appropriates space to make it a unique place according to cultural constructs, place "perpetuates culture," and thus influences the inhabitants in turn. Further, culture almost inevitably involves objects of shared transactional value that are created, used, exchanged, and transmitted from one group or generation to another group of people via social and individually-constructed places.

Place-based transactions thus enable travel through time and space, to explore, and to interact with people and objects, and share in the place-specific relations. Carla Cipolla (2004) would see this as travel as opposed to tourism: "The *relation* can begin when a traveller, without preconceptions or prefigured images, encounters a community, a monument, a natural environment or a cultural expression" (original emphasis).

## 16.2   Elements of Placemaking

Instead of focusing solely on creating spatially accurate digital reconstructions in virtual environments, we need ways to support a sense of place by drawing upon what we understand about representations and lived experiences. Based on the kinds of disciplines briefly mentioned above, we find some specific criteria emerging. For example, Yehuda Kalay and John Marx propose eight criteria for "cyber-placemaking" based on several concepts in architecture and town planning (2001, 230–240). These include place as setting for an event, that is engaging, provides relative location (i.e., orientation), provides authenticity, is adaptable, affords a variety of experiences, affords choice and control over transitions, and is inherently memorable. Although these criteria are useful, they do not help us to determine which of the above features are most important, necessary, or even desired, for different types of virtual environments. Further, they do not address several important features of place.

For instance, places are not just memorable but also evocative. Hartshorne said geography is a need to "fix the memory of the places which surround us" (Relph 1986, 5). Geography indirectly highlights our schemas of place—be they telluric, projected landforms, or urban. When triggering mental associations to these schemas, place is evocative, evoking remembered sensations of its previous self, of related activities or even of similar places.

Place gains unique character through time and use. Place is not just adaptable, but also markable, recordable, it leaves signs of its use; it reveals the way it was used and perceived by the way in which it has eroded (Massey 1993, 230, 240). In this sense one can argue that place is an artifact, as past events can often be inferred from it.

Place implies a certain type of setting, of occasion (Neumann 1996). A designer is required to invent cities, individual buildings and spaces and to provide them with a history, and with patterns and allusive traces of use in order to connect them to underlying narratives and to endow them with meaning. For while many see environmental designers as creating products, many actually create products that appear to have been eroded by use, in order to make them seem more familiar or appear more popular (Champion 1993, 81–84).

Place is also defined by its relation to other places, we often populate place with artifacts from other places. Hence some of its uniqueness is ironically formed in relation to *other places* where a traveler has been (Massey 1993).

To approximate reality requires settings for social transactions that are location-specific and task-specific. There is also a need for transition zones of perceived physiological comfort and discomfort. These features are often associated with thematic symbolism in architecture (Champion 1993). Architecture modifies behavior through symbolic cues, offers paths and centers so that we can navigate and orient ourselves, and suggests the passage of time as well as records the meetings of people.

These encounters are marked through the wearing of surface and the development of portable artifacts that locate and define social rituals.

Coyne (1999) argues one way of creating virtual space is where "cyberspace enables and constrains human interaction in ways similar to physical space." Architecture is also a filter of human-environment interaction, yet is not fully utilized in virtual environments as either affordance or as constraint. A feeling of place is dynamically impacted by environmental constraints. These constraints help define the place and suggest certain ways of acting inside that place. For example, we place or site and center ourselves optimally inside a field of forces that affect our task efficiency (e.g., path of least resistance), our social standing, and our feelings of comfort.

Place is also "artifactually" defined. We place artifacts in relation to our perception of how we appreciate or dislike environmental features. A bed may be close to the window but turned away from intense morning light. So our idea of place is identifiable as a locus between environmental features and personal or physical preferences. "Placed" (spatial) artifacts can indicate social relations between people and between artifacts, such as houses close to or far from each other (Schiffer and Miller 1999).

And artifacts are not just social, they are also cultural; they have a past meaning that informs a current use. The artifacts act like a library of memory cues to remind people how to behave according to certain events or locations.[4] Not just objects but also the wider environment can act as an artifact. Place is also a collection of symbolic cues for inhabitation and for territorial possession (Rapoport 1982).

As Edward Relph noted, "The identity of a place is comprised of three interrelated components, each irreducible to the other: physical features or appearance, observable activities and functions, and meanings or symbols" (1986, 25). So the placemaking criteria of Kalay and Marx address only two major types of environments addressed by Relph, environments that afford "physical features or appearances," and those that afford "activities." The Kalay-Marx criteria, being based on modes of reality, do not address virtual environments that attempt to offer interpretations of past and present cultures.

Partly this omission is due to the fact that it is difficult to simulate culture, virtually or otherwise. As Yi-Fu Tuan (1998) notes, "Seeing what is *not* there [our emphasis] lies at the foundation of all human culture," yet virtual environments by convention attempt to simulate what *is* there.

In order to create culturally evocative environments, we need to understand how elements disseminate cultural information. According to Schiffer and Miller (1999), we learn about a culture by dynamically participating in the interactions between *cultural setting* (a place that indicates certain types of social behavior), *artifacts* (and how they are used), and *social agency and contextual tasks*, i.e., people teaching others a social background and how to behave *along with* one's personal motives.

One way to approach this issue is to view (and design) digital environments which represent human cultures within a *hermeneutic* dimension or capacity, (that afford an actively engaged interpretation of the lives and intentions of past inhabitants). The hermeneutic features of place in digital environments are almost certainly more difficult to incorporate, but that does not negate their importance. Luckily for virtual environment designers, these hermeneutic features have been described by social scientists who maintain that people develop shared cultural perspectives of place in many different ways.

### Places and Functions

Instead of using the degree of visual correspondence between real and virtual worlds to discuss *place*, we propose a matrix (table 16.1) that correlates multiple dimensions of virtual environments in terms of purposes they serve, features they require, and experiential potential they offer.

Such a graduated categorization, on the one hand, allows us to correlate placemaking features to general aims of virtual environments (spatial visualization, entertainment, social and participatory to culturally immersive experience). On the other hand, it also suggests that a hermeneutic virtual environment (one that has to be actively interpreted by a participant) may be the most difficult to compose (table 16.1).

According to the proposed matrix, the most easily achievable kinds of places comprise visualization and manipulation of three-dimensional objects. A more advanced representation of such places may include their contextual settings (e.g., landscapes) and the ability to navigate through them. Although we can now adequately capture realistic detail and approximate believable physical behavior of objects, this type of digital environment, while achievable and useful for various scientific purposes, only represents spatial configurations and navigation through them.

The second type of virtual environment, the one that affords activity-based behavior, allows a more interactive form of empathetic insideness. Tasks can be accomplished inside the environment through interaction, supplemented with decision-making and navigation for a more immersed experience. Computer games and flight simulation applications perhaps best convey this type of digital environment.

**Table 16.1**
Graduation of Place and Cultural Functions

| Type of VE | Relph's categories | Features | Personal/Cultural Attachment |
|---|---|---|---|
| Spatial Visualization | Existential outsideness- (Objective) | Locational (links) | Locates setting |
| | | Navigational (orients) | Locates paths and centers |
| Activity-based | Vicarious-behavioral-empathetic insideness (Activity and Events) | Memorable (unique) | Has uniquely occurring events |
| | | Territorial (protects) | Locates shelter; repose in regards to dynamic environment |
| | | Modifiable | The artifacts and surrounds can be modified |
| Hermeneutic | Existential insideness (Symbolic) | Culturally coded | Supports an idea of agency-directed symbols that reveal secrets of the environment |
| | | Abandoned inhabitation | Evokes an idea of social agency and past inhabitation |
| | | Lived-in inhabitation | Supports interpersonal social behavior through human and/or computer agents |
| | | Home | Affords personal shelter, primary orientation, identification, possession and collection of artifacts |

However, only if the environment evokes a notion of others interacting with the environment in ways similar or dissimilar to us, does the digital world begin to form, or, to quote Heidegger, "worlds world" (the world around us unfolds as a world of possibility). While online game communities appear to create and record meaningful encounters (via "mods" and online forums and the selling of virtual roles or equipment), their add-on meaning is generally outside of the virtual environment. Only where the environment itself shapes and is shaped by interaction that is informed by appropriate and extensive social and cultural learning, can we begin to say that it is a "world" (Weckström 2004).

A hermeneutic environment requires the ability to personalize and communicate individual perceptions through artifacts, and the more deeply this cultural communication can be unselfconsciously expressed through our modification of our surrounds, the more this environment becomes a dwelling, a home, *a place*. The degree of complexity of such a virtual environment may range from merely believing people with a different worldviewpoint existed in an environment, to feeling that we are being rejected or assimilated by another culture, to feeling that we are "home" and that we "belong."

At the moment, we know of no virtual environment that can compare in emotional attachment to a real-world home, and hence we argue that this is the most difficult type of virtual environment to create (Weckström 2004). However, we can test for "mild" hermeneutic immersion in a virtual world, where a participant begins to use and develop the codes of other cultures in order to orient and solve tasks, and to communicate the value and significance of those tasks and goals to others.

The distinction between the three types of environments is determined by the degree to which the virtual environment can support tasks and activities through which one is able to form a mental model and understanding of another place and time. The particular type of virtual environment that might be required thus depends on the amount and intensity of cultural perspective that needs to be generated and communicated.

## *Place and Interpretive (Re)constructions*

For creating a virtual environment with a notion of a "place" (a region recognizable to a user as a culturally coded setting), we need to have more than merely identifiable or evocative virtual environments. A virtual environment must allow us to see as much as possible through the eyes of the original inhabitants. It must also suggest ideas of thematically related events, evidence of social autonomy, notions of territorial possession and shelter, and focal points of artifactual possession. In other words, the virtual environment must provide a perspective of a past culture to a user in a manner similar to that deduced by trained archaeologists and anthropologists from material remains (fossils, pottery shards, ruins, and the like).

In addition to goals for participants in virtual environments, interactive elements are needed to enable and encourage participants to reach those goals. We suggest that there are fundamentally three such interactive elements: social agency, modifiable artifacts, and dynamic environments.

There is a growing support for the view that physical space and engagements need to address perceptions of appropriate or believable social behaviors (Schuemie 2001, 182–202). If social behavior is an important way of transmitting cultural information in relation to artifacts, then we require some form of seemingly autonomous social agents, whether computer-based or other participants.

Designers of real and virtual environments also need to build on the relationship between patterns of inhabitation and usage of spatial artifacts, such as furnishings.[5] Even if the word "culture" is a noun and not a verb, cultures are intangible processes acting through tangible objects. Cultures can only exist socially through artifacts, labeled by Sauer as "agents of change" (Crang 1998). However, artifacts alone constitute only a fragment of the cultural process. To fully understand a cultural environment, one requires both artifacts and an idea of the task that motivates using them.

Some of the most effective constraints in both physical and virtual realms that offer and often dictate behavioral cues are derived from the dynamic nature of real-world environments. Modeling such dynamic environments can range from shelter and familiar territory, to a hostile world, depending on task direction, artifacts carried, and their impacts on users' abilities.

Such environments can be permanently modified by user interactions. Some parts of the environment may impede the progress of the user in order for the user to recognize trails and paths, and socially accepted ways of traveling through the environment. The other parts of the environment may be deleterious to the avatar's metaphorical health—in other words, they act as constraints. The dangers and opportunities of the environment could be contextually related to the local cultural perspective. Thus, to advance through the environment the user must develop an awareness of the cultural context as it supports or impedes his/her progress.

### Construction of "Place"

In the Collaboratory for Architectural and Environmental Visualisation (Dave 2001, 242–247), we are exploring and investigating whether digital environments that are *recordable, evocative, referential, and hybrid* contribute to a more engaging sense of "place" (figure 16.3).

One recent research project revolves around Palenque, Mexico, a major Mayan site full of rich details from geography to myths, from highly advanced scripts to ritualistically charged architecture. In this project, meaning is conveyed through the type of interaction and goal achieved, rather than through the quantity and quality of

Figure 16.3   Mythological sky snake (Palenque project).

photorealistic material. In future research we hope to also ascertain how cultural learning and a sense of authenticity is affected by the existence of others in the same virtual environment.

The digital information is purposefully designed to be abstract and suggestive rather than being perceptibly complete at a first glance. The digital environment employs tasks, events, artifacts, and interactors that are a function of place, time, and the user's understanding of Mayan beliefs and actions.

The underlying motive is to enable users to construct an understanding of another place and time (even if this may lead initially to erroneous choices and interpretations by the user). Further, in order to investigate various dimensions of how users develop such constructions of another culture, this project supports three separate interaction modes: observing (moving and looking around), being instructed (by scripted agents), and acting (manipulation of objects in order to accomplish tasks).

## Conclusion

A sense of place in virtual environments and real experiences is not just a consequence of being surrounded by a spatial setting but of being engaged in another place. A place is particular, unique, dynamic, and memorably related to other places, peoples, and events, *and* it is hermeneutic. The essay argues for and proposes a more graduated approach to understanding features of different kinds of virtual places, and the cultural and social functions they facilitate in order to guide the judicious selection of appropriate design elements and technologies.

## Acknowledgment

This work was supported by the Australian Research Council **SPIRT** grant in collaboration with the industry partner the Lonely Planet Publications.

## Notes

1. See Eiteljorg 1998 and Mosaker 2001.

2. See Benedikt 1991, Johnson 1997, Heim 1998, and Coyne 1999.

3. See Fred Lukermann, cited in Relph 1986, 3.

4. See Johnson 1997, Crang 1998, and Relph 1986.

5. See Rapoport 1982 and Beckmann 1998.

## References

Beckmann, John, ed. 1998. *The Virtual Dimension: Architecture, Representation, and Crash Culture*. New York: Princeton Architectural Press.

Benedikt, Michael, ed. 1991. *Cyberspace: First Steps*. Cambridge, Mass.: MIT Press.

Canter, Donald. 1974. *Psychology for Architects*. London: Applied Science Publishers.

Champion, Erik. 1993. "Scandinavian Architecture Redefined." In *Architecture New Zealand* (January/February): 81–84.

Cipolla, Carla M. 2004. "Tourist or Guest: Designing Tourism Experiences or Hospitality Relations?" *Design Philosophy Papers*, February. Available at ⟨http://www.desphilosophy.com/dpp/dpp_journal/paper2/dpp_paper2⟩ (accessed 7 January 2005).

Coyne, Richard. 1999. *Technoromanticism: Digital Narrative, Holism, and the Romance of the Real*. Cambridge, Mass.: MIT Press.

Crang, Michael. 1998. *Cultural Geography*. London: Routledge.

Dave, Bharat. 2001. "Immersive Modelling Environments." In *Proceedings of ACADIA 2001: Reinventing the Discourse*, ed. W. Jabi, 242–247, Buffalo, N.Y.: ACADIA.

Eiteljorg, Harrison. 1998. "Photorealistic Visualizations May Be Too Good," *CSA Newsletter* 11, no. 2 (fall). Available at ⟨http://csanet.org/newsletter/#fall98⟩ (accessed 29 March 2005).

Forte, Maurizio, and Alberto Siliotti, eds. 1996. *Virtual Archaeology: Great Discoveries Brought to Life Through Virtual Reality*. London: Thames and Hudson.

Hartshorne, Richard. 1959. *Perspectives on the Nature of Geography*. Chicago: Rand McNally.

Heim, Michael. 1998. "Creating the Virtual Middle Ground." *TECHNOS Quarterly for Education and Technology* 7, no. 3. Available at ⟨http://www.technos.net/tq_07/3heim.htm⟩ (accessed 29 March 2005).

Johnson, Steven. 1997. *Interface Culture: How New Technology Transforms the Way We Think and Communicate*. San Francisco: HarperEdge.

Kalay, Yehuda, and John Marx. 2001. "Architecture and the Internet: Designing Places in Cyberspace." In *Proceedings of ACADIA 2001: Reinventing the Discourse*, ed. W. Jabi, 230–240. Buffalo, N.Y.: ACADIA.

Lippman, Andrew. 1980. "Movie-Maps: An Application of the Optical Video Disc to Computer Graphics." *Computer Graphics* 14, no. 3: 32–42.

Lukermann, Fred. 1961. "The Concept of Location in Classical Geography." *Annals* (Association of American Geographers), 51: 194–210.

Massey, Dorothy. 1993. "A Global Sense of Place." In *Studying Culture: An Introductory Reader*, ed. A. Gray and J. McGuigan, 232–240. London: E. Arnold.

Mosaker, Lidunn. 2001. "Visualizing Historical Knowledge Using VR Technology." *Digital Creativity S&Z* 12, no. 1: 15–25.

Naimark, Michael. 1998. "Place Runs Deep: Virtuality, Place, and Indigenousness." *Virtual Museums Symposium*, Salzburg. Available at ⟨http://www.naimark.net/writing/salzberg.html⟩ (accessed 29 March 2005).

Neumann, Dietrich, ed. 1996. *Film Architecture: Set Designs From Metropolis to Blade Runner*. New York: Prestel-Verlag.

Norberg-Schulz, Christian. 2000. *Architecture: Presence, Language, Place*. Milan: Skira editore.

Rapoport, Amos. 1982. *The Meaning of the Built Environment: A Nonverbal Communication Approach*. Beverly Hills: Sage Publications.

Relph, Edward. 1986. *Place and Placelessness*. London: Pion.

Schiffer, Michael, and Andrea Miller. 1999. *The Material Life of Human Beings: Artefacts, Behaviour and Communication*. London: Routledge.

Schuemie, Martijn J., Peter van der Straatten, Merel Krijn, and Charles van der Mast. 2002. "Research on Presence in VR: A Survey." *Cyberpsychology and Behavior* 4, no. 2: 182–202.

Sutherland, Ivan. 1965. "The Ultimate Display." In *Proceedings of the IFIPS Congress*, 506–508, New York: IFIP.

Tuan, Yi-Fu. 1998. *Escapism*. Baltimore: John Hopkins University Press.

Weckström, Niklas. 2004. "Finding 'reality' in Virtual Environments." Helsingfors/Esbo: Arcada Polytechnic, Department of Media, Media Culture. Available at ⟨http://people.arcada .fi/~weckstrn/Degree_Thesis_NW_2004.pdf⟩ (accessed 29 March 2005).

Woodwark, J. 1991. "Reconstructing History with Computer Graphics." *IEEE Computer Graphics and Applications* 11, no. 2: 18–20.

# 17 *The Morphology of Space in Virtual Heritage*

Bernadette Flynn

---

## A Quest for Enchantment in Virtual Heritage

One of the central cornerstones of cultural heritage is authenticity. Its claims to the real and to some type of scientific truth or factuality are historically validated through the artifact as commodity and the aura of its unique presence in time and space. In this context, the main function of heritage is the preservation and exhibition of curiosities, with the monument as a tangible expression of permanence and historical authority.

Since the early twentieth century, heritage institutions have moved away from the collection of physical artifacts to photographic and filmic forms of reproduction, and more recently to digital creation and delivery. As the significance of digital images has grown, the form of the factual has become increasingly virtualized—that is to say, it has become separated from any real object. As Walter Benjamin (1968, 223) observed in discussing the replacement of the original by the copy, art is severed from its relationship to ritual and magic. Benjamin understood the mechanical reproduction of photography and film to involve a loss of aura associated with ritual and religious objects and a liquidation of the traditional value of cultural heritage.[1] He described how mechanical reproduction yielded new forms of perception that emancipated the work of art from its relationship with elite and private forms of consumption. Thus Benjamin's work brought into focus the notion of art as politics. It also brought into focus how enchantment, ritual, and other elements associated with the contextual integration of cultural heritage are displaced by technologies of reproduction.

In this era of digital technology and connectivity, access to heritage is increasingly mediated through the consumption of signs, electronic images, and simulacra. In virtual heritage, an algorithmically accurate large-scale 3D model of a cathedral or castle is taken as the hallmark of authenticity. However, the reduction of the monument or artifact to visual simulation disrupts its connection to material evidence and thus to history.[2] What is lost is the aura of the well-crafted object or the exquisitely designed monument that resonates with the memory trace of previous civilizations. The unique object removed from the place where it happens to be, takes it away from the fabric of tradition. The commodification of artifacts extracted from place and exhibited under spotlight and glass function as one form of separation. Plaques, accurate dating, and interpretive documentation act as another separation. As interpretation replaces direct experience of the cultural object, information replaces the presence of the past. As Benjamin noted, the separation of the object from history also erases a sense of enchantment and wonder. In its place is a move towards augmented reality pieces such as virtual tours, fly-through applications, and highly accurate data maps. In these digital landscapes, the emphasis is on the construction and mapping of space—for documentation, conservation, and visual realism. Spatial objects, terrain maps, and complex digital environments thus become the markers for a different, more spatially inflected heritage experience.

## The Shape of Space

The space of virtual heritage is not neutral ground. The application of digital media to cultural heritage privileges certain forms of spatial representation over others. Virtual heritage reconstructions such as the 3D modeling of Stonehenge and Amiens Cathedral, and the panoramic space of the Virtual Everglades take photorealism or its simulated geometric 3D equivalent as the standard, unquestioning model of vision. Characterized by an attention to mathematical accuracy and the placement of solids, the translation of the monument or artifact to geometry has become the hallmark of authenticity and the standard for a certain kind of realness. The result is a divided, directional space that reduces the complexity of spatial experience down to an XYZ grid of mathematical absolutes. The experience can be like wandering in a lifeless universe lacking in human scale, or social and cultural presence.

Take, for instance, the geometric perspective exemplified in software programs such as *3D Studio Max* or *Maya* (modeling programs used in constructing many 3D virtual models for heritage). *3D Studio Max*, developed from a CAD program for use in architecture and engineering, is designed to record the shape of concrete objects and the layout of concrete spaces. Working from a grid of lines, it forces a notion of space as mathematical division and a container for objects. Within this, mathematical perspective is a particular form of construction that splits space into object-horizon polarity (figure 17.1).

Figure 17.1    The Cartesian geometry of 3D Studio Max, in Corner of a room, from computer graphics class of Department of Computing and Information services, Kansas State University. Accessed from ⟨http://www.cis.ksu.edu/~fch6688/736/cis736.html⟩.

*3D Studio Max* was initially designed for the viewer looking at a painting from a fixed position, and Mani Kaul (1991, 415) suggests that mathematical perspective resists its restoration to a unified human space. He describes the geometric appropriation of nature within Renaissance art as an initiatory moment in the celebration of the isolated object and the division of space into distance and size. This can be seen most markedly in *The Flagellation of Christ*, painted by Piero Della Francesca in 1460 (figure 17.2).

In this painting, space is broken down into the separate components of the human form, the architectural form, and the background spatial matrix. In such paintings, space itself is the object of study, with structural distance and symmetry designed for an idealized eye. Mathematical perspective, as Kaul (1991) writes, is a space outside

Figure 17.2    Piero della Francesca, *The Flagellation*, ca. 1460, Oil and tempera on panel, Galleria Nazionale delle Marche, Urbino. © Mark Harden from Artchive ⟨www.artchive.com⟩.

the being, a projection of the unreal. Such an approach emerges from a longer history, based in Western cosmology, where space is understood as static and uninhabited. Part of the problem is the English word for space—conjuring up as it does the distance between two points, or the background containing objects. As Margaret Wertheim (1999, 153) notes, Rene Descartes' model of *res extensa* and *res cognitans* split space into the physically extended realm of matter in motion and the invisible realm of thoughts, feelings, and spiritual experience. As she points out, by the end of the eighteenth century, *res cognitans* was effectively eliminated with no place for mind/spirit/soul. As a result, the realm of the real was reduced to a self-contained mathematical machine with anything outside of physical matter categorized as empty space.

An understanding of space, of course, has multiple inflections and discourses relating to cultural context and perception. Outside of a European framework, the Cartesian model of spatial demarcations was barely noted except by Japanese thinking (Barnicol 1991, 313). Take, for instance, Indian cosmogony, which considers space as the prime factor as a "matrix of dimension." In this Indian framework, space encompasses what has been polarized in the West—that is, the outer and inner, spiritual and material, human and divine. The Sanskrit word for space, *akasa*, denotes multiple dimensions, not a void but a substance.[3] Aspects of *akasa*—*cittakasa* (mental space), *cidakasa* (consciousness), and *bhutakasa* (physical space)—include ritual space, human perception, and material space (Baumer 1991, 113). Within Yoga-Vasistha, *akasa* serves as the powerful symbol of pure consciousness, with *cidakasa* (consciousness), not the mental and physical aspects of space, as ultimately real. As Bettina Baumer notes, in Indian spatial cosmogony not only are things in space, space is also in things. In other words, we are active agents in spatial creation and create space as much as space creates us—a kind of "homo spatialis" (Vatsayana 1991, 20). Such an approach presents quite different ways of inhabiting space, as experienced in Hindu temples, Persian miniature painting, *yantras*, and classical dance.

In spite of its cultural specificity and limitations, virtual heritage remains resolutely bound to a cultural model that unquestioningly maintains mathematical perspective and its translation into Cartesian coordinates as standard practice. What the use of 3D laser scanning and 3D modeling programs then represent for virtual heritage is an embalming of the idea of space as the emptiness between three-dimensional solids, devoid of the realm of thoughts, feelings, and experiences. High levels of verisimilitude are taken as evidence of increased user presence or immersion in virtual environments. And, by implication, this increased density of object information is associated with a more engaged and enchanted user.

Imaging space as an extension of Cartesian mathematics as reflected in the 3D geometry of walkthroughs and modeling may have value for database and archival purposes, but it omits other, more useful forms of representation. While the simulation of the "real" through the XYZ dimensions of digital reconstruction remains standard practice, cultural heritage offers a limited range of ways to experience the past. Large-scale 3D models and algorithmic representations of a cultural site, while being "historically accurate," can only present history in an aesthetized, desocialized, and artificialized form. An over-emphasis on the production of "historically accurate" digital representations at the expense of other types of user engagement can result in a "dumbed-down" form of heritage devoid of cultural meaning. As many commentators have argued, an emphasis on detailed rendering and an exaggeration of the foreground-background axis limits the user's social presence (Penny 1993, 17–22). In the highly geometric space of 3D modeling, the space for sociality is denied along with the performative and ritual dimensions of heritage. Within it, the body remains disembodied, abstracted into the spectator, locked out from the expectation of individualized agency. A broader morphology can open up the field of virtual heritage to explore how past communities perceived and represented their physical and social environments.[4] As Anthony Pace (1996, 4) argues, this can include various nonmaterial aspects such as symbolism, ideology, and ritual; a concern with the social and abstract dimensions of heritage, ceremony, and the creative use of aesthetic objects and spaces. If virtual heritage is extended to take account of the complexity of interfaced being, there can be a return to a form of enchantment as a varied array of culturally and historically specific spatial encounters with the past.

## A User-centered Model of Interaction with the Past

One aspect of virtual environments is the user's ability to "walk through" digital screen space. Since the mid-1990s, laser scanning, 3D modeling, and photogrametry have provided astonishingly high levels of geometric verisimilitude. Such regimes of digital simulation undermine the indexical relationship to the object, replacing it with an emphasis on spatial modeling. So, while the art or archaeological object might still be visualized within the digital landscape, there is a privileging of spatial representation and navigable construction. Along with this, the role of the audience as merely a knowledge recipient is challenged. Yet, as researchers of virtual reality have observed, virtual environments lack the richness of association and the levels of user engagement that we find in computer games. As Mona Sarkis and Laetitia Wilson (1993) have identified, interaction has often been used to mean a form of interpassivity where the user has limited options within a preprogramed menu. In large-scale virtual heritage environments, user engagement is often restricted to a disembodied walk through a clean simulation, devoid of the marks of human habitation and endeavor. In being exposed to limited interaction, the user exercises little purposeful action that can generate a sense of "being there." Any approach that does not take user experiences as central will produce less than engaging spaces.

Rather than questing after the holy grail of greater realism, or forcing a narrative interpretation onto spatial experiences, it might be useful to ask different types of questions focusing on the role of navigation and the user's sense of agency—such questions as: What types of embodiment might be represented? How might these geographies then constitute an authenticity of experience? Can metaphysical or symbolic ritual experiences be simulated? In what ways can virtual environments take into account feeling, perception, and agency to convey cultural context?

In seeking to create forms of interaction and spatial engagements that convey the cultural context in virtual heritage, it is useful to explore the ways that games have brokered effective user models of interaction. Computer and console games offer the most sophisticated examples of user-driven navigation. Games from the massively multiplayer online computer role-playing games (MMORPGs), such as *EverQuest* and

*Ultima Online*, to PC games such as *Civilization* and PlayStation's *Final Fantasy* or *Grand Theft Auto* construct spatial landscapes which the player must navigate and interrogate. This approach substantially changes the role of the digital media consumer from knowledge recipient to active participant—so that, rather than data retrieval and access, there is aesthetic immersion—rather than narrative structure, there is interactive agency.

Computer games offer substantial attractions to the player through the provision of choices. These are too numerous to canvass fully here, but include challenges or tasks responsive to the user's level of knowledge; sociality with other players; discovery through exploration; and a sense of being there, or presence. In seeking to move away from costly supercomputers and CAVE or reality centers to more consumer-level applications, a number of virtual heritage developers are adapting game engines for heritage purposes. An early example of this is Victor DeLeon and Robert Berry's (1988) *Virtual Florida Everglades* project. Developed in 1998, it utilizes Epic's *Unreal* 3D games engine to introduce AI organisms, emergent behaviors, and environmental triggers. For DeLeon and Berry, the illusion of being there is constructed through high-end 3D visualization, detailed texture mapping, and atmospheric conditions. While these regimes of visuality are still privileged as the markers of authenticity and "realness," they are augmented by responsive ecologies that simulate the chaotic patterns of organic systems. Such an approach, resulting in random and self-generating environmental behaviors, can create a more complex and dynamic simulation that offers interactive options beyond representation.

### Virtual Heritage as Individualized Trajectories of Memory

In entertainment-based computer games, agency or the act of doing is constituted through diverse modes of spatiality. These modes of spatiality contribute to game-play through such strategies as constraint and concealment, challenging the player to negotiate terrain to access objects, meet avatars, find portals, and do battle (Adams 2003). One aspect that makes such games so interesting is that the environment—the roads, bridges, paths, and doorways—cannot be perceived as a whole all at once by the player. Building up knowledge requires commitment to a series of spectorial voyages —an extensive exploration of terrain through panoramas, cartographic maps, isometric diagrams, abstract expressionist planes, and geometric landscapes. Through these sets of spatial negotiations, players become involved in the sequential unfolding of a record of signposts and metaphors embedded in the landscape.

The particular ways that users inhabit computer game-space play an important part in rethinking the practice of virtual motion. Virtual motion provides ways of looking differently or more intensely. The close scrutiny of architecture required to find objects and decode the logic of the navigable world suggests geography—a sense of inhabiting a map—rather than filmic models of spectatorship. There is also a spatial pleasure gained through travel in a well-designed virtual landscape, which is quite separate from ludic or game-play pleasures. In games such as *Exile* and *Final Fantasy*, the movement of the player materializes picturesque space as spatial practice. In turn, this spatial practice offers up multiple habitations that afford ritual experiences or emergent narrative traces.[5] The designer structures a trajectory through game-space to regulate the practice of play in a specific direction. Often we disrupt these intended trajectories by adopting our own individual, idiosyncratic pathways or strategies of navigation. These individual trajectories—or, in Michel de Certeau's terms, "tactics"—transform place into space as a series of unfolding relationships. In this way, a process of labyrinthine meanderings invite re-collection of structure through "spectorial trajectories of wanderers who make up their own cultural maps along the way" (Bruno 2002, 154).

In navigation, we develop an internal map or associational relationship to place by recall and numerous visits, which becomes a type of architectonics of the memory theater.[6] A fragmentary aesthetics of space is joined together by the experience of disparate trajectories through space and the high points of memory associated with this traverse. As the Situationists highlighted in mapping the *dérive*, or the drift across Paris, our landscapes are fashioned by psychogeography—by the sphere of the emotional being moving through space. This affective traversing of space, which details the effect of the geographical environments on behavior and emotions, speaks to the experience of the meander or the wander in virtual space. As Giuliana Bruno (2002, 267) suggests, a form of psychogeographic mapping may periodically return to reinvent the measure of spatiotemporal configuration itself. Such an approach embedded within a montage of experience assumes that navigation is far from a primarily narrative configuration. Instead, it bypasses arguments about games telling stories, espoused by narratologists, or the primacy of conflict-ridden conquest, argued by ludologists, to suggest that navigation itself, as a mode of spatial engagement, activates an intersubjective terrain mobilized by aesthetics of fragments and discontinuities.[7]

An increasing number of game-players engage in navigating through landscapes that are alive with a simulated aura of place. Relationships between mastery of terrain, symbolic encounters, emergent patterns, and seductive objects transform the idea of a bounded place into a dynamic and resonant space. Michel de Certeau (1994, 117) writing about the art of doing, makes the distinction between space and place where place is a stable and distinct location, and space is composed of the intersection of mobile elements, taking into consideration vectors of direction, velocities, and time variables. Using de Certeau's terms, the user enunciates space through the practice of movement. Navigation as practice creates a sense of enchantment or wonder through the occupation of space (figure 17.3).

If, as de Certeau suggests, space becomes a practiced place only through the activity of walking, then by extension it could be argued that in virtual spaces it is only through navigation that the terrain has a language and can bring meaning to the experience. This can be considered a different type of aura—an aura of spatial enchantment associated with the practice of virtual motion. In the atmosphere of these emotionally moving situations, space becomes an affective landscape inhabited by an intense form of subjectivity.

### Embodied Movement

What, then, of a body moving in virtual space? The mathematical realism of virtual mobility has tended to assume a passive viewing subject. In this legacy, the virtual body is rendered invisible or translated into a visual element of screen design. However, spatial aesthetics, ways of moving through space, structures of navigation, and perceptions of the virtual self or avatar are not only forms of the visual. A too-narrowly inflected approach to spatial design obscures the fact that virtual forms/embodied beings need a space. In *Architecture from the Outside: Essays on Virtual and Real Space*, Elizabeth Grosz (2001, 116) argues for space that takes the body into consideration: "Space like time, is emergence and eruption, orientated not to the ordered, the controlled, the static, but to the event, to movement or action."

She takes up Henri Bergson's concern for spaces sensitive to the emotion and actions that enfold in them suggesting that perhaps space has a materiality itself rather than the materiality resting only with the contents (116). For Bergson's space is an emerging phenomenon through specific motions and specific spaces where motion unfolds and actualizes space (128). Rather than motion as distance or space over time, space is unfolding and defined by the arc of movement, and thus is a space open to becoming:

Figure 17.3   Wonder and enchantment in Exile: Myst 111.® © UbiSoft Entertainment Inc. All rights reserved. Exile: Myst 111® UbiSoft Entertainment Inc. Accessed from ⟨www.mypage.bluewin.ch/myst3/screenshots/22.htm⟩.

To remember (to place oneself in the past, to relocate, to cast oneself elsewhere) is to occupy the whole of time and the whole of space, even admitting that duration and location are always specific, always defined by movement and action. It is to refuse to conceptualise space as a medium, a container, and instead to see it as a moment of becoming, of opening up and proliferation, a passage from one space to another, a space of change, which changes with time. (Grosz 2001, 119)

These notions move us beyond a limited, divided, directional space and have a particular relevance for cultural heritage as an embodied way of evoking the presence of the past. This evocation of space represents a shift away from a Cartesian model of space as mathematical division and space as a container for objects to a notion of space as mobile, dynamic, and active, and as reflective of the marks of human endeavor. Space then emerges from techniques and forces before representation, suggesting that the morphological scheme of virtual spaces might be responsive to the act of doing rather than seeing. That is, it might be responsive to process as becoming, to identity formation, and to an embodied subjectivity.

Games make space between virtual architecture and the virtual body so that the player is engaged in multiple transitions between space and the body. In their radical examination of architecture, Arakawa and Madeline Gins theorize how the body will seek out perceptual landing spaces, sites where the viewer's perception lands here or there.[8] A mobile and structured series of events is based on the "heres" of position and the "theres" of perceptual landing places. For Arakawa and Gins, space is active and dynamic and occupied by a sensate body. This approach bypasses subject/object distinctions and conceives of agency in a manner common in many games (figure 17.4).

In *Shenmue*, *Final Fantasy*, and *Grand Theft Auto*, as well as many other first-person exploratory and simulation games, agency emerges from the player's planning of architectural landing sites. In *Grand Theft Auto*, this is negotiated through shifts between top-down, first-person, third-person points-of-view. The player maps the landscape through the filling in of the gaps between an imitation of position and imaged landing sites. Such an approach includes intention and an invitation to perception, and thus presupposes a mobile and a thinking body. Space has a trajectory—imagined, planned, and made real through user action. In the presence of an embodied player, terrains become paths, paths become perceptual landing places, landing places become habitable environments. Maurice Merleau-Ponty, discussing the limits of scientific thinking in the ideology of cybernetics, makes distinctions between geometric space and anthropological space—that is, between the static position of the body in a self-enclosed feedback loop and the mobile body in a socially embodied spatiality.[9] His understanding of

Figure 17.4   Landing-site studies of an ordinary room, 1993–1994. Reproduced from Reversible Destiny: We have decided not to die. Arakawa/Gins. Published by The Guggenheim Museum, New York.

phenomenology suggests that space is not positioned at a distance in front of us, but surrounds the body and emerges from a spatiality of situation. Game-designers strive towards such a sense of space—space as an affect of the body's own motility—so that players form an understanding of the world and situate themselves within it. In the games *Grand Theft Auto* and *Splinter Cell*, the body is such an experiential body, developing skills and engaging in any number of bodily situations. The virtual body of the player is expressed as a portion of space, with frontiers and vital centers, defenses and weaknesses. The player's spatial domain is literally (technologically) created in real-time through the action of driving, running, crashing, and numerous other spatial acts (figure 17.5).

Representing space as embodied perception moves us away from understanding space as numerals and geometry. While Cartesian geometry underlies much game design, a number of perceptual and experiential tricks are used to enhance the composition of the landscape and its modes of reception. Immersion in play is also based on transference between the haptic feedback at the controls of the console and the gover-

Figure 17.5    The mobile body of the player in Grand Theft Auto 111.® © Take Two Interactive Software, Inc. All rights reserved, Grand Theft Auto 111® Take Two Interactive Software, Inc. Accessed from ⟨http://back.boom.ru/WallPapers/26/0306.jpg⟩.

nance of the onscreen avatar. For the player, kinetic journeys across a fragmented terrain generate kinesthetic feelings. The feelings of sweat, excitement, frustration, and anxiety are evidence of the visceral engagement of the body. Any trip to a games arcade shows the level of spatial bodily engagement as players physically jump, rotate their necks, and duck to avoid onscreen attackers. Pursuing, conquering, devouring spaces, and the spatial bodily contents translate directly to physical bodily affects. Players find this visceral environment exciting to the point where players remain in virtual space long after the body has reached its limit.[10] This internal kinesthesia and the dynamic feedback between machine and player indicate a phenomenological subjectivity whereby the player is simultaneously immersed in screen and physical worlds.

The state of virtuality is a matter of this dual immersion between the virtual world of the interactive and the external physical world. There is also a parallel oscillation between the experience of presence in simulated screen space and the reflection required for interpretation. Roger Caillois (1962) described the simultaneous presence in simulated and external worlds as vertigo characterized by a state of dizziness and disorder. Andrew Murphie (2004) argues for the centrality of vertigo in the experience of game-play—in the experience of new affects in the body. As Murphie puts it, this state of distraction emerges from identifying with your own actions and perceptions at the same time as simulating these perceptions and actions. He reminds us that, for Caillois, mimesis (role-play) can be as much about losing oneself to space, and thus the pleasures of vertigo, as finding oneself. For Caillois, mimicry and vertigo can operate to destabilize more competitive and regulated forms of play. Arguing that pantomime or role-play assures intensity and social cohesion in more collaborative societies, he suggests that mimicry and vertigo can be productive in terms of shaping and expressing culture.

### From Knowledgeable Tourist to Active Agent

The cultural tourist engages in a particular form of vertigo refined through codes and cultural patterns associated with the mobilization of the gaze (Solnit 2002). Losing oneself, the art of wandering, placing oneself as the subject in a landscape, capturing history—these are all modes of the discourse of tourism. As Scott Lash and John Urry (1994, 256) suggest, the objects of the tourist gaze are functionally equivalent to the objects of religious pilgrimage. In the traditional pilgrimage, the pilgrim followed an itinerary dotted with stages and high spots, the completion of which was a rite of passage or an initiation. From this perspective, tourism to the historical monument is not just a form of sightseeing; it is also an opportunity for a form of spiritual contemplation, direct experience, and immersion in the past.

Opportunity for direct experience through personal observation is disappearing. In an effort to safeguard cultural sites, UNESCO and other conservation groups are increasingly opting for reduced public access.[11] Extreme examples include the closing of Stonehenge to public access and the construction of a simulacrum of the Laseaux Caves next to the original. Access to heritage is increasingly mediated through digital simulations of the original or digital photographic representations. These simulations, as well as being accurate representations of material culture, can also offer forms of simulated travel adopted by the virtual tourist. In the simulation of "site seeing," the aura of the object as understood by Benjamin may be lost. Instead of Benjamin's cult object imbued with a mysterious power, we may have gained the experience of navigation, immersion, or vertigo as another form of "losing oneself."

The contemporary museum has itself moved away from being solely defined through the exhibition of objects or curatorship of physical sites. In a recent charter, the International Council of Museums (ICOM) outlined its commitment to public service—to society and its development (Weil 2002, 34). This extends the role of the museum to a place not bound by material culture, but "a place for inquiry into the memories of the past, a forum for consideration of the present, and a site from which to inspire, instruct, and inform" (Weil 2002, 111). Lash and Urry, reflecting on network culture, describe the heritage industry as an emblem of "aesthetic reflexivity" (1994, 256). By aesthetic reflexivity they mean a reflection on the social conditions of existence. For them, aesthetic reflexivity is characterized by an informed and knowledgeable tourist/visitor, openness to others, a willingness to take risks, an ability to reflect on and judge aesthetically between different nature, places, and societies.

Reinstating an informed and knowledgeable tourist opens up the possibility of a form of subjectivity and the reconstitution of community and the particular. In a simulation environment, presence and excitement are privileged over the reflection and contemplation associated with direct experience at the monument. As has been shown, there is much in game design that can be applied to virtual heritage aspects of user exploration and aesthetic immersion. Virtual movement has the potential to create a simulated spatiality that extends the real to a more imaginative, enchanted, lyrical relationship with spatial immersion. As users pass through and inhabit virtual space, the creation of a simulated reality moves beyond the purely visual. As the player inhabits screen worlds—takes up space—the relationships between mastery of terrain, kinetic engagement, and experiences of space transform data space into place. Like geography, virtual traverse is a metaphoric and associative experience owing much to our memory of garden promenades, speeding in a car, the country meander, and other personal recollections (Solnit 2002). The sense of being able to mark territory and impose an idiosyncratic trajectory on its lines of action leads to the sense of a private cartography. Metaphors of space are used which construct a discursive structure or arrangement. This structure demands kinetic journeys across a fragmented terrain, allowing users to apprehend the world and use it for their own ends. Like geography, virtual traverse is associative and transforms what might be initially understood as a mainly visual experience into a kinetic one. In this way, virtual heritage can become a space for tactics—for contextual, personal adaptation of cultural heritage. Rather than the authority of the object, we can take pleasure in a series of games and the paradox of choices—where not just objects or monuments, but interfaced being and space itself are there to be interpreted.[12]

Reconfiguring navigation as a cultural act allows for "diverse geographies of curiosity" that have the potential to reanimate virtual heritage spaces (Bruno 2002, 154). In large-scale virtual walkthroughs, a fascination for views and the physical hunger for space lead subjects from vista to vista in an extended search for spatial emotion. Adopting the strategies of immersion by incorporating the user into the action can increase the spatial pleasures of presence and embodiment. These strategies are more than the graphic immersion and organized sensuousness in computer games. Applied to cultural heritage, navigation structures can operate as significant elements in the creation of engaging and "authentic" experiences for visitors. As Caillois notes, the destabilization associated with mimicry and vertigo—the experience of dizziness, of "losing oneself"—can be productive in terms of shaping and expressing culture. In this way, spatial immersion becomes central in the animation of the past through the movement of a perceptive body incorporating memory and personal experiences. The user, as a knowledgeable tourist, occupies a cultural and dynamic space through a responsive haptic engagement with the past.

The adoption of quests with inventories, records of interaction history, and gameplay responsive to the user's skills can create emotional purchase through the vehicle of responsive interaction. In adopting interaction strategies from games, a virtual environment has the potential to intensify sensation. Task-directed artifacts, advances in complexity over time, and real-world conflict can add to a heightened sense of agency and achievement through the act of doing. Through the subjectivity of the lived body, spatial environments can function precisely at the level of metaphysical or symbolic ritual, offering different cultural perspectives and creating a sense of being.

To evoke the presence of the past relies on a different treatment of space that creates cultural and social presence. Applying constraints, affordances, and challenges within the architecture of virtual reconstruction can convey a direct negotiation with specific cultural context. This can lead to the imagined presence of the organically social rather than the ordeals of solitude. In this way, heritage is restored not only as a spatial representation onscreen, but also as a place for habitation—as a kinesthetic sense of presence in the past.

*Notes*

1. As an example of the aura, Benjamin gives the example of the elk portrayed on the prehistoric cave wall painted as an instrument of magic rather than an exhibit.

2. This paper uses the term *simulation* to refer to a dynamic rule-bound system whose mode of representation, particularly in cultural heritage, has adopted the tradition of verisimili-

tude. Frasca and other ludologists have taken another approach in arguing that simulation is not necessarily a mimetic or representational system. See Gonzalo Frasca, at ⟨http://www.ludology.org/articles/sim1/simulation101.html⟩ (accessed 18 April 2005).

3.  See Kramrisch 1991, 103. The term *akasa* is from the Upanishads, which are the final part of the Vedas and the philosophy of Vedanta written in 8–4 BC.

4.  This idea draws on Rupert Sheldrake's work on morphic resonance, where he proposes that memory is inherent in nature and these morphic fields occupy "structures of probability."

5.  See Jenkins 2004, 118–130. Jenkins proposes the term *emergent narrative* to suggest that narrative is not predefined in virtual environments, but is embedded in the architecture and emerges from the user's navigation through space and interaction with spatial objects.

6.  Bruno 2002, 352. The memory theatre was an imaginary theatre used during the Renaissance as a rhetorical aid for memorising concepts and knowledge. For more on the memory theatre, see Yates 1966.

7.  For more on the debates about narratology and ludology, see Janet Murray, Brenda Laurel and Henry Jenkins on the significance of narrative; and Markuu Eskelinen, Jesper Juul, and Gonzalo Frasca on the importance of game rules and play action.

8.  See Arakawa and Gins 1994. Thanks here to Andrea Blundell, who drew my attention to Gins's work on architecture and navigation.

9.  See Merleau-Ponty 2000. Cybernetics is the self-regulating feedback loop between humans and machines.

10.  In extreme cases, this can lead to players continuing game-play to exhaustion, collapse, and even death. Two cases that have been given much media attention are the death of a man in China who died after playing the popular online game *Saga* for twenty hours in 2004 and the death of a South Korean man who played *Binge* for eighty-six hours: ⟨www.theage.com.au/articles/2004/03/09/1078594344830.html⟩ and ⟨www.geek.com/news/geeknews/2004Mar/bga20040309024200.htm⟩.

11.  ICOMOS Burra Charter (International Council of Monuments and Sites).

12.  See Lash and Urry 1994, 276. Lash and Urry propose a notion of post-tourism as "a series of games with media texts and no single authentic experience."

## References

Adams, Ernest W. 2003. "The Construction of Ludic Space." Paper delivered at the Digital Games Research Association conference, Utrecht, Holland.

Ange, Marc. 1995. *Non-places: Introduction to an Anthropology of Super-modernity*. London: Verso.

Arakawa, Shusaku, and Madeline Gins. 1994. *Architecture: Sites of Reversible Destiny*. London: Academy Editions.

Barnicol, Hans Fisher. 1991. "Body as Measure in Architecture (Criteria for Human Living)." In *Concepts of Space: Ancient and Modern*, ed. Kapila Vatsayana. New Delhi: Abhinav Publications.

Baumer, Bettina. 1991. "From *Guha* to *Akasa*: The Mystical Cave in the Vedic and Saiva Tradition." In *Concepts of Space: Ancient and Modern*, ed. K. Vatsayana, 112–22. New Delhi, Abhinav Publications.

Benjamin, Walter. 1968. "The Work of Art in the Age of Mechanical Reproduction." In *Illuminations*, ed. H. Arendt, trans. H. Zohn, 217–251. London: Jonathan Cape.

Bruno, Giuliana. 2002. *Atlas of Emotion: Journeys in Art, Architecture, and Film*. New York: Verso.

Caillois, Roger. 1962. *Man, Play and Games*. London: Thames and Hudson.

De Certeau, Michel. 1994. *The Practice of Everyday Life*. Berkeley: University of California Press.

DeLeon, Victor J., and Robert Berry. 1998. "Virtual Florida Everglades." In *Proceedings of the 4th International Conference on Virtual Systems and Multimedia* (VSMM '98), Gifu. Japan: Ohmsha.

Eskelinen, Markku. 2001. "The Gaming Situation." *Game Studies* 1, no. 1 (online), *Game Studies Journal*. Available at ⟨http://cmc.uib.no/gamestudies.0101/eskelinen⟩.

Flynn, Bernadette. 2002. "Factual Hybridity: Games, Documentary and Simulated Spaces." *Media International Australia Incorporating Culture and Policy*, no. 104: 42–54.

Frasca, Gonzalo. 2001. "*Simulation 101: Simulation Verses Representation*" (online). Ludology .org, videogame theory. Available at ⟨http://www.ludology.org/articles/sim1/simulation101 .html⟩ (accessed 18 April 2005).

Gins, Madeline, and Arakawa. 1994. *Architecture: Sites of Reversible Destiny*. London: Academy Editions.

Grosz, Elizabeth. 2001. *Architecture from the Outside: Essays on Virtual and Real Space*, Cambridge, Mass.: MIT Press.

Hazan, Susan. 2001. *The Virtual Aura—Is There Space for Enchantment in a Technological World?* (online). Museum and the Web. Available at ⟨www.archimuse.com/mw2001/papers/hazan/ hazan.html⟩ (accessed 18 April 2005).

Jenkins, Henry. 2004. "Game Design as Narrative Architecture." In *First Person: New Media as Story, Performance, and Game*, ed. N. Wardrip-Fruin and P. Harrigan, 118–130. Cambridge, Mass.: MIT Press.

Juul, Jesper. 2001. "Games Telling Stories." *Game Studies* 1, no. 1 (online). *Game Studies Journal*. Available at ⟨http://cmc.uib.no/gamestudies/0101/juul-gts⟩.

Kaul, Mani. 1991. "Seen from Nowhere." In *Concepts of Space: Ancient and Modern*, ed. K. Vatsayana, 415–428. New Delhi: Abhinav Publications.

Kramrisch, Stella. 1991. "Space in Indian Cosmogony and in Architecture." In *Concepts of Space: Ancient and Modern*, ed. K. Vatsayana, 101–104. New Delhi: Abhinav Publications.

Lash, Scott, and John Urry. 1994. *Economies of Signs and Space*. London: Sage.

Laurel, Brenda. 1991. *Computers as Theatre*. Reading, Mass.: Addison-Wesley.

Manovich, Lev. 2002. *The Language of New Media*. Cambridge, Mass.: MIT Press.

Merleau-Ponty, Maurice. 2000. *Phenomenology of Perception*. Trans. Colin Smith. London: Routledge.

Murray, Janet H. 1997. *Hamlet on the Holodeck: The Future of Narrative in Cyberspace*. Cambridge, Mass.: MIT Press.

Murphie, Andrew. 2004. "Vertiginous Mediations: Sketches for a Dynamic Pluralism in the Study of Computer Games." *Media International Australia Incorporating Culture and Policy*, no. 110: 73–95.

Pace, Anthony, ed. 1996. *Maltese Prehistoric Art 5000–2500 BC*. Fondazzjoni Patrimonju Malti in association with The National Museum of Archaeology, Malta: Patrimonju Publishing.

Penny, Simon. 1993. "Virtual Bodybuilding." *Media Information Australia*, no. 69: 17–22.

Poole, Stephen. 2000. *Trigger Happy: The Inner Life of Videogames*. London: Fourth Dimension Publications.

Sarkis, Mona. 1993. "Interactivity Means Interpassivity." *Media Information Australia*, no. 69: 13–16.

Solnit, Rebecca. 2001. *Wanderlust: A History of Walking*. London: Verso.

Shinkle, Eugénie. 2003. *Gardens, Games and the Anamorphic Subject: Tracing the Body in the Virtual Landscape* (online). Fine Art Forum Inc. Available at ⟨http://www.fineartforum.org/Backissues/Vol_17/faf_v17_n08/reviews/reviews_index.html⟩ (accessed 18 April 2005).

Vatsayana, Kapila. 1991. *Concepts of Space: Ancient and Modern*. New Delhi: Abhinav Publications.

Wardrip-Fruin, Noah, and Pat Harrigan, eds. 2004. *First Person: New Media as Story, Performance, and Game*. Cambridge, Mass.: MIT Press.

Weil, Stephen E. 2002. *Making Museums Matter*. Washington, D.C.: Smithsonian Institution Press.

Wertheim, Margaret. 1999. *The Pearly Gates of Cyberspace: A History of Space from Dante to the Internet*. Sydney: Doubleday.

Wilson, Laetitia. 2003. *Interactivity or Interpassivity: A Question of Agency in Digital Play* (online). Fine Art Forum Inc. Available at ⟨http://www.fineartforum.org/Backissues/Vol_17/faf_v17_n08/reviews/reviews_index.html⟩ (accessed 18 April 2005).

Yates, Frances. 1966. *The Art of Memory*. Chicago: University of Chicago Press.

# 18 *Toward Tangible Virtualities: Tangialities*

Slavko Milekic

## Introduction

The ease with which it is possible to make any information accessible through the World Wide Web has led to an explosion of repositories of digitized information. For example, large art museum digital libraries have collections frequently containing more than 100,000 digitized artifacts.[1] However, apart from cataloging a museum collection, the value of such repositories for any kind of knowledge transfer is highly questionable. Seeing a thumbnail of a Jackson Pollock painting, or even a "large" image measuring a whopping $800 \times 600$ pixels on a computer screen can hardly convey the essence of Pollock's technique. The same is true of other kinds of museums where, for example, a photograph of Galileo's telescope accompanied by an abstract diagram of its optics conveys much less information than what one would get by playing with various lenses and a paper tube. Playing with telescope parts creates a unique experience that, in view of many cognitive sciences, is a basic building block for any knowledge acquisition. In this chapter I will focus on two characteristics of virtual environments: the absence of support for meaningful (experiential) interactions with virtual information, and the fact that currently the emphasis in virtual environments is placed on the *quantity* of information rather than its *quality*. Following this premise, the designers of virtual environments have to meet two challenges:

*1.* to support user interactions that contribute to information transfer and retention; and

*2.* to make the quality of virtually presented information meet or exceed a real-life experience.

In this chapter I would like to make the case that one can make knowledge transfer (learning) more efficient by tying abstract information to (tangible) experience.

Although this is currently not a common practice, the paradox is that the building of such tools is possible, even at the current technological level. Before giving suggestions on how to make virtual information tangible, I will present a very brief overview of the properties of the digital medium and problems inherent to using this medium for communication.

## Making the Virtual Tangible

There is a growing consensus among human-computer interaction (HCI) researchers that the description of interaction between humans and computers in terms of *action* is more adequate than the description in terms of information processing. This is especially true in the area of HCI design for children.

## Experiential Learning

Long before they are capable of understanding it, children are capable of acting within and upon their environment. Numerous studies indicate that these actions are biologically and contextually appropriate. Early emphasis on computational and/or representational explanations of "cognitive" functions could not account for the smart and context-appropriate behaviors of organisms with severely limited computational resources (like insects), or explain many features of children's cognitive development (Thelen and Smith 1994). In his monograph, *Being There*, Andy Clark writes:

> Cognitive development, it is concluded, cannot be usefully treated in isolation from issues concerning the child's physical embedding in, and interactions with, the world. A better image of child cognition (indeed of *all* cognition) depicts perception, action, and thought as bound together in a variety of complex and interpenetrating ways. (1997, 37)

Clark uses the example of a puzzle assembly task. A possible approach to putting the puzzle together would simply be to look at each piece and figure out mentally where its place would be. However, both children and adults often use the strategy of "trying out" the fit of various pieces, and rotating the pieces themselves rather than trying to perform the same operation mentally. These actions, labeled as "epistemic" by David Kirsh and Paul Maglio (1994, 513), have the purpose of making the task easier by reducing the cognitive effort necessary to achieve the goal.

Some authors go so far as to suggest that human cognitive abilities are an extension of bodily experiences created by our early activities. Mark Johnson (1987) calls these *kinesthetic image schemas*. An example of such a schema would be the *container schema*, which can provide conceptual structural elements (such as *interior*, *exterior*, or *boundary*), basic logic (something can be either *inside* or *outside*), and allow for numerous complex metaphorical projections (for example, one gets *in* and *out* of a relationship, is a member of a certain group or not, etc.).

## The Concept of Tangialities

The meanings associated with the adjective *tangible* in the online version of the Merriam-Webster dictionary include:

> **1 a**: capable of being perceived especially by the sense of touch : **PALPABLE; b**: substantially real : **MATERIAL.**
>
> **2**: capable of being precisely identified or realized by the mind (her grief was tangible).[2]

The listed synonym for *tangible* is "perceptible," which, in turn, has the following synonyms: "sensible," "palpable," "tangible," "appreciable," "ponderable." The evolution of the term, starting with sense percepts related to the sense of touch and ending with precise mental identification and realization of abstract concepts (like "grief" in the definition above) corresponds, more or less, to my view on this topic. I argue that association of virtual and abstract information with multimodal sensory experiences creates a new layer of knowledge and action spaces that are more natural and efficient for humans. These in-between domains, where interactions with virtual data produce tangible sensations, I have dubbed *tangialities* (figure 18.1). Please note that my definition of tangiality (2002) includes all sensory modalities and is not reduced to those related only to the sense of touch (haptic, cutaneous, tactile).

Our bodies may be considered the first interface between the real world and ourselves. Our most basic interactions are guided by our goals (intentions), carried out through actions, and repeated or corrected based on our perceptions of the consequences of actions (observations). Over time, body actions have been enhanced through the use of tools (artifacts). D. Norman (1991) introduced the concept of a cognitive artifact as a tool that enhances cognitive operations. Although the enhancement of body actions is sometimes achieved by sheer magnification (using a lever, or

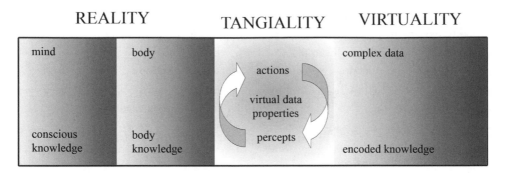

Figure 18.1    Representation of different action/knowledge domains.

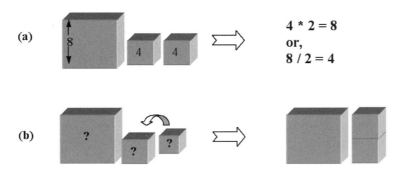

Figure 18.2    *(a)* Getting the result using abstract operation or *(b)* direct manipulation. Note that for the direct manipulation solution it is not necessary to know the actual dimensions of objects in order to arrive at the correct conclusion.

inclined plane), the enhancement of cognitive operations is most often the consequence of *changing the nature of the task*. An example of a cognitive artifact that enhances our ability to memorize and recall events is a personal calendar. Instead of trying to rehearse and memorize all of the events for weeks to come, one has to remember only to write them down on the calendar, and to consult the calendar every day. In the context of tangialities, cognitive enhancements are also the consequence of changing the nature of the task. Most often, a change occurs in the shift from relying on formal, abstract operations as a means of gaining knowledge, to directly manipulating data (properties) to produce instantaneously observable results. For example, in order to answer the problem illustrated in figure 18.2, "are the dimensions of the smaller cubes exactly one half of the larger one?" we may use conventional knowledge of algebra to solve the problem. Alternatively, using direct manipulation would allow one simply to position two smaller cubes next to the larger one and the answer would become self-evident. Note that the formal solution can be made harder by choosing different dimensions (for example, the height of the big cube could be 8.372914), but this does not influence the direct manipulation solution.

In this chapter, however, I will not address a very fruitful area of research often referred to as "tangible interfaces" (Ullmer and Ishii 2000, 915). The cornerstone of this approach is in using real objects with desired physical (manipulable) properties as data representations/containers. These objects embody computation regardless of whether they are connected to a computer or not. Although the areas of development of tangible interfaces and tangialities overlap, in this chapter I will focus on procedures that endow *virtual data representations* with tangible properties and thus make manipulations carried out on data available to our senses. The reason for this is economical and practical in nature. Building tangible interfaces is costly and out of reach for the average user. However, even with the currently available (and cheap) hardware and software, it is entirely possible to support multimodal exchanges and dramatically enhance the quality of interactions in virtual environments. I hope to be able to demonstrate that making virtual information tangible, even in the current virtual environments, makes the information more accessible and enhances knowledge-transfer.

## Historical Overview

The first human-computer interfaces were abstract, efficient, and accessible only to expert users. They involved learning the vocabulary and syntax of a command language, which was used to initiate some operations on the digitally stored data, and often one needed to issue a separate command to see the results of the previous one. There was no continuity of interaction—once the command was issued there was no way of interfering with the process, short of aborting it. There was also no sensory feedback that would provide relevant information about the operation on an experiential level.

## Direct Manipulation

One of the first examples of a tangiality domain was the introduction of Graphical User Interface (GUI) and the concept of direct manipulation. "Direct manipulation," a somewhat misleading term, was introduced by Ben Shneiderman in 1983 (57) to describe what we take today to be an integral part of human-computer interaction—the use of a mouse (cursor) for pointing and manipulation of graphically represented objects. The crucial characteristics of direct manipulation are: a) continuous visibility of the manipulated object; b) all the actions carried out on the objects are rapid, incremental, and reversible; and c) the consequences of actions are immediately visible (Shneiderman 1998).

What makes direct manipulation a tangiality-domain is the fact that it provides continuous sensory input (visual and kinesthetic feedback from hand-on-mouse positions) while acting on abstract parameters (like location coordinates, adjacency, parallelism) of digitally represented data. Despite the fact that the output in direct manipulation depends on a single sense (vision) it truly revolutionized human-computer interaction. Suddenly, anyone who could see and make hand movements could use the computer. However, it is the very success of the direct manipulation paradigm that is now one of the obstacles to creating even more efficient interfaces.

## Problems with the Traditional Interface

As Malcolm McCullough (1996) aptly observed in his book *Abstracting Craft*, one of the problems with the traditional "point-and-click" interface was the increased separation of the hand and the eye. In performing operations on digital data, the eye was given the major role of identifying, focusing, monitoring, and interpreting, while the hand was reduced to performing simple repetitive gestures. The fact that the rising number of repetitive motion injuries is associated with individuals working with computers indicates that this analysis is not just a handy metaphor. Shifting the control to the eye did not bring any benefits either. Besides its physiological role in finding and interpreting visual clues, in the traditional GUI the eye is forced to play the role of the white cane of the blind for the hand. All of the cursor guidance and positioning, often demanding a single-pixel precision (in graphics programs), is done under the guidance of the eye. This leads to overstraining of this sensory channel to the point that computer operators "forget" to blink, ultimately developing chronic conjunctivitis.

## Multimodal Interaction

In our other daily activities we almost never depend on a single-sense feedback. Take, for example, the simple act of putting a pencil on the table. Although the eyes are involved, their role is more general—seeing whether the surface of the table is within our reach, and if it is clear of other objects. The actual act of putting the pen on the surface is guided more by cutaneous and proprioceptive clues, and augmented by discrete but definite auditory cues. This multimodal and complementary feedback is the reason why our daily actions do not cause over-straining of any particular sense.

In this chapter I will focus on ways of introducing multimodal interactions in virtual environments that can be implemented using readily available, low-cost computer accessories and software. The examples will cover the following modalities:

- tactile (how does something "feel"?)
- kinesthetic (gesture, movement)
- verbal/auditory (speech recognition and synthesis)

### "Touching" the Virtual

One of the commercially available devices that provides haptic feedback is a mouse[3] that allows the user to "feel" different objects (for example, folders) and actions (dragging, scrolling) in the traditional GUI interactions. Haptic feedback is provided by a vibrotactile unit in the mouse and can be finely tuned to fit individual preferences. Although additional information initially seems trivial and discrete it becomes evident very quickly that it significantly increases the comfort of interactions (figure 18.3).

When a user receives complementary information about cursor location or action he or she does not have to rely as heavily on an already overstrained sense of vision. Experimental results presented by Robert Hughes and Robin Forrest (1996, 181–186) indicate that adding haptic information to mouse manipulations increases the

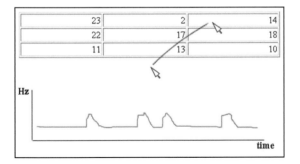

Figure 18.3   Movement of the iFeel mouse (cursor) over the table cells is augmented by haptic output produced by vibrotactile unit in the mouse. Although the stimulus is fairly discrete it significantly reduces the load of the visual sense during table input or selections using pull-down menus.

efficacy of accomplishing certain tasks. The difference in the experience can be compared to the difference between typing on a standard keyboard where every keypress is accompanied by tactile, kinesthetic, and auditory cues, and typing using keyboards where the keys are outlined on a touch-sensitive surface and provide no specific feedback. Numbers seem to confirm the benefits to the user of haptic technologies—in the first year they were introduced, Logitech sold a quarter of a million iFeel™ mice (as reported on the Immersion TouchSense™ Web site).[4]

The noticeable increase in the quality of interaction experience is making the devices that use haptic or force-feedback a standard in the area of computer games. A quote from James Biggs and Mandayam Srinivasan (2001) illustrates the use of haptic interfaces in PC game-playing:

> Active haptic interfaces can improve a user's sense of presence: Haptic interfaces with 2 or fewer actuated degrees of freedom are now mass-produced for playing PC video-games, making them relatively cheap (about US$100 at the time of this writing), reliable, and easy to program. Although the complexity of the cues they can display is limited, they are surprisingly effective communicators. For example, if the joystick is vibrated when a player crosses a bridge (to simulate driving over planks) it can provide a landmark for navigation, and signal the vehicle's speed (vibration frequency) and weight (vibration amplitude).

However, although there are more than nine commercially available haptic-enabled interaction devices (see ⟨www.ifeelpixel.com⟩ for more information), with prices comparable to traditional computer mice, trackballs, and joysticks, I am not aware of any instance of their use in museum Web environments. This is unfortunate since there is abundant evidence from learning theory that demonstrates that providing information through different modalities increases information retention.

Besides making information search and retrieval more efficient, haptic devices can play an important role in making information accessible to populations of blind and visually impaired users. Several companies have already developed products that allow the users with special needs to "feel" the information on their screens by using haptic feedback devices (iFeelPixel, VirTouch Ltd.).

An important area where multimodal sensory feedback plays a crucial role is the area of affective computing. The pioneer of research in affective computing, Rosalind Picard (1997) suggests a number of possible applications where the affective state of a human user becomes accessible to a computer, or another remote user. One of the applications is TouchPhone, developed by Jocelyn Sheirer (Picard 2000, 705), where the pressure that a participant in a phone conversation applies to the headset is transmitted to the other party's computer screen as a color range—blue corresponding to slight pressure and red corresponding to the maximum pressure value. Inspired by TouchPhone, I dared imagine a more natural model where haptic information is transmitted back and forth by holding the other party's simulated "hand" while carrying on the conversation. Of course, the extended "tactile" hand could be as easily attached to the personal computer, adding a very personal, nonverbal layer to our communications over the Internet.

A number of researchers in the field recognized the potential benefit of adding a tactile component to remote communication. Chang et al. (2002) proposed, for lack of a better phrase, a "squeezable mobile phone" device that would also convey finger-pressure information about communicating parties.

Introduction of tactile feedback in virtual environments has already led to explosive development of new areas of research like telemedicine. Adding a realistic haptic dimension to digitized data makes it possible to build sophisticated simulators for complex medical procedures. Using this technology, medical students can practice performing different surgical procedures in a virtual environment before applying their skills to real patients. Indeed, a whole new science—nanoscience—is being developed based on literally touching the surfaces of individual atoms, using Scanning Tunneling Microscopes since their dimension falls well below the half-wavelength of lightwaves necessary for "seeing" them (Gimzewski and Vesna 2003, 7–24).

## Using Gestures

It is impossible to imagine everyday communication without the use of gestures. We use them for pointing, as descriptors, to indicate agreement or disagreement, or to convey an emotional state. Although it is possible to capture gestures, even in traditional digital environments, using a computer mouse, they are neglected as a mode of interaction. An illustrative example is the navigational mechanism used for browsers and many digital documents. It usually consists of two buttons that allow the user to go to the next, or the previous screen. As mentioned before, moving the mouse cursor to the small target involves a certain level of eye-hand coordination, and constant visual attention, since there are no other clues to indicate that the target area has been reached. The interaction would be much more natural and would not require any visual attention if one could just "hit" the edge of the screen ("left" to go to the previous page, "right" to go to the next one) with a mouse cursor. This idea has already been implemented as an alternative navigational mechanism for a CD-ROM-based text containing more than 1200 screens (Weisler and Milekic 1999). A cheap, commercially available program is available (StrokeIt, ⟨www.tcbmi.com⟩) that allows users to define their own "mouse gestures" in Windows OS.

Another way of capturing natural gestures is by using touch-sensitive surfaces that can be integrated with computer monitors or by using large-scale projection surfaces, (like SmartBoard from SmartTechnologies.)[5] Some of the touchscreens (such as Surface Acoustic Wave technology) are capable not only of detecting the location of touch but also the amount of pressure, allowing the creation of pseudo-3D environments. Unfortunately, most touchscreen applications are used as oversized virtual buttons. An example of gesture interaction using a touchscreen is the "throwing gallery," a part of an interactive installation at the Speed Art Museum in Louisville, Kentucky.[6] This application allows even very young children to browse the museum collection by "grabbing and throwing" away a digital painting which is then replaced by a new item.

Currently available technologies allow fairly sophisticated gesture recognition even beyond the screen or projection surface. There are a large number of free and commercially available programs that allow the use of a cheap Webcamera as a "motion detection" system. Several years ago Intel put the me2cam on the market, a small camera with a program that could place the user directly "into" an application and supported interactions with objects on the screen (for example, playing virtual musical instruments, or juggling balloons).[7]

An example of a gesture-based museum installation is the "Gesture Gallery" at the Phoenix Museum of Art (Milekic 2001). Using simple gestures (similar to flipping a newspaper page), a visitor controls a large-scale digital projection (figure 18.4). Projected images correspond in size to the original paintings and are color corrected to be perceptually as close as possible to the originals. The appearance of each painting is accompanied by audio narrative that can be easily changed to fit a particular audience, from young children to Art History students.

## Using Voice

Although the software for continuous speech recognition has reached a level of sophistication that allows its use for numerous commercial applications (such as directory assistance, airfare booking, medical, and legal transcriptions), it is not commonly used in personal applications, for Web-based interactions, or in cultural heritage archives. An example of a possible use of speech recognition in the cultural heritage domain has

Figure 18.4   Gesture Gallery in a museum/gallery environment. Unlike a museum kiosk, the interactions with Gesture Gallery are large scale and visible to other visitors, creating a shared social context.

been described[8] where a visitor to a Web page can request more information about anything displayed on the screen by using simple voice requests, (for example, "more about Rauschenberg" while viewing a Web page mentioning Rauschenberg).

However, although the current situation seems bleak, it is probable that voice interactions will be the first modality that will be added to current interactions with the digital medium. The World Wide Web Consortium has recently approved a number of standards[9] for building Internet-based communications between humans and computers using voice.

## Implementation Problems

While it may seem that all arguments are in favor of building tangialities, it is worth investigating the barriers and problems associated with this approach. As I see them, there are some closely related general problems, which are listed here separately only for the sake of clarity. These are:

- success of "drag-and-drop" and "point-and-click" interface (the current "desktop" metaphor)
- failure to realize that the addition of another interaction device calls for redesign of the GUI
- need to unlearn established (imposed) conventions

Although it may seem paradoxical that an early solution should be blocking the introduction of more advanced ones, historically this is a common occurrence. For instance, the QWERTY keyboard arrangement, which became standard for typewriters, continues to be used for computer keyboards despite the fact that any conceivable keyboard layout is equally accessible. In the same way, the success of the mouse/cursor point-and-click interface made it a de facto standard closely associated with the very concept of what it means to interact with a computer.

This standardized notion of what an interface should look like adversely affects the introduction of any new interaction device. For example, using a touchscreen as an interactive device while preserving the traditional GUI creates a problem for the user because of the discrepancy in scale of objects necessary for comfortable interaction. This is because certain elements, such as a window closing box—an icon with dimension of $8 \times 8$ pixels—is perfectly acceptable for a single-pixel active tip of the mouse cursor, but it is definitely inappropriate for finger interaction. In environments adequately scaled for finger interactions, touchscreens have shown to be superior pointing devices (Sears and Shneiderman 1991, 593). The same holds true for other interactive devices: introduction of continuous speech recognition calls for adequate feedback that verbal information has been successfully transmitted; the use of haptic mice and joysticks introduces the texture as a GUI design element, etc.

Another general problem that has to be taken into account when introducing new ways of interaction is the need for "unlearning" adopted conventions. This can be a very slow process and a transitional stage should be part of the plans for any new design (a lack thereof is often the reason for failure).

More specific problems with the introduction of multimodal feedback come from our lack of knowledge of the complexities of multimodal interaction. Just adding another channel to human-computer interaction is not going to be, by default, beneficial. A logical and commonsensical analysis tells us that additional information presented through another channel, like any other information, may fall into the following categories:

- conflicting
- competing
- redundant
- complementary

In the above list, only the last two categories may have a beneficial effect on interaction. In the following paragraphs I will provide examples of different types of multimodal information.

*Conflicting information* is often the consequence of hardware/bandwidth limitations, as is the case with a lack of audio/video synchronization streamed in Web-based videos. Sometimes it is a product of poor design, such as when an animated character's mouth movements are inappropriate for actual utterances. Conflicting information is sometimes purposefully designed into an application, as is the case in some Web site designs that try to keep the user "glued" to their site in order to artificially boost their ratings.

An example of *competing* multimodal information is a voice overlay that is not synchronized to the printed text that one is trying to read. Animated graphics (gifs) on Web sites are another example of information competing for the same sensory modality (visual) and claiming a part of cognitive resources. An example from everyday life would be the effect that carrying on a phone conversation has on driving ability.

A definition of *redundant* multimodal information would be that of information conveyed through another sensory channel, which does not increase the total amount of information but may distribute it across different modalities, thus reducing the strain of one channel.

*Complementary* multimodal information is information conveyed through another sensory channel that does increase the total amount of information received and has a beneficial effect on interaction. This effect can manifest as an increase in efficacy of interaction, or a decrease in the number of errors. This is the only instance where the bandwidth of a human-computer information channel is increased by engagement of another channel.

### Implications for Cultural Heritage Information Access and Dissemination

Tools that support user interactions in virtual environments can be classified broadly as those that make the environment, as a whole, more conducive to information foraging, and those that serve as the actual instruments of knowledge acquisition.

### Making Cultural Heritage Virtual Environments Visitor-friendly

One of the problems associated with virtual environments is that they are most often accessed using a traditional browser (MS Explorer, Netscape, etc.). As the very term "browser" implies, the browser architecture and navigational elements were not specifically designed to promote structured access to information. This, combined with bandwidth limitations, often has deleterious effects on knowledge acquisition. The paradox is that it is entirely possible to bypass browser-defined navigational elements, even while using the browser engine for accessing the Internet. It is also possible, and not even a big programming challenge, to design a browser that would be better-suited for accessing and interacting with cultural heritage information. As described in previous sections, the technology and the software that would make it possible to support multimodal interactions (haptic information, gesture, and voice recognition) already exist. Brewster (2001, 1–14) notes that just the introduction of haptic information to museum Web sites would have the following benefits:

· Allow handling of rare, fragile, or dangerous objects.
· Allow long-distance visitors.

- Improve access to cultural heritage information for visually impaired and blind individuals.
- Increase the number of artifacts on display.

In addition to these characteristics, such an environment should also provide support for activities that are associated with knowledge acquisition in the real world. These would be, at the very least:

- Support for observation and comparison by allowing juxtaposition of different items.
- Support for visual and textual annotation of Web-based content.

Unfortunately, current browsers are far from allowing a user to freely and easily manipulate or annotate the content of Web pages.

## Digital Tools

A good example of a virtual exploratory/learning tool is a digital "magnifying glass." Building this tool only requires the acquisition of high-resolution scans of desired artifacts and simple software that would allow the user to explore the artifact at will (for example, a user could explore a [digital] painting simply by moving a digital lens over its surface). Levels of magnification can range from standard magnifying glass to microscopic examination ($\times 30$, $\times 120$, $\times 200$, $\times 600$ magnification). The tool can also be modified to provide different "views," such as the examination of a painting under ultraviolet light or using the magnifying glass to "feel" the texture of different artifacts. Note that, in this case, the digital tool provides *more* information about the artifact than what one would get from free observation. This is in accord with what Nielsen (1998) claims should become a fundamental Internet principle: that *it is better than reality*. Making a described tool available on the World Wide Web (through a browser plug-in or as a standalone application) would not only make individual learning more efficient, but would potentially lead to general knowledge-building by making the artifacts available for detailed inspection by interested experts.

An example of how the availability of such tools promotes discovery is illustrated in figure 18.5. The Whitney Museum of Art holds a number of Edward Hopper's sketchbooks. The documents are extremely valuable and cannot be subjected to handling by the general public. Displaying them in a traditional manner, in a glass case, would allow the viewing of only several pages. I was asked by the Directors of the Art Museum Image Consortium (AMICO) to suggest possible ways of making artifacts of this kind

Figure 18.5    An example of the use of "digital magnifying" glass that led to a discovery during the demonstration of the tool itself. For details see text. (Page of Edward Hopper's Sketchbook reproduced by courtesy of the Whitney Museum of Art. Digital tool demonstrated at the AMICO users group meeting, "Enabling Educational Use of Museum Media," New York, 2002).

accessible to museum patrons. One of the tools I presented was the "magnifying glass" that allowed the user to "stretch" an entire page of the manuscript or to magnify particular details. I was given a random selection of high-resolution scanned pages from one of Hopper's sketchbooks. While moving the magnifying glass across the page I discovered a note by Hopper describing the painting he submitted to the Whitney Museum for the First Biennial Exhibition! And this was happening at the meeting hosted by the Whitney Museum at the end of the Thirteenth Biennial. It is easy to imagine the possible results of making such a tool available to the scholars and the interested public worldwide.

Currently, there are a number of "zooming" tools associated with different museum Web sites. However, for the general audience, these tools do not provide an experience that is "better than reality." Very often there is a compromise between traditional graphic interface and new function—for example, using traditional scrollbars to move the "zoomed" image. A notable exception in terms of wealth and quality of information is the Web site Visual Collections from David Rumsey.[10]

## Conclusion

In this chapter I have tried to make a case for the introduction of multiple modalities in interaction with virtual environments. This is especially relevant for digital archives containing cultural heritage information. Cultural heritage artifacts tend to be "conserved, preserved, and protected," which dramatically reduces the possibility for personal interaction. Introducing an experiential component to digital archives enhances discovery-based knowledge building and makes even abstract concepts more accessible. In summary, the advantages of adding a meaningful experiential dimension, tangialities, to virtual representations are:

- widening the human-computer communication channel bandwidth.
- adding an affective dimension to interaction.
- making grasping and interactions with complex concepts possible without the need for explicit formalization.
- reducing cognitive load by use of intuitive body (biological) knowledge.
- reducing the strain to one sense—vision—single sense fatigue.
- possibly adding another dimension to metadata—"how does it feel?"

Current browser-supported interactions are not conducive to active exploration and learning. Thus, there is a need to build another environment that would be more learner-friendly. Apart from supporting interactions through interface design (for example by allowing easy manipulation and annotation of displayed content), this environment should take advantage of cheap, commercially available hardware (vibrotactile mouse, force-feedback joystick) to provide support for multimodal interactions.

Another significant use of tangible descriptions of data and results of data manipulations is in making the digital domain more accessible for populations with special needs, especially for visually impaired and blind users. There are already some promising results in this area (Yu and Brewster 2001, 41–51).

Currently, we are lacking a satisfactory theory of multimodal interaction and most often arrive at usable results through a process of trial and error. It is evident that this theory has to come from interdisciplinary efforts bridging disciplines as diverse as learning theory, neurophysiology, telerobotics and computer science.

### Notes

1. For example, Art Museum Image Consortium (AMICO) contains more than 100,000 digitized artifacts, San Francisco Museum of Fine Arts more than 82,000.

2. Merriam-Webster Dictionary, at ⟨http://www.m-w.com/home.htm⟩ (accessed 18 April 2005).

3. One of the first was iFeel mouse produced by Logitech. Unfortunately, after selling a quarter of a million of these peripherals, Logitech discontinued the production. However, a number of "haptic-enabled" devices produced by other companies are still available.

4. Immersion TouchSense™ technology, at ⟨http://www.immersion.com/products/ce/generalmice.shtml⟩ (accessed 24 March 2002).

5. See ⟨http://www.smarttech.com/⟩ for more information.

6. The installation is described at ⟨http://www.uarts.edu/faculty/smilekic/kiddyface/jbspeed/⟩ (accessed 18 April 2005).

7. Unfortunately, Intel discontinued production and support of the me2cam camera and other Intel-play products. Although I am not familiar with the reasons for this decision, that Intel used proprietary software and was unwilling to allow third-party developers to create other applications played a significant role in relative market failure of the product.

8. In the original paper the application is described for searching and getting more information from a newspaper Web site but the principle remains the same for any digital archive.

9. See Gonsalves 2004.

10. See ⟨www.davidrumsey.com⟩.

## References

Brewster, S. A. 2001. "The Impact of Haptic 'Touching' Technology on Cultural Applications." In *Proceedings of EVA 2001*. Vasari UK. Available at ⟨http://www.dcs.gla.ac.uk/~stephen/papers/EVA2001.pdf⟩ (accessed 18 April 2005).

Biggs, J., and M. A. Srinivasan. 2001. "Haptic Interfaces." In *Handbook of Virtual Environments*, ed. K. M. Stanney. Mahwah, N.J.: Erlbaum.

Clark, A. 1997. *Being There: Putting Brain, Body, and World Together Again*. Cambridge, Mass.: MIT Press.

Chang, A., S. O'Modhrain, R. Jacob, E. Gunther, and H. Ishii. 2002. "ComTouch: Design of a Vibrotactile Communication Device." *DIS*. London: ACM. Available at ⟨http://www.cs.tufts.edu/~jacob/papers/dis02.pdf⟩ (accessed 18 April 2005).

Généreux, M., A. Klein, I. Schwank, and H. Trost. 2001. "Evaluating Multi-modal Input Modes in a Wizard-of-Oz Study for the Domain of Web Search." Presented at HCI-IHM 2001, September 10–14, Lille, France. Available at ⟨http://www.ai.univie.ac.at/~michel/publications.html⟩ (accessed 18 April 2005).

Gimzewski, J., and V. Vesna. 2003. "The Nanomeme Syndrome: Blurring of Fact & Fiction in the Construction of a New Science." *Technoetic Arts: A Journal of Speculative Research* 1, no. 1: 7–

24. Available at ⟨http://vv.arts.ucla.edu/publications/publications_frameset.htm⟩ (accessed 18 April 2005).

Gonsalves, Antone. 2004. "W3C Approves Internet-Based Voice Technologies." *TechWeb News* (March 16). Available at ⟨http://www.techweb.com/wire/story/TWB20040316S0014⟩ (accessed 18 April 2005).

Gouzman, R., I. Karasin, and A. Braunstein. 2000. "The Virtual Touch System by VirTouch Ltd: Opening New Computer Windows Graphically for the Blind." *Proceedings of "Technology and Persons with Disabilities" Conference*, Los Angeles, March 20–25. Available at ⟨http://www.csun.edu/cod/conf2000/proceedings/0177Gouzman.html⟩ (accessed 18 April 2005).

Hughes, R. G., and A. R. Forrest. 1996. "Perceptualisation Using a Tactile Mouse." *Proc. Vis. '96*: 181–186.

Hendriks-Jansen, H. 1996. *Catching Ourselves in the Act: Situated Activity, Interactive Emergence, Evolution, and Human Thought*. Cambridge, Mass.: MIT Press.

Immersion TouchSense™ technology. Available at ⟨http://www.immersion.com/products/ce/generalmice.shtml⟩ (accessed 24 March 2002).

Johnson, M. 1987. *The Body in the Mind: The Bodily Basis of Imagination, Reason, and Meaning*. Chicago: University of Chicago Press.

Kirsh, D., and P. Maglio. 1994. "On Distinguishing Epistemic from Pragmatic Actions." *Cognitive Science* 18: 513.

Kirsh, D. 2001. "Changing the Rules: Architecture in the New Millennium." *Convergence*.

Logitech Web site, ⟨http://www.logitech.com⟩.

McCullough, M. 1996. *Abstracting Craft: The Practiced Digital Hand*, Cambridge, Mass.: MIT Press.

Milekic, S. 2001. "Gesture Gallery." An interactive installation at the Phoenix Art Museum, presented at "Arts & Science Labs Exhibit," ICHIM, Berlin. Available at ⟨http://www.ichim.org/jahia/Jahia/pid/339⟩ (accessed 18 April 2005).

Milekic, S. 2002. "Towards Tangible Virtualities: Tangialities." In *Museums and the Web 2002: Selected Papers from an International Conference*, ed. D. Bearman and J. Trant. Pittsburgh: Archives & Museum Informatics.

Mitsuishi, M. 2003. "Robot Surgery Expanding the Possibilities of Telemedicine." *Nature Interface* 4: 40. Available at ⟨http://www.natureinterface.com/e/ni04/P040-043/⟩ (accessed 16 April 2004).

Nielsen, J. 1998. "Better Than Reality: A Fundamental Internet Principle." Alertbox for March 8. Available at ⟨http://www.useit.com/alertbox/980308.html⟩ (accessed 22 February 2004).

Norman, D. 1991. "Cognitive Artifacts." In *Designing Interaction: Psychology at the Human-Computer Interface*, ed. M. J. Carroll. Cambridge: Cambridge University Press.

Patten, J., H. Ishii, J. Hines, and G. Pangaro. 2001. "Sensetable: A Wireless Object Tracking Platform for Tangible User Interfaces." In *Proceedings of the SIGCHI Conference on Human Factors in Computing Systems 2001, Seattle, Washington*. New York: ACM Press.

Picard, R. W. 1997. *Affective Computing*. Cambridge, Mass.: MIT Press.

Picard, R. W. 2000. "Toward Computers that Recognize and Respond to User Emotion." *IBM Systems Journal* 39: 705.

Sears, A., and B. Shneiderman. 1991. "High Precision Touchscreens: Design Strategies and Comparisons with a Mouse." *International Journal of Man-Machine Studies* 34: 593.

Shneiderman, B. 1983. "Direct Manipulation: A Step Beyond Programming Languages." *IEEE Computer* 16, no. 8: 57.

Shneiderman, B. 1998. *Designing the User Interface: Strategies for Effective Human-Computer Interaction*. Reading, Mass.: Addison-Wesley.

Thelen, E., and L. Smith. 1994. *A Dynamic Systems Approach to the Development of Cognition and Action*. Cambridge, Mass.: MIT Press.

Ullmer, B., and H. Ishii. 2000. "Emerging Frameworks for Tangible User Interfaces." *IBM Systems Journal* 393, no. 3: 915.

Weibel, P. 1996. "The World as Interface: Toward the Construction of Context Controlled Event-Worlds." In *Electronic Culture: Technology and Visual Representation*, ed. T. Druckrey, 338–351. New York: Aperture Foundation.

Weisler, S., and S. Milekic. 1999. *The Theory of Language*. Interactive CD ROM & introductory linguistics text. Cambridge, Mass.: MIT Press. Available at ⟨http://mitpress.mit.edu/catalog/item/default.asp?ttype=2&tid=4009⟩.

Yu, W., R. Ramloll, and S. A. Brewster. 2001. "Haptic Graphs for Blind Computer Users." In *Haptic Human-Computer Interaction* (LNCS 2058), ed. S. A. Brewster and R. Murray-Smith, 41–51. Berlin: Springer.

# 19 Ecological Cybernetics, Virtual Reality, and Virtual Heritage

Maurizio Forte

## The Question

The revolution of digital technologies in the past has focused attention mainly on its technical power rather than at the semantic level—the informative and communicational aspects. In the field of virtual heritage the enhancement of the aesthetic features of virtual worlds can come at the cost of informative/narrative feedback and cognition. How much information can one get from a virtual system? How does it communicate? What is its economic impact? How can we process this kind of interactive information?

The importance of virtual reality systems in the applications of cultural heritage should be oriented towards the capacity to change ways and approaches of learning. The Virtual communicates, the user learns and creates new information; typically we define a linear mode of learning—tools and actions, such as books, audio guides, catalogs and so on, and reticular learning VR systems where the user is immersed within reticules of information and visual data. An environment can be recognized in relation to others of which it also forms a part. VR is ontology by virtue of the relations that are created with the actor/observer of the system. The significance (information) is not the form that the mind imposes through acquired or innate schemes; it is generated in the relational contexts of the surrounding world. In this chapter I embrace the philosophy of ecological thinking for virtual reality applications, interpreting virtual worlds as we would ecosystems. As Ellen Levy and Philip Galanter argue,

> Complex systems are those that include large numbers of components interacting in nonlinear ways, often leading to unexpected behaviour. Complexity sciences explore how parts are related to wholes, describing the interactions between environment, system, and observer. In common language one is reminded of the saying that "the whole is greater than the sum of its parts." Understanding art as a language, the historical

development of art, and the creative process itself are all key areas in which art and complex systems have a common ground. (2003, 259–267)

## Introduction to the Ecology of the Virtual

In this contribution we discuss the concepts of virtual reality and cultural heritage according to an ecosystemic perspective. The Virtual represents a complex of relations. And the virtual translation of heritage is explainable according to a connectivity of information able to create a system. Therefore we will address these issues according to an ecological and cybernetic viewpoint, considering virtual reality as an ecosystem and subject to biodigital epistemology.

Epistemological discussion would seem to be a theoretical speculation but, on the contrary, is fundamental for projecting and communicating the virtual and for understanding the relations between digital ecosystems and cultural heritage. A virtual reality system is a complex environment and, because of this, has to be considered according to a cybernetic approach as well as a technological one. The transmission of knowledge through VR applications is one of the great challenges of the future, not least because of its ability to engage cultural nomadism and innovations in learning.

The virtual is an environment and an ecosystem: in fact there are important relations between cybernetics, system theories, and the cognitive sciences. According to certain instances of ecological thought, the living being must be comprehended in terms of relations. The living being is an autopoietic system because it is self-produced and self-organized.[1] In short, one can define virtual reality as an autopoietic system reflecting processes of mutual interaction (Maturana and Varela 1980 and 1987). In autopoietic theory, cognition is a consequence of the circularity and of the complexity in the shape of each system, the behavior of which includes the maintenance of that exact shape.

Ecological thinking takes as its starting point the condition of a whole system: organism-person thought as mind-body unit. In autopoietic theory these states are called descriptions and an organism operating within the domain of its descriptions is an observer. The existing relations between organism and environment dominate the knowledge of the Real and the Virtual and constitute the central theme of biological research. According to Humberto Maturana and Francisco Varela (1980), the Observer is the system itself: the interactions with the environment are instructive, they are part of the definition of the organization of the system and lead the course of the transformation. This typology of interaction exists in a virtual reality system, even if with some limitations. Behaviors inside and outside the system create other, unpre-

Figure 19.1    E-SPARKS (By M. Annunziato, P. Pierucci, and Plancton ⟨www.plancton.com⟩), an example of artificial life in a VR system. The goal of this project is an audio-visual interactive installation to explore the creative content and suggestions that artificial life (alife) environments have in their potentialities. The suggestion we want to communicate is the evocation of an artificial society able to self-develop over time and interact with humans. The basic idea is a vision of future digital worlds as a way to better explore the mechanisms which are the basis of the formation of our societies, languages, psyche.

dicted, behaviors, causing the concatenated generation of events and actions (figure 19.1).

In his recent book, *Digital Biology*, Peter Bentley imagines a future where computers can create universes:

> Imagine a future world where computers can create universes—digital environments made from binary ones and zeros. Imagine that within these universes there exist biological forms that reproduce, grow, and think. Imagine plant-like forms, ant colonies, immune systems, and brains, all adapting, evolving, and getting better at solving problems. Imagine if your computers became greenhouses for a new kind of nature. Just think what digital biology could do for us. (2001, 9)

According to this sense the Virtual is an ontology not opposed to the Real (Levy 1998 and 2001). Rather, it justifies the existence of other ecosystems with other rules, completely digital.

The abovementioned premises introduce the sense of an ecology of the Virtual for defining the study of interconnected relations in a developed digital system, a perceptible ontology of a biological system (figure 19.1). Therefore Real and Virtual are not opposed ecosystems, but parallel ontologies in the perception and interrelations of information (Levy 1995 and 2001).

## Virtual Reality

A virtual reality application is defined by the following features: immersion, 3D interactivity, interactivity in real time, and three dimensions (figures 19.2 and 19.3). It

Figure 19.2    Digital metaphor of the Scrovegni's Chapel: the vault of the crypt is painted with stars like the vault of the chapel. The charm of this representation is to imagine two chapels in Giotto's project; the first one related to the earthly world, the second one to the celestial world. So, in this "virtual tale," the user-visitor imagines walking on a sky, contemplating the frescoes of the chapel through the transparent vault of the crypt.

means that each action or interaction has to be in an inclusive space, within a 3D world where the navigator, rather than following preordered paths, can be free to move, exploring in real time all the available space or creating new spaces. Therefore the navigator can follow visual paths, activate behaviors, or create new worlds during the exploration. In any case, the cybernetic relations to be generated are many and unpredictable, because they embrace a personal approach to space and manner of interaction. The software calculates in real-time perspectives, and shapes dimensions, geometry of the polygons, direction of the light, and movement, according to the path initiated by the navigator-operated camera.

The user-navigator acts within the scene, choosing the behaviors, driving the action, perceiving the information and the informative echo surrounding him/her. The interaction is immersive at the interface and at the visualization level (according to the hardware). In this domain, two categories of relations and behaviors exist—the visual

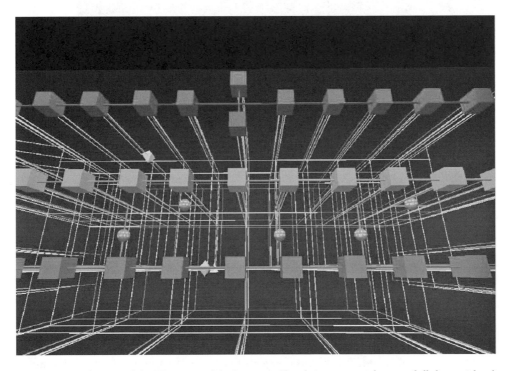

Figure 19.3   Cybermap of the VR project of the Scrovegni Chapel: it represents the map of all the spatial and thematic cyber relations of the Scrovegni Chapel Project (see Forte et al., 2002), a grammar between real and virtual.

Figure 19.4    Mindscape: the Aksum Project. The VR installation allows the user to interact with the landscape and its spatial relations.

immersion (relation of human-to-frame), and interaction in the virtual world (auto-poietic relation). An autopoietic organization is an autonomous and self-maintaining unity which contains component-producing processes. Visual immersion involves both haptic and visual interaction between the actor and the system, between observer and frame (see below and figure 19.4).

Interaction in the virtual world also involves the behaviors of the navigator within the digital ecosystem. Visual immersion and interaction in the virtual world produce differences, in a cybernetic horizontal sense, between navigator and frame and in a vertical sense, between navigator and behaviors.

The difference between navigator and frame represents the feedback deriving from the observation of the virtual world through the interface of the system (frame). The difference between navigator and behaviors is given from the feedback of each action within the virtual world. It is in this domain of virtuality that one seeks to position virtual heritage in a correct spatial connection between observer and observed world, between place and space. The spatiality of information increases the cognitive impact and, mainly in archaeology, is a fundamental premise for obtaining the maximum effect of alphabetization from a virtual ecosystem.[2]

Is it possible to think of a near future in which virtual worlds of cultural and heritage significance are accessible to everyone? According to Hal Thwaites, Professor of Communication Studies at Concordia University, Montreal:

> Something very important in the world of art happened at the end of the 20th century. The time and space axes of information are warping the content and form of an artistic piece—in the mindscape of the information consumer. . . . The perceiver of an artistic work is more and more freed from the conventions of audio-visual form. These start to move freely within his/her imagination due to the awesome capacity of the human brain to combine the carefully designed elements of subject, contents, and forms into a most fantastic and invisible pattern of thoughts, dreams and imagination. (2001, 268)

The cybernetic manifesto from which I draw inspiration is that of the ecological school of Gregory Bateson (1904–1980),[3] who anticipated the development of cybernetic theories now applicable even in the field of virtual reality software for personal computers. The ecosystemic approach of Bateson, whether in terms of informative fluxes, feedback, difference, or communication-contextualization, is fundamental to the correct evaluation of the Virtual as system. According to Bateson, "without context there is no communication" (1972 and 1979). Hence without the creation of contexts through rules of difference there cannot be an informative exchange.

The rules of difference manage the activities of interaction and learning between us and the environment, real or virtual; the more we create difference between ourselves and the ecosystem, the more we learn. In the process of learning we use a reconstructive-symbolic and perceptual-motor approach (Antinucci 2004). The latter is based on the activities of doing, touching, moving, and seeing; that is, physical interaction. The first one, the reconstructive-symbolic, works on symbolic and logic representation and reconstructs in the mind "objects"—meanings and their mental representations. In both approaches one is able to learn by creating difference and rules of difference between the subject (mind and body) and the environment, between input (perception) and output (motor).

According to this methodological approach the information processing correlated to heritage is able to modify character, interpretation, and value. We could say that the last goal of digital processing is the perceptive and cognitive enhancement of cultural heritage through digital access to virtuality with an ability to feed back cultural information.

The translation of heritage into virtual heritage recapitalizes heritage as physical object, disseminating it, diffusing its message and content. Cybernetic relations in the Virtual constitute the basis of the system, and the threshold of the exploration of the psychovision of learning. The goal is to reread the complex relations between cultural heritage and virtual reality according to an ecological approach. I will now convey the wider context of an ecology of the virtual, the description of a set of relation-interconnections that create a cybernetic epistemology.

The relations between transmission and perception of cultural heritage are largely unknown. We do not know the exact nature of what happens in the virtual reality domain, nor which relations interact between VR systems and cultural translation/alphabetization of heritage (Loftin 2002, 145–148). One could say that in virtuality we look at the deobjectification of the Real. Significations are opened up, so that new meanings, or different cognitive geometries, can be tabulated and explored. This theme is relevant because it involves future generations for whom cultural communication and new collective memories will be generated via a nomadic and deterritorialized relationship with information.

Atemporality of the virtual in the field of virtual heritage regards only the actual dynamics of the interaction (interactive action itself, is a nonplace). In artificial space, in the map of the memory (place) and in time (diachronic factor of the cultural information), navigation is fourth-dimensional.

## Map and Territory

"The map is not the territory" is the basic principle of cybernetics.[4] This fundamental difference (figure 19.3) allows for informative exchange, and the further identification of difference. The learning of information is always determined by a difference. According to the linguist Alfred Korzybski, "when there is thought or perception or communication of perception there is a transformation, a codification, between what is communicated, the *Ding an sich*, and its communication."[5] According to Bateson, the map is "a sort of effect which sums up the differences, which organises the information about the territory's differences." The differentiation, for example, between map and territory, which the semanticists insist that scientists respect in their writings must, in cybernetics, be located in the very phenomena about which the scientist writes. Predictably, communicating organisms and badly programmed computers will mistake map for territory; and the language of the scientist must be able to cope with such anomalies."

In the Virtual the map represents the cyber-geography, the alphabet, the cartography of learning, whether in the perceptive act or in the information metabolism (memory). Hence, the map is the digital code of the information we infer from the virtual environment. If, in Bateson, the map is constituted from the connectivity of the information, in the Virtual the map is represented by the spatial connectivity of the information (figure 19.3). In virtual reality all information is interconnected in a 3D space. An ontology of the connectivity involves a mutual causality—actor and environment modify each other, creating new information. In the relation between map and territory—presuming the Virtual, the map, and the territory as items of knowledge—the transfer of information between map and territory and vice versa can be seen in the circular relation of the interaction "map-territory," between coded and uncoded information.

Knowledge and learning of the "map" can produce a new knowledge of the territory and, as a consequence of this, knowledge of the territory will produce a newer knowledge of the map. The informative exchange between map and territory is bidirectional and it represents very well the relations between real (territory) and virtual (map). Bateson writes:

> The original statement for which Korzybski is most famous, the statement that the map is not the territory. . . . We know the territory does not get onto the map. . . . What gets onto the map, in fact, is difference, be it a difference in altitude, a difference in vegetation, a difference in population structure, difference in surface, or whatever. Differences are the things that get onto the map. . . . But what is the territory? . . . What is on the paper map is a representation of what was in the retinal representation of the man who made the map; and as you push the question back, what you find is an infinite regress, an infinite series of maps. The territory never gets in at all. The territory is *Ding an sich* and you can't do anything with it. Always the process of representation will filter it out so that the mental world is only maps of maps of maps, ad infinitum. All "phenomena" are literally "appearances." (1972, 460)

The map in this context is the visual pattern able to recompose a comprehensible and completely perceived image. One could say that the map creates a pseudoframe for rationalizing the information and checking the content. The bewilderment that can be induced by a work of art, a collection, or an object, is derived from the evocation of multiple and/or hidden contents. The complexity of cultural patterns pertaining to a single work of art, a museum, an archaeological landscape, needs a high level of information connectivity in order to reach an adequate cognitive representation, a contest of transmissible and narrative communication (Antinucci 2004).

A cultural heritage not perceptible and not perceived is denarrative, destitute of connections and contexts. It is informative wreckage left in virtual space, hence it is destitute of place, of transmissible memory. The information in virtual reality should correspond to a passage of condition: from territory to map. If all these passages of perceptive condition create differences, the feedback is the effect of the map. Hence the signifier of the virtual, which produces difference, is the map and not the territory.

In cybernetics we could summarize this discussion with these homologies: territory = space (the Real); map = place; map = Virtual. The Virtual becomes the place and, simultaneously, the map. Territory, like space, is a "not-place," a denarra-tivized environment, atemporal.

In this context what role does the museum play? In theory it could be either map or territory: map because it exhibits didactic criteria of visual communication, a place in itself, because it tries to construct newer contexts. It is even territory because it extrapolates objects/items from their original communicative context (or "signification") for representing them in a decontextualized frame, out of their informative medium. Therefore, the museum can be defined as a metaterritory (territory of a territory), because it deterritorializes cultural information from its original contexts/content/communication systems, in order to recreate (when it can) new meta-alphabets able to generate partial information codes. Hence the museum is not strictly a territory, neither can it be considered a map, having lost the original function of maps.

Although spatial metaphors are the prevailing ones to support interaction, it is actually a notion of place that frames interactive behavior. According to Steve Harrison and Paul Dourish (1996), in contrast to space, place is the desired notion; it constitutes a set of common and shared cultural understandings about behavior and action; a place is a social space.

In the museum field it is evident that problems of learning, disorientation, or confusion between contexts and contents, come from the superimposition of "map" and "territory" and from the erroneous relations that ensue.

## The Issue of the "Frame"

The frame in cybernetic terms represents the interface between ontologies. In the picture the frame distinguishes the interpretation, the visual difference between what is observing and what is observed, the focal direction of the context. As Thwaites argues, "the human animal, because it interposes a semiotic screen between mind and external environment . . . can drive from inside the perception, getting free from the direct influence of the external environment" (Cimatti 2000, 246). In the relationship between interaction and virtual contexts the frame is typically constituted from a display, which ontologically separates the perception or informative feedback that is generated by the differences between real and virtual, map and territory. The crossing of the frame constitutes the metaphor of the passage from the territory to the map and of the consequent feedback. It means that we distinguish the digital world because of the frame, whether a simple display or a virtual theater, verifying the cybernetic signal from the physic arrangement of an image or of a metacontext. In any case, a frame is a condition of perceptual passage, from a territory to a map. A new cognitive challenge for the future is the following: when the physical or identifiable frames disappear, will we be able to identify a spatial threshold between real and virtual?

In Bateson, the distinction between picture and ground, and what is out of the frame and the picture, is a demonstration of difference (a difference creating a difference):

> Psychological frames are exclusive, i.e., by including certain messages (or meaningful actions) within a frame, certain other messages are excluded. . . . Psychological frames are inclusive. . . . The frame around a picture, if we consider this frame as a message intended to order or to organize the perception of the viewer, says, "Attend to what is within and do not attend to what is outside. Figure and ground, as these terms are used by gestalt psychologists, are not symmetrically related as are the set and nonset of set theory. Perception of the ground must be positively inhibited and perception of the figure (in this case the picture) must be positively enhanced. . . . The picture frame tells the viewer that he is not to use the same sort of thinking in interpreting the picture that he might use in interpreting the wallpaper outside the frame. . . . Any message, which either explicitly or implicitly defines a frame, ipso facto gives the receiver instructions or aids in his attempt to understand the messages included within the frame. . . . Every metacommunicative message is or defines a psychological frame. (Bateson 1972, 188)

This circularity of subject, frame, meaning, and visual perception, is shown by the metaphor represented from Escher (figure 19.5) of the observer inverting the perspective of a scene (the observer is out or in?) A visitor is observing a print within a museum gallery; the print represents a city but the city itself is included as part of the representation of the prints gallery. The picture draws a perceptive illusion of feeling simultaneously inside and outside that which is represented. In this world drawn by Escher the map lies in the relation between observer and observed objects: the city is the object of the picture and subject of the scene, hence it is inside even as it is outside.

Figure 19.5    The "Print Gallery" of Escher, 1956, lithograph.

If the picture ends where the frame begins, in the same way is the Real interrupted/ended when one enters a virtual space? In the ecosystemic approach the issue is solved according to a logic of relations and the link between different information: Real and Virtual are two ontologies not opposed but exchanging contexts in continuous communication. An iconic and informative universe doesn't end in a frame, but it catalyzes the meaning, it multiplies the content, it attracts the surroundings.

In Bateson the word *context* is always in relation to the idea of difference and of significance: the frame allows the viewer to recognize and to learn the context. Therefore, the frame of the display permits the identification of the digital cartography, the place where interaction occurs. In relation to the Virtual, is it possible, therefore, to deconstruct the frame? And, if so, what would happen to the relation between map and territory?

In a broad sense, the frame can be interpreted in two ways. That is, frame as iconic interface and metaphor of the threshold between real and virtual or as a technological device of visualization (display) and interaction (haptic interfaces). Immersive virtual reality tries to delete the frame, substituting the reality we perceive in the surrounding with another reality. If the frame constitutes a metaphor for the interface of learning, it seems likely that the multiplication of frames would allow for greater observational perspectives.

Let's do the example of the landscape or, better, what I define as mindscape (Forte 2003), archaeological landscape perceived through mental maps. If we imagine the landscape as chaos, an interconnected structure of finished elements, but too many and too complex and interrelated to be easily translated, the input of one or more frames would define paths of learning or visible and classifiable layers (figure 19.4).

The issue of the frame concerns even the principle of observation for interpreting reality; therefore, what do we need for perceiving? The perception depends on the relations we construct with the environment/context of interaction (ecosystem) in the unit mind-body. In detail, the cybernetic relations can be summarized thus:

In-In       = relations between different elements interacting within the virtual environment.

Out-In      = relations between user/observer (out of the environment) and the behaviors of the virtual elements existing in the digital ecosystem (in).

Out-Out   = relations between users/observers out of the system.

Out-In-In = relations between users/observers out of the system and behaviors or autopoietic events interacting with each other (*in*) and regenerating within the system (still *in*).

It is evident that the relations of cybernetic exchange can multiply and create new differences. Therefore, the relations between observation, environment, and observed things are very complex.

### Contexts and Rules of Learning

According to the Bateson's cybernetic approach, we learn through encounters with difference. Difference represents the continuous interaction between us and the ecosystem, between us and the relations we produce with the surrounding environment

(Deleuze 1967). To receive, to process information, means to always acquire news of difference, and the perception of the difference is always limited; differences too smooth or showing too slowly are not perceived (Gibson 1979).

Bateson's theory of knowledge explains the mechanism of information processing. The data are neutral objects and the knowledge of a spatial system is attained through interaction by encountering the difference between components and interconnected events. The more difference is increased in a virtual interaction, the more learning is enhanced. As Bateson (1972, 458) writes:

> A difference is an abstract matter. In the hard sciences, effects are, in general, caused by rather concrete conditions or events—impacts, forces, and so forth. But when you enter the world of communication, organization, etc., you leave behind that whole world in which effects are brought about by forces and impacts and energy exchange. You enter a world in which "effects" are brought about by differences. That is, they are brought about by the sort of "thing" that gets onto the map from the territory. This is the difference. . . . And within the piece of chalk, there is for every molecule an infinite number of differences between its location and the locations in which it might have been. Of this infinitude, we select a very limited number, which become information. In fact, what we mean by information, the elementary unit of information, is a difference which makes a difference, and it is able to make a difference because the neural pathways along which it travels and is continually transformed are themselves provided with energy.

In virtual reality, learning follows informative geometries of reticular types (information spatially connected inside a net); the user is immersed within networks of information and visual data (Forte 1997 and 2000, 247–263). A feedback or retroaction is the property of learning and knowing the digital ecosystem through actions, interactions, and reactions. The perceptive phenomena involve a level of interaction or, better, an exchange or absence of behavior between action and reception. In this field one can identify multiple levels of interaction in real time (Forte 2000). According to Bateson,

> The hierarchy of contexts within contexts is universal for the communicational aspect of phenomena and drives the scientist always to seek for explanation in the ever larger units. . . . without context, there is no communication. . . . In the general semantics system, lived experience can fall within a number of terms: Territory, silent level, nonverbal level, "facts," "unnameable level," object-level. Even at a cognitive level, "seeing" an object means creating an image of it, an inside map which is not the external object.

To communicate, we use signs that *belong* to something else, they are abstractions, "pretenses." The "facts," the "phenomena" are thus "appearances" since they are cognitive elaborations that, during communication are elaborated still further. (1972, 408)

Therefore, the virtual hierarchical space is recontextualized and hierarchically restructured in order to allow the identification of logic units of information in the geometry of the models (Forte 2000).

## Conclusion

In a virtual ecosystem how much and what information can I obtain? How do we communicate with the system? How do we learn in the Virtual domain? The connectivity of the information and its relations require an ecosystemic approach, and in this direction we will address our efforts. The cybernetic frame deobjectifies and derelates the Real, bringing us to other dimensions of knowledge, the virtual reality system: the cultural and psychocognitive value of this new map is yet to be explored. In ecological thinking it is impossible to separate mind and body and, in the same way, the epistemology of the virtual is positioned between production of cultural information (mind) and communication/transmission of information (the VR system, body); the relations between production and communication create additional universes of knowledge. The study of these processes will benefit from interaction with other disciplines such as neuroscience, cognitive psychology, digital technology, philosophy, and epistemology. Currently, the mind is still an unknown object (Horgan 1999). While it is important to research the epistemological aspects of systems and of virtual reality applications, it is important to remember that a system can be studied and described, the mind only in part. The learning process with regard to data virtualization can be described as a series of communicating phases: the information transmission passes from an unstable phase, the interaction with the system to a more stable phase, the feedback and the threshold; at this point the information becomes a narrative and mnemonic process (hence the experience can be renarrated) (Forte 2002). A good cognitive impact corresponds to a good memory. A good memory starts a narrative process constituted by an essential quality of cultural transmissibility. In a museum exhibition, for example, when interpreting an object we compare old mental maps with newer maps. This mutual interaction can create an aesthetics of fruition, a new context.

In the domain of virtual reality the integration of art, science, and technology is fundamental for understanding and planning the envisioning of cultures of tomorrow:

dynamic phenomena, processes, rather than static definitions. Virtual reality opens different perspectives for a collective alphabetization. In the near future we expect to plan virtual environments based on neural networks and artificial intelligences. Worlds where avatars and artificial organisms will learn information from the digital ecosystem. Interacting with observers, they will describe their environment, their mindscape.[6] New artificial life will create possible artificial societies (Annunziato and Pierucci 2002). In a postmodern era, bodies, organisms, and communities have to be retheorized because they will be composed from elements born in three different domains, with fleeting boundaries: the organic, the technical-virtual, and the textual-cultural. In this way we would have interactions between humans, nature, and machines (Escobar 1994, 211–231).

An environment can be known in relation to the organisms hosting it; virtual reality is ontology by virtue of the relations created by the actor/observer of the system. The ecological information is what is generated in the relational contexts with the surrounding cyber system. We define autopoiesis as the cumulative feedback of innovation and use of the innovation.[7]

The discussions presented in this chapter on the epistemology of virtual heritage show in ecological and cybernetic dimension the complex and unexplored relations between psychovision, learning, information, and feedback. The early popularization of the phenomenon of the Virtual in the 1980s and 1990s, in science and literature, has overestimated some evolutionary processes not yet effectively instigated or concluded. This overvalue has created a slackening of expectations and, in consequence of this, a reappraisal of the phenomenon at the level of media and research, without an adequate epistemological consideration.

If we imagine an evolution of postmodern humans based on capacities adaptive to the ecosystem one could describe this sequence: homo *legens*, homo *videns*, homo *communicans*, homo *ludens*,[8] and, lastly, homo *"virtualis."* If we are really contextualizing in the era of *homo virtualis*, as Gianfranco Pecchinenda (2003, 47) argues, "the importance of the new technologies has to be evaluated not only in relation with the creation of new space-time realities beyond the screen, but even in relation with the capacities of stimulating new and original space-time perceptions of the reality also on this side of the screen, in daily life." In this context, we think that an anthropology of knowledge defined as ecosystemic can represent the cybernetic challenge of virtual heritage, for the cognitive modeling and virtual transmissibility of the culture—a "cybernetics of cultural heritage."

*Notes*

1. From Principia Cybernetica Web, at ⟨http://pespmc1.vub.ac.be/ASC/AUTOPOIESIS
.html⟩ (accessed 18 April 2005):

> Autopoiesis is the process whereby an organization produces itself. An autopoietic
> organization is an autonomous and self-maintaining unity which contains component-
> producing processes. The components, through their interaction, generate recursively
> the same network of processes which produced them. An autopoietic system is opera-
> tionally closed and structurally state determined with no apparent inputs and outputs. A
> cell, an organism, and perhaps a corporation are examples of autopoietic systems.

2. See Forte 2002, 81–94, and Forte 2003, 95–108.

3. See Bateson, quoted in Manghi 1998, and Cotugno and DiCesare 2001.

4. See Korzybski 1941 and Bateson 1972.

5. See Bateson 1972 and 1979 and Korzybski 1941.

6. See Thwaites 2001 and Forte 2003.

7. See Castells 2002 and Pecchinenda 2003.

8. See Huizinga 1971 and Pecchinenda 2003.

*References*

Annunziato, M., and P. Pierucci. 2002. "Experimenting with Art of Emergence." *Leonardo* 35,
no. 2 (April).

Antinucci, F. 2004. *Comunicare Il Museo*. Laterza: Rome.

Bateson, Gregory. 1972. *Steps to an Ecology of Mind*. San Francisco: Chandler Press.

Bateson, Gregory. 1979. *Mind and Nature: A Necessary Unit*. New York: Dutton.

Bentley, P. J. 2001. *Digital Biology*. New York: Simon & Schuster.

Bowen Loftin, R. 2002. "Psychophysical Effects of Immersive Virtual Reality." *Proceedings of
the IEEE Virtual Reality Conference*, 145–148. Los Alamitos, Calif.: IEEE.

Castells, M. 2002. *La Nascita Della Società in Rete*. Milan: Università Bocconi Editore.

Cimatti, F. 2000. *La Scimmia Che Parla. Linguaggio, Autocoscienza E Libertà Nell'animale Umano*.
Turin: Bollati Boringhieri.

Cotugno, A., and G. Di Cesare. 2001. *Territorio Bateson*. Rome: Meltemi.

Deleuze, Giles. 1967. *Différence et répétition*. Paris: Puf.

Deriu, M. (A Cura Di). 2000. *Gregory Bateson*. Milano: Bruno Mondatori.

Escobar, A. 1994. "Welcome to Cyberia: Notes on the Anthropology of Cyberculture." *Current Anthropology* 35, no. 3: 211–231.

Forte, Maurizio, ed. 1997. *Virtual Archaeology*. New York: Harry Abrams.

Forte, Maurizio. 2000. "About Virtual Archaeology: Disorders, Cognitive Interactions and Virtuality." In *Virtual Reality in Archaeology* (Bar International Series S843), ed. J. Barceló, M. Forte, and D. Sanders, 247–263. Oxford: Archaeopress.

Forte, Maurizio. 2002. "The Remote Sensing Project for the Archaeological Landscape of Aksum (Ethiopia)." In *The Reconstruction of Archaeological Landscapes through Digital Technologies* (Bar International Series 1151), ed. M. Forte and P. R. Williams, 81–94. Oxford: Archaeopress.

Forte, Maurizio. 2003. "Mindscape: Ecological Thinking, Cyber-Anthropology, and Virtual Archaeological Landscapes." In *The Reconstruction of Archaeological Landscapes Through Digital Technologies* (Bar International Series 1151), ed. M. Forte and P. R. Williams, 95–108. Oxford: Archaeopress.

Forte, M., E. Pietroni, and C. Rufa. 2002. "Musealising the Virtual: The Virtual Reality Project of the Scrovegni Chapel of Padua." In *VSMM 2002, Proceedings of the Eighth International Conference on Virtual Systems and Multimedia— "Creative and Digital Culture."* Gyeonggju, Korea, 25–27 September.

Gibson, J. J. 1979. *The Ecological Approach to Visual Perception*. Boston: Houghton Mifflin.

Harrison, S., and P. Dourish. 1996. "Re-Place-Ing Space: The Roles of Place and Space in Collaborative Systems." *Proceedings of the ACM 1996 Conference On Computer Supported Cooperative Work*, Boston. New York: ACM.

Horgan, J. 1999. *The Undiscovered Mind: How the Human Brain Defies Replication, Medication, and Explanation*. New York: Free Press.

Huizinga, Johan. 1971. *Homo Ludens*. Boston: Beacon Press.

Korzybski, Alfred. 1941. *Science and Sanity*. New York: Science Press.

Levy, Ellen K., and Philip Galanter. 2003. "Complexity." In *Leonardo* 36, no. 4.

Levy, P. 1995. *Qu'est-ce que le virtuel*. Paris: Editions La Decouverte.

Levy, P. 1998. *Becoming Virtual: Reality in the Digital Age*. New York: Plenum.

Levy, P. 2001. *Cyberculture*. Minneapolis: University of Minnesota Press.

Manghi, S. (A Cura Di). 1998. *Attraverso Bateson: Ecologia della mente e relazioni sociali*. Milan: Raffaele Cortina Editore.

Maturana, H., and F. Varela. 1980. "Autopoiesis and Cognition: The Realization of the Living." *Boston Studies in the Philosophy of Science*, vol. 42, ed. R. S. Cohen and M. W. Wartofsky. Dordrecht: D. Reidel Publishing Co.

Maturana, H., and F. Varela. 1987. *The Tree of Knowledge: The Biological Roots of Human Understanding*. Boston: Shambhala (rev. ed., 1992).

Pecchinenda, Gianfranco. 2003. *Videogiochi e cultura della simulazione*. Rome: Editori Laterza.

*Principia Cybernetica Web*, ⟨http://pespmc1.vub.ac.be/ASC/AUTOPOIESIS.html⟩ (accessed 18 April 2005).

Thwaites, Hal. 2001. "Fact, Fiction, Fantasy: The Information Impact of Virtual Heritage." In *VSMM 2001. Virtual Systems and Multimedia. Enhanced Realities: Augmented and Unplugged*, University of California, Berkeley, 263–270. Los Alamitos, Calif.: IEEE.

# 20 Geo-Storytelling: A Living Archive of Spatial Culture

Scot T. Refsland, Marc Tuters, and Jim Cooley

## Introduction

While standard archaeological and historical documentation methodologies serve humankind well in accurately preserving and indexing cultural and natural heritage, next generation emerging technologies are now challenging current archival, presentation, and historical versioning practices. While first and second generation attempts at creating a "virtual heritage" experience through technologies like virtual reality have typically fallen dismally short of their much publicized claims, current versions are now succeeding in accurately presenting and documenting history. An excellent example is the University of Birmingham's seismic data visualization of a spectacular prehistoric landscape previously unknown to science, where early man roamed more than 10,000 years ago, deep beneath the North Sea (figure 20.1) (Stone 2005).

The stabilization of inexpensive and highly usable technology has accelerated the focus upon more sophisticated content development and presentation, adding new multi-viewpoints, new layers of storytelling, and creating dynamic, living archives of spatial culture. These emerging technological paradigms are pushing for new, transforming notions of how historical documentation and indexing might be both constructed and maintained in the near future.

Due in large part to the "off-the-shelf" commercial availability and highly inexpensive usable technologies such as data-capable cellular phones, Geographical Positioning Systems (GPS) receivers, and Smart Personal Objects Technology (SPOT), the notion of location-encoded media has begun to seep into the public consciousness. GPS technologies now appear regularly in Hollywood movies, such as "Mission Impossible," as well as in mobile, location-aware computing games such as "MOGI," which utilize GPS to enable players to see each other's locations (Terdiman 2005). Soon we will

Figure 20.1    Meso House near the Shotton River Stream in a recreation of the prehistoric landscape over 10,000 years ago that is now the North Sea. Credit: R. Stone and E. Ch'ng.

see a sweeping ubiquity of location-aware technology become a standard feature embedded in the next generation of wireless devices.

Location-aware wireless devices will hypothetically permit an immersive experience in which users will be able to browse layers of digital information encoded to a particular place, just as they now surf the Internet, invoking potential models of interaction, that appear "as limitless as the possibilities of reality" (Fisher 1991, 101–111). In its darker manifestation, however, pervasive location-awareness also brings with it the possibility of a user being tracked wherever there is a service signal. This sense of "big brother" has been the subject of much debate among media artists, who are often early adopters/explorers of emerging media, where the field is frequently referred to as "locative media."

Although cultural producers have not been using locative media for long, an unconventional theater group from the UK known as Blast Theory has developed a series of award-winning locative projects modeled on networked interactive entertainment. Blast Theory projects such as Roy All Around You (figure 20.2), allow participants equipped with mobile location-aware devices to interact with other players online. Yet, while these early experiments have proven remarkably popular with audiences, their "command and control" model of interaction is not perhaps as well-suited to cultural heritage asset production.

Perhaps a more appropriate example of locative media for the purposes of virtual heritage can be found in the concept of Geograffiti, which proposes an open-access spatial authoring system for mobile, network-enabled, location-aware devices (Tuters 2004, 78–82). Etymologically derived from the Roman practice of scratching political messages onto public walls (graffito), graffiti has a long political history, throughout many cultures, as a form of public expression, yet today it is widely considered a violation of some basic principle of consensual social order. Geograffiti, however, can be conceived of as a kind of "virtual graffiti" that allows one to interact with a space without visibly altering it [1] Indeed, what is interesting about locative media in the context of virtual heritage is that it makes possible the notion of a collaborative mapping of space, and the intelligent social filtering or "narrowcasting" of that space, so that it is only experienced by those who so desire, and are so equipped. This has the potential to significantly impact the dominant modes of representation, most notably the linear, expository narrative.

## *Flexible Storytelling*

Storytelling is one of the oldest art forms of human beings; initiated in the oral tradition, its form has evolved with changes in society and available media with which to work. The oral storyteller often altered the tale as it was told, choosing words based on response from the listeners to build the collective moment. Of particular relevance here is the Australian aboriginal storytelling tradition in which territory is not perceived as a piece of land enclosed within borders, but as "an interlocking network of 'lines' or 'ways through'" (Chatwin 1987). Sung into existence by the ancestors, these stories actually function as maps of their terrain that can be augmented by the accounts of travelers. Interactive in the beginning, with the advent of the written medium, storytelling evolved into a noninteractive narrative style.

Figure 20.2    A street player using a handheld computer to search for Uncle Roy, guided by an interactive map and messages from online players. Courtesy Blast Theory: Mixed Reality Lab.

The digital era has provided a more flexible medium for storytelling. In the primitive digital sense technology, like hyperlinking, allows weaving of myriad paths through an otherwise linear presentation, and also allows inclusion of side-bar material without disrupting the story for those uninterested in its detail. Stories or articles can be authored statically, ahead of the reading, or dynamically generated "on-the-fly." This provides a medium that can capture more of the richness of the interactive storytelling present in the original oral tradition than previously available. In a mobile context within a cultural heritage landscape, technology can provide rich scenes upon which to set a tale or narrative, allowing the viewer to see details that could previously only be imagined.

Heritage content is typically pieced together from "objective" scientific records, making it almost impossible to intelligently translate a coherent first-person narrative from all the facts. In the context of a museum or heritage display, the physical artifacts are arranged in a collection of linear stories by a curator in order to convey a prepackaged slice of some larger set of holistic events. Yet each artifact has its own story and those stories have different prevailing accepted and controversial depictions.

Collaborative cartography potentially causes a crisis for those areas which had been defined by single dominant narratives. With the anticipated accessibility of locative-media authoring systems, a flood of amateur content producers, eager to contribute to the collaborative map, will inevitably be among the early adopters. With a multitude of voices telling different stories in any given space, one might consider this a kind of ecosystem, which produces a decentralized vision of virtual heritage.

## Heritage Ecology

Media ecology is a term coined by the famed media scholar Marshall McLuhan for considering the whole field of production when engaging with a problem regarding media (Haynes 2004, 227–247). In 1977, McLuhan said that media ecology means:

> arranging various media to help each other so they won't cancel each other out, to buttress one medium with another. You might say, for example, that radio is a bigger help to literacy than television, but television might be a very wonderful aid to teaching languages. And so you can do some things on some media that you cannot do on others. And, therefore, if you watch the whole field, you can prevent this waste that comes by one canceling the other out.

While having identified a problem regarding the absence of multiple points-of-view in the traditional field of heritage, a media ecologist would argue that if anything, there is an overabundance of "first-person" points-of-view in the broader field of contemporary media production. Perhaps the most obvious example of this is the Web-logging, or blogging phenomenon. While the majority of "blogs" are simply personal diaries, written for niche audiences, the sheer number of blogs means there is something out there for everyone. Because of the fact that blogs tend to be written in machine-readable format, that can be reused across application, enterprise, and community boundaries, bloggers are able to syndicate their data in the form of "feeds," which readers can, in turn, subscribe to. With the emergence of applications that allow one to associate feeds with given locations and the migration of blogging to mobile platforms (moblogs), it may only be a matter of time before these merge (Kellner and Petersen 2004). Such locative media systems would allow tomorrow's mobloggers to interact with as well as to author a multitude of different points-of-view on a particular location.

When considering locative media, perhaps the field of virtual heritage could highly benefit from this emergent public interest in archiving spatial culture. Endorsing and supporting a program of community review of cultural heritage documentation might facilitate the emergence of a common resource for local knowledge-sharing. As with the online encyclopedia Wikipedia, this would be constructed and maintained by an army of hobbyist volunteers. Such a distributed effort could prove invaluable in maintaining an informative, unbiased presentation of the often-disputed facts of history.

A potentially relevant future direction for this global project to collaboratively map spatial culture might be that of the Geoscope, a project proposed by the visionary philosopher of science, R. Buckminster Fuller (Fuller and Kuromiya 1981, 133–140). Geoscope was to be a fully computerized globe, a tool for planning and stewardship that was supposed to make the entire world "dynamically viewable . . . to all the world, so that common consideration . . . of all world problems by all world people would become a practical everyday, hour and minute event." According to Fuller, "With the Geoscope humanity would be able to recognize formerly invisible patterns and thereby to forecast and plan in vastly greater magnitude than heretofore. The consequences of various world plans could be computed and projected, using the accumulated history." The beauty of drawing on Fuller's science fiction utopia as a pedigree for the virtual heritage archiving project is that it does not necessitate the destruction of the old in order to construct the new. To the contrary, in fact, in the service of heritage, locative systems preserve space and culture.

Channeling the current zeitgeist around online collaborative development into a project for mapping cultural heritage as a shared resource would help the field of virtual heritage steward the construction of a definitive archive of spatial culture. Having recognized the validity of this decentralized model of cultural asset production, virtual heritage could greatly improve its efficacy by developing user-centered and dynamic systems for nonlinear storytelling. Such a system would give users the sensation of being able to navigate beyond the official story of heritage into a web of interconnected complexity. Crucially, however, the objective of virtual heritage might also be to help simplify and organize the chaos of "Geograffiti" so as to facilitate a degree of situated awareness among the supertechnologically equipped citizens of the locative future. And while plans to steward such a project might indeed sound like science fiction, in closing we might simply consider the words of Fredrick Jameson (1982) for whom "the 'science fiction' utopia serves not to give us 'images' of the future . . . but rather to defamiliarize and restructure our experience of our own present" (147–158). We need these new visions of the technological future in order to better manage the present.

## Note

1. The Geograffiti project developed a wireless Web-client based on a public waypoint-sharing database known as GPSter.net. Anyone with an Internet connection could add to the database, as well as search for personal points of interest, encoded with positional information via a system known as WhereFi. Rather than use GPS to obtain such positional awareness, the project sought to transform the wireless landscape of Wi-Fi access points, GSM towers, or bluetooth beacons into a public location awareness infrastructure for collaborative cartography.

## References

Chatwin, Bruce. 1987. *The Songlines*. London: Jonathan Cape.

Fisher, Scot S. 1991. "Virtual Environments, Personal Simulation, & Telepresence." In *Virtual Reality: Theory, Practice and Promise*, ed. S. Helsel and J. Roth, 101–111. San Francisco: Meckler.

Fuller, R. B., and K. Kuromiya. 1981. *Critical Path*. New York: St. Martin's Press.

Haynes, L. W. 2004. "Original Sin or Saving Grace? Speech in Media Ecology." *Review of Communication* 4, nos. 3–4: 227–247.

Jameson, F. 1982. "Progress Versus Utopia; or, Can We Imagine the Future?" *Science-Fiction Studies* 9, no. 2: 147–158.

Kellner, S., and F. Petersen. n.d. "plazes.beta," by Plazes. Available at ⟨http://new.plazes.com/⟩ (accessed 24 February 2004).

Stone, Robert J. "Scientists Reveal a Lost World Discovered under the North Sea." *Virtual Heritage Network*. Available at ⟨http://www.virtualheritage.org/news/Article.cfm?NID=1543⟩ (accessed 18 April 2005).

Terdiman, D. 2004. "Making Wireless Roaming Fun." *WIRED News* (online). Available at ⟨http://www.wired.com/news/games/0,2101,63011,00.html⟩ (accessed 18 April 2005).

Tuters, Marc. 2004. "Locative Media as the Digital Production of Nomadic Space." *Geography* 89, no. 1: 78–82.

# 21 Urban Heritage Representations in Hyperdocuments

Rodrigo Paraizo and José Ripper Kós

---

## The Problem of Representing an Urban Heritage Environment

The representation of urban environments has always been a challenge for those who want to portray a city or a town. Several authors agree that even the term "city" is an artificial creation to facilitate theorizations and representations of urban spaces and the lives performed within them.[1] In the past, cities were enclosed by walls and their territories could be easily determined. Once they became larger than their walls—and even walls were forever removed—city limits could be determined by abstract State regulations, in many cases without a distinct physical counterpart. According to Rob Shields, we "classify an environment as a city and then 'reify' . . . that city as a 'thing.' The notion of 'the city,' the city itself, is a representation" (1996, 227–252).

José Barki (2003) reminds us that people can only understand reality through representations, and that representations are constituted by selected elements put together for a partial and focused understanding of reality, given the complexity of real objects. Each person elaborates a different representation of the same object and even the same person represents it differently in distinct moments. Since many different individuals inhabit cities, they generate a very complex set of social relations, and are constantly elaborating different representations of the place in which they dwell. As a result, there is no ultimate representation of an object—let alone of a city.

When representing an urban heritage environment, two other issues should be addressed: one of them is the need to rely upon limited historical documents to support representation, and the other is the lack of documentation of urban tissues—and ordinary people's lives—in order to depict urban dynamics.

On the other hand, urban spaces, and their buildings, are characterized not only by their forms, but also by their dynamics. If an abandoned church can still be praised for its form, in spite of change (or lack) of use, the same thing seldom happens when dealing with the urban scale—lack of use usually leads to the creation of some sort of ghost town. Context—social and economic among others—and the very dynamics of daily or past life, are essential rather than circumstantial for the understanding and representation of urban heritage. Therefore, considering spaces that are also stages for social events, hyperdocuments can be powerful tools for displaying not only physical urban structures but also the connections that create the urban spaces people dwell in.

Throughout this chapter, we aim to explore some features of hyperdocuments according to existing and elaborated classifications. To illustrate these we will refer to projects developed by our research group in Brazil at the Federal University of Rio de Janeiro, in addition to other authors' projects. We do not intend to present a definitive classification or rules to reveal urban heritages through electronic tools, but to discuss different methodologies that can illustrate paths for future investigations in this area.

## Informational Devices and Urban Heritage Representation

This study focuses on one of the several ways to represent, analyze, and communicate heritage—the hyperdocument—and locates characteristics that are appropriate for the representation of urban heritage.

Pierre Lévy defines hyperdocuments or networked messages as a specialization of one of the dimensions of communication, namely the informational device, that is, the way in which information is arranged. Other informational devices include linear narrative, the most traditional way of telling a story; virtual worlds or spatial narratives, the spatial arrangement of information to deliver a message; and information flow, facilitated by database filters (1999, 61–66). Fold-out maps, for instance, work as virtual worlds, and travel books rely heavily on informational flow to organize hotels and attractions indexes. The mnemonic technique called "memory palace" is used to build a virtual world in one's mind to organize information (Malamud 2000). All of these informational devices have digital instances that can be used to depict urban heritage.

They do not constitute a strict classification of representations, however, since we often find different informational devices in any document, rather, they offer a way to analyze representation. In order to clarify the application of each informational device, we shall examine some of the academic works of digital representations of urban heritage.

The system used to build the *Virtual Reality Notre Dame*, from Digitalo,[2] borrows technologies from the electronic games industry in order to achieve faster rendering and allow multiple-user interaction inside a virtual Notre Dame. It exemplifies the virtual world narrative, where information has a spatial distribution.

The research group ABACUS, located at the University of Strathclyde, developed two noteworthy Web-based projects using historical databases. The *Glasgow Directory* binds virtual worlds to informational flow (databases) in order to either filter or provide additional information on buildings in a 3D map (a VRML model) of central Glasgow. The model, with the geometry of the topography, roads-network, and 10,000 buildings, was divided into twenty-eight neighboring city "chunks" to be interactively explored on the Internet.[3] This project connects data related to buildings and streets to the 3D model, and information is organized by different categories in a database and linked to individual buildings in the 3D model.

The other project is called *TheGlasgowStory* (ABACUS 2005). It is part of a great initiative from the UK government to provide learning and research material online, and is being carried out by a large consortium including two universities and all of the public libraries in the city. Its aim is to digitize archives from different institutions relating to the city of Glasgow, collate the material, and provide free online public access to medium-size resolution images.

All institutions agreed to give away these digitized versions of their collections to be freely accessed through *TheGlasgowStory* database over the Internet. This is remarkable, particularly to educators, students, and researchers, whose access will be extremely facilitated. The number of people who have access to these historical archives will probably increase exponentially, so that similar initiatives ought to change the way history is communicated and apprehended. Unlike *The Glasgow Directory*, *TheGlasgowStory* is not associated with a 3D model of the city, and due to the lack of spatial reference of the historical documents, even those who are familiar with Glasgow may have difficulty locating some of the historical documents which refer to locations in the city.

Projects like *Ename 974*, in Belgium,[4] and *Projeto Missões* (Missions Project), from the University of Vale do Rio dos Sinos, Unisinos, use immersive-oriented electronic modeling to involve the reader (Rocha and Danckwardt 2000, 191–193). The first is basically a kiosk with one of its windows facing the excavation of a medieval abbey. Into the window is projected the corresponding perspective of a model from the abbey in its diverse stages of existence, helping visitors localize the ruins. The latter evolved from a project of image restoration of the detailed electronic modeling of a church and the corresponding settlement of São Miguel Arcanjo, in Rio Grande do Sul, Brazil, made by the Computer Graphics Center of Unisinos.

In addition, digital panoramas constitute tools for virtual world narratives, using actual photographs, digital modeling, or both. Examples include the project of digital reconstruction of the *Synagogue Neudeggergasse*, in Vienna (Martens et al. 2000, 165–170), and the *Edinburgh Royal Mile Project* of the University of Edinburgh (Wright et al. 1999, 217–223), that conveys nineteenth-century panoramas. The Brazilian project *Visorama* uses a binocular telescope with a rotation and tilting feedback system, installed in specific belvederes of Rio de Janeiro, to display computer-generated images of past times from the same point of view of the observer in those places (Parente and Velho 2000, 257–260).

As we see in these examples, informational devices hardly appear isolated in a document. They are combined to get the most out of their ability to point out some specific characteristic. Thus, for example, when exposing the spatial relationship among elements, virtual worlds are to be preferred. When the urban scale and the effects of time are involved, or there is a multiplicity of resources of different natures, it is often useful to adopt the hyperdocument as the main narrative.

### Digital Representation and the Urban Scale

As previously stated, the informational device is not necessarily bound to computers. However, a more strict definition of hyperdocuments from Lévy assumes that they are networked messages conveyed in digital media, or, more comprehensively, digital interactive multimodal documents. This means hyperdocuments take into account user intervention to reorient the flow of information in real time and involve many of the user's senses in their interpretation. The qualitative leap of this approach is the displacement of the linking operation from the subject navigator, who does it manually, to the navigated object, which does it automatically; also, the destination point is displayed within the space of origin. As a result, following the links is much faster, and can be more easily done by the reader. At the same time, the differences between each piece, whether text or image, become blurred, ultimately leading to some interesting questions on authorship and publishing rights, especially on the Net.[5]

Robert Laurini is even more specific, defining "hyperdocument" as ". . . a modern version of nonlinearly organized materials (2001, 124). That is, they are electronic documents with direct access to information of diverse form by means of window presentation and mouse clicking on important words or other displayed information."

Digital hyperdocuments have some important features for the representation of urban heritage. There is a tendency to make the most of the image-processing capability of computers (more restricted in printed publications), which makes it economically viable to use several images and even to use them as the basis for navigation. They work with nonlinear narrative, allowing the development of parallel structures. Also, they are composed of fragments linked through highlighted remissions. Besides, they usually have some sort of index, readily accessible from most parts of the document—a meta-remission that helps understand the structure of the message—or at least the presence of other forms of metaremission such as navigation buttons.

If the representation of isolated architectural objects does profit a lot from 3D digital models, these advantages are not fully carried from the architectural to the urban scale. One of the reasons is the ongoing difficulty for ordinary PCs to manipulate complex models in real-time, as in the case of detailed urban reconstructions. The limited size of computer screens poses another problem, as urban plans tend to be quite large, and even the zoom feature can be disorienting when the whole figure is not simultaneously available. This leads to the search for alternatives to sheer virtual worlds for depicting urban scale narratives, and the need to use them in conjunction with other devices.

Unlike architectural heritage, usually apart from the daily life of the city, and therefore with a "limited" or "finished" history to be told, urban heritage is usually a dynamic space where people still live, work, and rest. Therefore, to be successfully represented, it should afford links to its current and ever-changing character.

The hyperdocument offers an extremely useful way to portray different, even contradictory discourses, such as those found in the city. Its very nature stimulates juxtaposition and comparisons. The hyperdocument, however, needs to link the different pieces of information, whether images, text, or audio, in a way that helps users to make sense of each piece and its connections, that is, providing some perspectives of the material. Otherwise, it can be very confusing for the purpose of analysis and comprehension of the urban environment, no matter how poetic the new kind of *dérive* and *detournement* that results.

### Urban Tales

Since 1995, the Brazilian research group at the Laboratory of Urban Analysis and Digital Representation at the Federal University of Rio de Janeiro (LAURD-PROURB/UFRJ) have explored hyperdocuments, both as CD-ROM and Web sites, that aim to make evident the symbolic structures of Latin American cities. Our first attempts focused on Rio de Janeiro and Havana, as part of a research called "Evolution of the

Latin American cities symbolic systems". Both "Rio Colonial" and "Havana Colonial" CD-ROMs, as their titles state, deal with the Colonial period of each city. Professor Roberto Segre, one of the LAURD coordinators, also conducted a similar project in Buenos Aires.

Typological classifications used here are nothing more than an attempt to establish a vocabulary for the hypertext concerning urban heritage, in a similar way to that proposed by Bernstein for fictional narratives, with the same objective: naming them to provide a common ground for their exploration (2002, 83–97). Narrative types here described—augmented linear narrative, tree structure, array of categories, rules of combination, virtual world array, and historical virtual world databases—were defined following an analysis of the projects developed by our research group at LAURD. Thus, they do not represent the only possibilities for historical hyperdocuments. This classification should be seen as a contribution to the creation of a theoretical framework for the understanding of hyperdocument projects in this field. Therefore, it should be completed by other authors' contributions.

### Augmented Linear Narrative

These projects work with two main types of narratives: linear narratives on thematic sections, or modules, that make reference to the symbolic structures; and navigation through prerendered images of the 3D model. The latter is limited to a few viewpoints comprising the main buildings and the urban tissue of the period, in a catalog-like virtual world derived from the buildings depicted on screen.[6]

The 3D modeling also provides images to support the animated and interactive analysis, such as the influence of religion on the establishment of plazas in downtown Rio, or the association between military power and the polycentrism of Havana. In such topics, the main informational device is the linear narrative (figure 21.1).

However, these systems should not be taken for traditional linear narratives. Indeed, they contain a few aspects that alter considerably the experience of reading: hyperlinks to small additional text and images and highlights on the main image, usually a full-screen rendering of the 3D model. The main image assumes the role of conducting the linear narrative along with the text. In essence, these elements, supported by a timeline control, help the user customize the amount of information onscreen. In terms of informational devices, these are small-scale networked messages that do not overrule the main linear narrative.

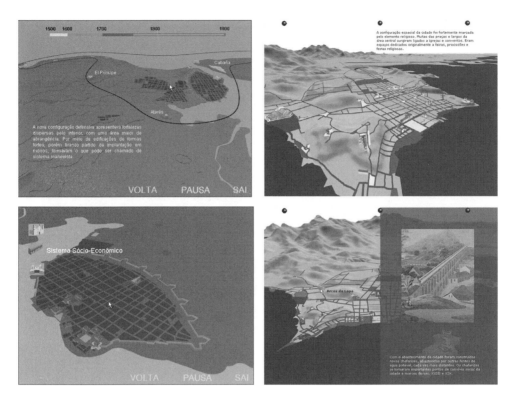

Figure 21.1    Linear narratives supported by virtual world devices in "Havana Colonial" and "Rio Colonial." Hyperdocuments from the Laboratory of Urban Analysis and Digital Representation (LAURD-PROURB/ FAU-UFRJ).

The continuous use of images from the city 3D model as a base for the screen interface helps to establish spatial relationships between elements through time. They were used both in linear narratives and in subsection menus, constituting a virtual world narrative subjacent to the main linear narrative. These images also set the ground for image-based links leading to text, or to new images. As an intellectual apprehension of the space, the emphasis on interaction does not need to be immersive—the realm of virtual worlds—but should rather allow connection between elements that are not necessarily spatially close—which is the main characteristic of the hyperdocument.

These examples may look a bit modest as hyperdocuments, especially when compared to works in fiction or even to games. But then it is worth remembering that the problem we faced then—as now—was the representation of the city evolution that could provide a comprehensive urban analysis.

The analysis people produce still tends to be linear, making it difficult to move away from traditional narratives even in applications of this nature. How can one be sure that the reader is following the line of thought of the authors, and reaching the authors' conclusions, if the argumentation is made of random-visited pieces? Nevertheless, in traditional narratives of analysis we also divide them into several sections and subsections, and they often can be read in different orders. This might indicate that the linearity of the whole argumentation should come more from habits drawn from our traditional writing devices than from an absolute premise, for analysis to take place.

One solution to enable the application of writing habits and new informational device possibilities is to take the tendency towards subdivisions to an extreme, splitting up the discourse as much as possible into discrete parts of information, so they still remain related to each other, but can be presented at random. That path is corroborated by the limited size of computer screens and the pragmatic observation that long texts on the screen are uncomfortable to read. Of course, it still means that writing should be done specifically for a particular informational device. Nevertheless, much can be transposed from traditional devices, such as books, with relative facility, or at least more easily than attempting to directly write in hypertext, given also the difficulty of finding adequate tools.

### Tree Structure

In 1996, the team of LAURD engaged in a short-term project focused on a single building. It was an inversion of our usual perspective—a building, the Catete Palace, was used as a point of departure for the understanding of some parts of the city. The Web site "Palácio do Catete" was also an opportunity to refine some of the propositions presented in "Havana Colonial" (Segre et al. 2005).

This project takes advantage of the essential elements of the hyperdocument. Links are differentiated by whether they lead to the page itself, to an external frame (a photo, for instance), or to an external searching tool—at that time we used altavista .com. It offers an example of navigation through dataflow that despite all parameterization is dynamically updated and often surprising, and that became even more common after the popularization of google.com.

Urban heritage objects are likely to be examined from several points of view. Hyperdocument structures can easily accommodate this, as this Web site shows, ranging from explanations on the site to the excellence of interior decoration.

The visual appeal that is characteristic of hyperdocuments on heritage—after all, they deal with objects that are part of the visible city—stimulated us to profit from navigation through the model in parallel-to-text-based buttons. It also led to the development of a few animations—linear graphic narratives operating at the same time as discrete components of the networked message.

The structure is still made of sections and subsections, decomposing the linear argumentation into units organized in a two-level tree structure that allows independent and random reading of modules. On the other hand, we avoided excessive depth levels, keeping information close to "surface" in pages as rich in information as possible (figure 21.2).

An even greater level of fragmentation is present in the article "The City That Doesn't Exist" (Kós et al. 1999, 1–10). Here, the index shows some theme sections composed by titles of minimum text units; and this index, with all its components, is always present on the left side of the screen. Again, the text was written with the fragmentation and random access in mind. The tree is only one level deep, so it weakens

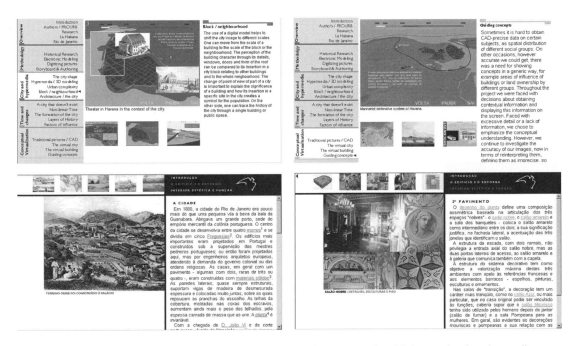

Figure 21.2    Network narrative using tree structure in "Palácio do Catete" and "The city that doesn't exist." Hyperdocuments from LAURD-PROURB/FAU-UFRJ.

the hierarchy—in fact, if the themes were followed, the arrangement of the parts could be random without loss of intended meaning. The index is one of the elements of this discourse, an orientation, but not the only path provided. Its constant presence only reiterates that it allows instant access to any part of the document.

## Array of Categories

The CD-ROM "Havana Colonial" presented a section named "Symbolic Structures Relationships," that consisted of a 3D-modeled plan of the city associated to a matrix of urban elements over three periods of one hundred years. Each urban element is represented by schemes on the map, being selected independently from the others, thus allowing the superimposition of schemes; the mouse pointed over an icon brings a short explanation of the role played by that particular element in that period. The timeline makes it possible to observe the development of the selected items through time. This matrix scheme was later used in another CD-ROM on the Ministry of Education and Public Health Building.

The current project of LAURD-PROURB aims to narrate the history of Rio de Janeiro during the twenty-first century through selected architectural icons. The analysis of the historical context that generated these architectural icons, following the Web site on the Catete Palace, is the starting point for an understanding of the city. The analysis of the first icon is being concluded as an individual CD-ROM and presents the former headquarters of the Ministry of Education and Public Health (MESP), in Rio de Janeiro (Segre et al. 2002, 75–78). The building was designed in 1936 by a group of young architects who became some of the most recognized Brazilian Modern architects—Lucio Costa, Oscar Niemeyer, Carlos Leão, Ernani Vasconcellos, Affonso Reidy, and Jorge Moreira, with the advice of Le Corbusier. The "MESP" CD-ROM is divided into three main parts: previous context, the building, and posterior influences. The section on the building comprises, besides an analysis of the architectural object, subsections on the evolution of the design and the competition for the building design.

The public architectural competition section and the one dealing with the evolution of the project use a networked narrative in the form of a bidimensional array as the basis for interaction. In the first case, it allows comparisons among five different relevant entries related to the building analyses: typologies, volumes, circulation schemes, fenestrations, and relationships with the surroundings. The interface was designed with similar screen layouts for each intersection of the array to help with comparisons, either by these approaches or by the different building proposals (figure 21.3).

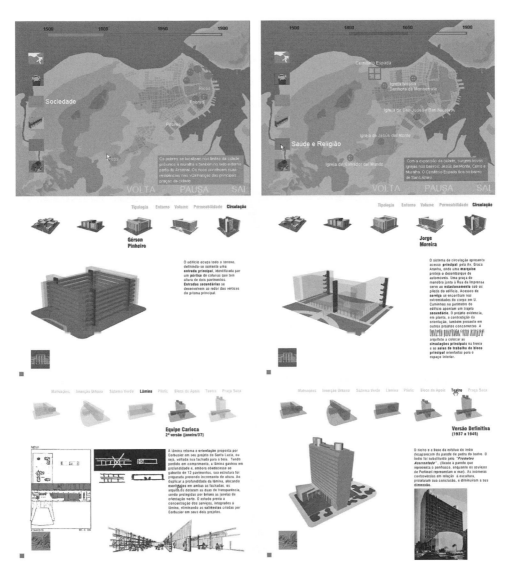

Figure 21.3    Network narratives by array in "Havana Colonial" and in "MESP." Hyperdocuments from LAURD-PROURB/FAU-UFRJ.

The section on the evolution of the design combines five key moments of the design process materialized in five different proposals and the main components of the project—items such as the theater, the columns, the green system, and the vertical slab—as a similar combined menu. The major difference is that the analysis uses different—freer—screen layouts, and the comparisons also take place in the models that are part of the menu.

The section dealing with historical context assumes a one-level tree hierarchy of fragmented pages. The same structure is followed within the typology context section, except that it groups the pages by subject and relies more heavily on the images of each page, since they are an essential tool for establishing the comparisons.

If in "Havana Colonial" we had three to four interfaces to analyze a city, the Ministry of Education and Public Health demands no fewer options to be understood within the city. In "Havana," however, analysis occurs mainly through linear narratives and virtual world narratives with subjacent network narratives; in "MESP" it shifts to specializations of network narratives—topics, groups, arrays—aided by virtual worlds and linear narratives. One of the reasons for this is the experience gained in manipulating hyperdocuments. Another important reason is the capacity hyperdocuments have to easily confront different approaches of a single object, which is of great value when it comes to observing the influences of a building on a city.

### Rules of Combination

The analysis of the city of Rio de Janeiro through architectural icons was also the theme of postgraduate courses in 2001 and 2002 at PROURB. The students were instructed to create representations of iconic buildings or elements of Rio in ways that depicted the reasons for them to be considered elements of convergence of meaning in the city.

One of the examples analyzed an office building, located in a central square, named "Centro Empresarial Internacional Rio," or "Rio Branco 1," or simply "RB1" (Paraizo 2005). We chose it as both a way to understand the changes that occurred in the square where the first high-rise building in the city was constructed, and as a way of demonstrating where the Municipality intends to initiate the waterfront revitalization (figure 21.4).

We tried to explore the possibilities of interaction as elements of representation— just like the texts and illustrations—from the start, believing that this macrostructure and the interaction it made possible could communicate as much as the pages themselves.

Figure 21.4   Network structure by domino-like association in "RB1" (left) and in "The Monroe Palace" (right). Author: Rodrigo Cury Paraizo.

The master guidelines for this structure came from one of Italo Calvino's books, *The Castle of Crossed Destinies*, where a set of narratives is established from a deck of tarot cards, each one recombined to form different stories (1991). The combination of minimum and highly complex semantic units, using some specific previously stated rules, forms each narrative. The restriction imposed by these rules, as Calvino himself recognizes, stimulates creativity rather than restrains it.

The metaphors of cards and the domino game helped us to develop another urban heritage hyperdocument for analytical purposes. It resembles an average network structure, where pages have links to other pages of related subjects. The pages/cards are divided into five suits or subjects. All cards of a suit are accessible from one of their own. In order to establish crosslinks between suits, we arbitrated a maximum of three connections in each card.

As we departed from a hypothetical cardboard version of the work, to be played like an actual game with the ultimate objective of observing the final domino-like composition, we had from the start the intention to somehow record and display the paths followed by users/players. This is a feature yet to be implemented, and not only in our own work. As Steven Johnson notes (2001, 80–101), the potential of links in browsers is not yet fully realized, because one of the still-missing features is precisely the possibility to record and share the several jumps from link to link, just as foreseen by Vannevar Bush (1945) in his Memex.

We strongly believe this kind of network structure, based on some rules, can enhance the experience of navigating through analytical hypertext. In fact, we are developing another hyperdocument, based on the same principles of interaction, for the representation of the Monroe Palace, in Rio de Janeiro (Paraizo 2003).

The structure may also be of assistance in the production of such documents, since one of the problems associated with arrays is the need to organize the subject in equivalent categories with a page for each combination of the sets. Besides the amount of work required to produce all the resultant pages, sometimes it requires an intellectual effort to conceptually fit some aspects to this grid. The possibility of limiting links through some sort of criteria displaces emphasis from comparisons to the actual linking between messages. On the other hand, it avoids the problem of excessive linking leading to meaningless connections.

An analog situation happens in games, where rules coordinate interactivity to produce similar and yet altogether different results each time they are played. Johnson (2001, 159) suggests that we should evaluate electronic representations, namely electronic games, not using the same parameters we employ to appraise novels or even movies, but for their interface, that is, for the relationships they establish with users. We should add that traditional games are already evaluated that way: it is the process, and being part of it, that counts. When applying this logic to hyperdocuments, the suggestion to take games as inspirational models for further types of interactivity comes as no surprise.

### Virtual World Array

Currently the "MESP" CD-ROM uses QTVR panoramas as elements of discourse, by means of hotspots and the use of similar panoramas combined in a one-dimensional array. Each black and white panorama highlights a few elements in such a manner that when the user changes from one theme to another, the point of view of the active panorama is the same; only the highlighted elements change. Each highlighted element leads to additional text and images that explain their importance.

The idea was to bring actual elements and details of the building and its context closer to the user, as active parts of the analysis, an implementation of some form of augmented reality for the desktop—even if it relies on the simplification (black and white, highlights) of reality. Also, we wanted the building to speak for itself as much as possible, with the electronic model playing a secondary role, but nonetheless establishing a virtual world as part of the narrative.

### Historical Virtual World Databases

Electronic databases are powerful tools for representing the complexity and different versions of urban heritages. In addition, Web-based documents allow access, for a wider range of the public, to a number of historical images and documents related to the history of the city, which a traditional, printed vehicle could not afford.

The Web-based system "Rio-H" was developed as a form of digital alternative for the representation of the city history (Kós 2003). The tool is in the prototype stage, and is grounded on 3D models representing different periods of the city, linked to a database containing a great diversity of historical documents. Thus, the city history is accessed through images of the significant sites from the 3D models. The prototype development is based on the assumption that this process of retrieving historical information related to city spaces facilitates the understanding of the past culture. Furthermore, when readers associate the space they know in the city with the historical information, they gain a greater understanding of the past culture that shaped it, strengthen their identities, and intensify their relationships with the place in which they dwell (figure 21.5).

"Rio-H" was an outcome of a PhD thesis (Kós 2003), which investigated, among other issues, how historians structure their narratives. Its structure, therefore, might be viewed as an alternative for the traditional narrative historians use to represent the outcome of their research. Historical documents such as photographs, paintings, newspaper articles, and articles from historians are associated, through the database, to places within the city. Users retrieve the documents through 3D representations of historical settings of the city of Rio de Janeiro. Each hotlink within the 3D image is associated with keywords, and each keyword lists documents that are linked to both the place and the keyword. Thus, users make their choices selecting *Time*, *Place*, and *Subject*.

Figure 21.5    Rio-H interface, displaying the selection of an architectural drawing after choosing a place within a historical 3D image of the city and a keyword. Author: José Ripper Kós.

The importance of the *Place* selection through the image is critical for "Rio-H." The place is usually familiar to users because most of them know the city in its current status. Thus, this connection facilitates the understanding of the historical document though a relationship between past and present.

The structure of navigations through databases allows more flexibility for users who usually have more control over the results of their choices. Although the authors of the database systems are responsible for providing content and defining how it is organized and retrieved, users of these interactive systems have more opportunities to construct a unique sequence of navigation. This means that those users usually have greater possibilities for structuring individual narrative through the hyperdocument.

## Conclusion

We hope that this structure-oriented analysis can enhance the discussion and the production of hyperdocuments on urban heritage. Our intention has not been to establish closed and strict definitions, or to restrain creativity with rules, but to provide enough vocabulary for grouping, as a tool for description and the recognition of common issues among works. In spite of the fact that informational devices can be combined among themselves and presented in several ways in the final product, we need to name its different developments in order to compare and evaluate their effects on the deployment of the message.

## Acknowledgments

The authors would like to offer thanks to CNPq (Brazilian Ministry of Science and Technology) and CAPES (Brazilian Ministry of Education), for the research funding that made this work possible; to PROURB, for hosting our research team; to professors José Barki, Andrea Borde, and Naylor Villas Boas, and especially to Roberto Segre, for providing orientation of the research team; and to our fellow colleagues of LAURD who are responsible for the works presented here.

## Notes

1. See King 1996, 1–19, and Deriu 2001, 794–803.

2. See DeLeon 1999, 484–491, and Digitalo 2002.

3. See Ennis and Maver 2001, 423–429, and ABACUS 2003.

4. See Pletinckx 2000, 45–48, and Provinciaal Museum t'Ename 2004.

5. We would also like to point out another major change that digital media brings that arises when dealing with the preservation of documents, that is, the need to preserve not only the physical media, but also the software that enables reading those specific files. See Brand 1998.

6. The CD-ROM "Rio Colonial" enables user interaction with 3D models suggesting a real-time navigation. Prerendered still images are programmed to give the user the illusion of actual manipulation of the 3D object.

## References

ABACUS. 2005. *TheGlasgowStory*. Available at ⟨http://www.theglasgowstory.com/⟩ (accessed 25 February 2005).

Architecture and Building Aids Computer Unit (ABACUS). 2003. *The Glasgow Director*. Available at ⟨http://www.vrglasgow.co.uk/⟩ (accessed 25 February 2005).

Barki, José. 2003. "O risco e a invenção: um estudo sobre as notações gráficas de concepção no projeto." Ph.D. Thesis, Universidade Federal do Rio de Janeiro.

Bernstein, Mark. 2002. "Padrões do hipertexto." In *Interlab: Labirintos do pensamento contemporâneo*, ed. L. Leão. São Paulo: Iluminuras.

Brand, Stewart. 1998. "Purpose." *The Long Now Foundation: Library*. Available at ⟨http://www.longnow.com/10klibrary/library.htm⟩ (accessed 25 February 2005).

Bush, Vannevar. 1945. "As We May Think." *Atlantic Monthly*, July. Available at ⟨http://www.theatlantic.com/unbound/flashbks/computer/bushf.htm⟩ (accessed 24 July 2002).

Calvino, Italo. 1991. *O castelo dos destinos cruzados*. São Paulo: Companhia das Letras.

DeLeon, Victor J. 1999. "VRND: Notre-Dame Cathedral—A Globally Accessible Multi-User Real-Time Virtual Reconstruction." In *Proceedings of the 5th International Conference on Virtual Systems and Multimedia 1999*, vol. 1, ed. The Scotland Chapter of the International Society on Virtual Systems and Multimedia. Dundee: VSMM Society.

Deriu, Davide. 2001. "Opaque and Transparent: Writings on Urban Representations and Imaginations." *Journal of Urban History*, 27.

Digitalo. 2002. "Virtual Reality Notre Dame Cathedral." *FilePlanet*, 17 June. Available at ⟨http://www.fileplanet.com/22623/20000/fileinfo/VRND---Notre-Dame-Cathedral⟩ (accessed 25 February 2005).

Ennis, Gary, and Thomas W. Maver. 2001. "Visit VR Glasgow—Welcoming Multiple Visitors to the Virtual City." In *Architectural Information Management: Proceedings of the 19th eCAADe Conference*, ed. H. Penttilä. Helsinki: eCAADe.

Johnson, Steven. 2001. *Cultura da interface: como o computador transforma nossa maneira de criar e comunicar*. Rio de Janeiro: Jorge Zahar.

King, Anthony D. 1996. "Introduction: Cities, Texts and Paradigms." In *Representing the City: Ethnicity, Capital and Culture in the 21st Century Metropolis*, ed. A. King. New York: New York University Press.

Kós, José. 2003. "Urban Spaces Shaped by Past Cultures: Historical Representation Through Electronic 3D Models and Databases." Ph.D. Thesis, University of Strathclyde.

Kós, José. 2003. *Rio-H*. December. Available at ⟨http://www.kos.med.br/thesis⟩ (accessed 25 February 2005).

Kós, José Ripper, et al. 1999. "The City That Doesn't Exist: Multimedia Reconstruction of Latin American Cities." In *Proceedings of the 5th International Conference on Virtual Systems and Multimedia 1999, vol. 1*, ed. The Scotland Chapter of the International Society on Virtual Systems and Multimedia. Dundee: VSMM Society. Available at ⟨http://www.fau.ufrj.br/prourb/cidades/vsmm99/⟩ (accessed 25 February 2005).

Laurini, Robert. 2001. *Information Systems for Urban Planning: A Hypermedia Co-operative Approach*. London: Taylor and Francis.

Lévy, Pierre. 1999. *Cibercultura*. Rio de Janeiro: Editora 34.

Malamud, Carl. 2000. "Memory Palaces: A Millennial Metaphor?" *Mappa.Mundi Magazine*, 1 January. Available at ⟨http://mappa.mundi.net/cartography/Palace/⟩ (accessed 25 February 2005).

Martens, Bob, et al. 2000. "Synagogue Neudeggergasse: A Virtual Reconstruction in Vienna." In *Construindo (n)o espaço digital: Anais do IV Seminário Ibero-Americano de Gráfica Digital*, ed. J. R. Kós, A. P. Borde, and D. R. Barros. Rio de Janeiro: SIGraDi/UFRJ/PROURB.

Paraizo, Rodrigo. 2002. *RB1—Rio Branco 1* (January). Available at ⟨http://www.fau.ufrj.br/prourb/cursos/icones/alunos/rb1/index.html⟩ (accessed 25 February 2005).

Paraizo, Rodrigo. 2005. "A representação do patrimônio urbano em hiperdocumentos: um estudo sobre o Palácio Monroe." MA Thesis, Universidade Federal do Rio de Janeiro, 2003. Available at ⟨http://www.fau.ufrj.br/prourb/dissertacoes/rparaizo⟩ (accessed 25 February 2005).

Parente, André, and Luiz Velho. 2000. "Visorama: a Arte do Observador." In *Construindo (n)o espaço digital: Anais do IV Seminário Ibero-Americano de Gráfica Digital*, ed. J. R. Kós, A. P. Borde, and D. R. Barros. Rio de Janeiro: SIGraDi/UFRJ/PROURB.

Pletinckx, Daniel, et al. 2000. "Virtual-Reality Heritage Presentation at Ename." *IEEE Multimedia: Virtual Heritage* 7, no. 2: 45–48.

Provinciaal Museum t'Ename. 2004. *Ename 974 Project* (11 August). Available at ⟨http://www.ename974.org/⟩ (accessed 25 February 2005).

Rocha, Isabel, and Voltaire Danckwardt. 2000. "Projeto Missões, Computação Gráfica." In *Construindo (n)o espaço digital: Anais do IV Seminário Ibero-Americano de Gráfica Digital*, ed. J. R. Kós, A. P. Borde, and D. R. Barros. Rio de Janeiro: SIGraDi/UFRJ/PROURB.

Segre, Roberto, et al. 1998. *Um palácio na cidade*, August. Available at ⟨http://www.fau.ufrj.br/prourb/catete/⟩ (accessed 25 February 2005).

Segre, Roberto, et al. 2002. "Investigação Digital dos projetos do MESP: a busca de vestígios do modernismo brasileiro [Digital exploration of MESP projects: the search for the Brazilian footprints]." In *Proceedings of the 6th Ibero-American Seminar of Digital Graphics*, ed. G. L. Sánchez and G. V. Jahn. Caracas: SIGraDi.

Shields, Rob. 1996. "A Guide to Urban Representation and What to Do About It: Alternative Traditions of Urban Theory." In *Re-presenting the City: Ethnicity, Capital and Culture in the 21st Century Metropolis*, ed. A. King. New York: New York University Press.

Wright, Mark, et al. 1999. "Edinburgh: 200 Years of Heritage Through Image-based Virtual Environments." In *Proceedings of the 5th International Conference on Virtual Systems and Multimedia 1999*, vol. 1, ed. The Scotland Chapter of the International Society on Virtual Systems and Multimedia. Dundee: VSMM Society.

# 22 Automatic Archaeology: Bridging the Gap between Virtual Reality, Artificial Intelligence, and Archaeology

Juan Antonio Barceló

## The Nature of Archaeological Problems

What is archaeology? In contrast to most usual descriptions of the field, archaeologists do not study artifacts as mute witnesses of the past, but analyze social actions performed sometime in the past. We archaeologists are historians using archaeological data to understand the dynamic nature of present society. We need ancient pottery, prehistoric tools, garbage accumulated during centuries to understand what actions were performed in those places at those times. But all such objects are the consequence of social actions. We are not looking for objects, but actions which produced objects with specific features. These are our data: social acts, causal processes relating to people and the products of their work. Therefore, the goal of archaeology is to discover what cannot be seen (*social causes*) in terms of what is actually seen (*material effects*).[1]

Archaeological artifacts have specific physical properties because they were produced so that they had those characteristics and no other. They were produced that way, at least in part, because those things were intended for a particular use: they were tools, or consumed waste material, or buildings, or containers, or fuel, etc. If objects appear in some locations and not in others, it is because social actions were performed in those places at those moments. Therefore, archaeological items have different shapes, sizes, compositions, and textures, and they appear at different places at different moments. That is to say, the changes and modifications in the form, size, texture, composition, and location that experience nature as a result of human action (work) are determined somehow by these actions (production, use, distribution) having provoked its existence. This is the proper goal of archaeological research.

We should describe differences in those features and explain the sources or causes of that variability. Production, use, and distribution are the social processes that

# Problem Solving in Archaeology

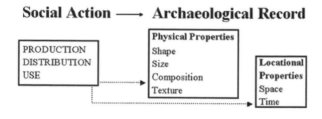

Figure 22.1    A schematic view of what archaeology really is.

in some way have produced (*cause*) observable differences and variability (*effect*). Why do stone axes have different shapes and sizes? Why do graves have different contents? Why do pottery shards have different textures? All of these are *archaeological problems*. In general we are looking at *how* perceptual properties (shape, size, composition, texture, location) allow us to indirectly solve the archaeological problem and attain our goal: finding the social *cause* (production, use, distribution) of what we "see" (figure 22.1).

For instance, some tools have different use-wear texture because they were used to cut different materials; some vases have different shapes because they were produced in different ways; graves have different compositions because social objects circulated unequally between members of a society and were accumulated differentially by elites. Furthermore, shape, size, content, composition, and texture values vary from one location to another, and sometimes this variation has some appearance of continuity. This can be understood as variation between social actions due to neighborhood relationships. Why have stone axes in a specific location in space and time different shapes and sizes? It is impossible to answer why "stone axes" have different shapes and sizes, because there is no general law that applies to all stone axes in the world. But we can ask why stone axes from this location have shapes, sizes, compositions, and textures, when compared to stone axes from another location.

Interpretations of this kind typically constitute what we may call *inverse engineering*.[2] Inverse problems refer to problems where one has observations on the response, or part of the response, of a system and wishes to use this information to ascertain more detailed properties. It entails determining unknown *causes* based on observation of their *effects*. This is in contrast to the corresponding direct problem, whose solution involves finding effects based on a complete description of their causes. When the relevant properties of the causal process are assumed, as well as the initial and boundary conditions, a model then predicts the resultant effect. When we know how a house, a castle, a burial, a tool have been made, under which social and economic conditions, we have a model which predicts the resultant shape and stylistic characteristics given as input into a specification of production context.

Finding solutions by inverse engineering is a particularly difficult task because of the "nonuniqueness" difficulties that arise. Nonuniqueness means, in effect, that the true solution cannot be selected from among a large set of possible solutions without further constraints being imposed (Thornton 2000). This undesirable behavior is due to noise in the measurements, and insufficient numbers of measurements. That is to say, we generally do not know the most relevant factors affecting the shape, size, texture, composition, and spatio-temporal location of material consequences of social action. Instead, we have sparse and noisy measurements of perceptual properties, and an incomplete knowledge of relational contexts. From this information, we need to resolve the causal distribution of nonperceptual properties to adequately interpret archaeological observables. We also must take into account the circumstances and contexts (social and natural) where actions were performed and the processes (both social and natural) that acted upon that place after the original cause, because they may have altered the original effects of primary actions.

## Computers, Robots, and Archaeology

Is it possible to build a machine to practice archaeology by virtue of internal mechanisms? Such a machine has to be able to *solve* archaeological problems, by using relevant knowledge, by detecting the lack of well-organized knowledge, and by "automatically" acquiring knowledge. According to the previous definition of the archaeological problem, this "automatic" archaeologist would be able to solve inverse problems, that is, to infer causal processes and mechanisms, given some observable effects of the causal process.

A formal definition for such an automaton is then "a device with defined inputs, outputs and structure of inner states."[3] This formal definition includes a human being in the class of automata. A human has input senses, a muscular output, and an ability to use this to great effect (excavating old sites, teaching archaeology, etc.). A human

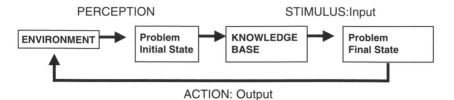

Figure 22.2    A schematic view of an automatic archaeologist functioning.

has a state structure of wondrous complexity, which is the seat of consciousness and intelligence. This complexity distinguishes archaeologists from computers, but mathematically even the best of us can be described as an automaton.

Imagine this "automatic archaeologist" as a *computer* agent (figure 22.2). It can be viewed as a mechanism which perceives its *environment* through sensors and acts upon that environment through *actuators* or *effectors*. The environment provides the problem to be solved, and the solution is the behavior which brings the problem to an end.

In a sense, our automatic archaeologist may be seen as a *cognitive robot*. By cognitive robotics, I refer to the multidisciplinary approach of using techniques from the field of cognitive science to enable the creation of more intelligent, dynamically autonomous robots. We are dealing with a cognitive robot because it is a "physically instantiated . . . system that can perceive, understand . . . and interact with its environment, and evolve in order to achieve human-like performance in activities requiring context-specific (situation and task) knowledge."[4] Research in cognitive robotics is concerned with endowing robots and software agents with higher level cognitive functions that enable them to reason, act, and perceive in changing, incompletely known, and unpredictable environments. Such robots must, for example, be able to reason about goals, actions, when to perceive, what to look for, the cognitive states of other agents, time, collaborative task execution, etc. In short, cognitive robotics is concerned with integrating reasoning, perception, and action within a uniform theoretical and implementation framework. That is to say, when robots use reasoning mechanisms and representations that are similar to those used by humans, we can enable better, more effective human-robot interaction and collaboration.[5]

Robots are computer agents which *act*, that is, they *do* things. In our case, we are speaking of *epistemic actions*:[6] namely, actions whose purpose is not to alter the world so as to advance physically toward some goal (e.g., excavating in order to unearth some archaeological material), but rather to alter the world so as to help make available information required as part of a problem-solving routine. In that sense, *explanations* are for our automatic archaeology machine a form of *behavior*. Expressed in its most basic terms, the task to be performed may be understood in terms of *predicting* which explanations should be generated in the face of determined evidence:

GIVEN a set of observations or data,
FIND one or more general principles that explain those data.

This task is performed by the *function* that maps any given percept sequence to an action. The computer agent correlates evidence and explanation adequately in order to *generate* an adequate action. An automatic archaeologist is then capable of solving archaeological problems through a sequence of epistemic actions or cognitive behaviors which are described by the *function* that maps any given percept sequence to an action.

Let us look at a trivial example. Imagine you have some kind of device able to classify archaeological material, based on its original function. There is a conveyer belt on which the artifacts are loaded. This conveyer passes through a set of sensors or detectors, which measure three properties of artifacts: *shape, texture, and size*. These sensors are somewhat primitive. The shape sensor will output a 1 if the prehistoric tool is approximately round and a − 1 if it is more elliptical. The texture sensor will output a 1 if the surface of the artifact is smooth and a − 1 if it is rough. The size sensor will output a 1 if the artifact is big (greater than 20 cm.), and a − 1 if it is small (less than 20 cm.).

The three sensor outputs will then be put in to a computer program. The purpose of the program is to execute a function deciding which kind of artifact is on the conveyor, so that archaeological materials can be correctly classified. An input pattern is determined to belong to class *knife* if there is a function which relates in some way to incoming inputs with an already defined concept "knife," and a *scrapper*, otherwise. As each artifact passes through the sensors, it can be represented by a three-dimensional vector. The first element of the vector will represent shape, the second element will represent texture, and the third element will represent size.

$$P = \begin{pmatrix} \text{Shape} \\ \text{Texture} \\ \text{Size} \end{pmatrix}$$

Therefore, a prototype knife would be represented by

Figure

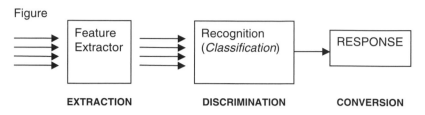

EXTRACTION          DISCRIMINATION          CONVERSION

Figure 22.3    The task of the automatic archaeologist.

$$P_1 = \begin{pmatrix} -1 \\ 1 \\ 1 \end{pmatrix}$$

and a prototype scrapper would be represented by

$$P_2 = \begin{pmatrix} 1 \\ -1 \\ 1 \end{pmatrix}$$

The task of the automatic archaeologist would be to assign any artifact, represented by a vector $\mathbf{x}$, to one of $c$ meanings, or concepts. This can be done according to the schema depicted in figure 22.3.

Our "automatic" archaeologist "reasons" because it is able to infer the values of features that cannot be sensed directly. In this sense, "inference" is only a mechanical way of creating a well-formed formula.

Essentially the idea is to set up appropriate, well-conditioned, tight feedback loops between perception and action, with the external world as the medium for the loop. The actions are gated on perceptual inputs and so are only active under circumstances where they may be appropriate.[7] Because the robot behavior is the cumulative product of an extensive event history, it cannot be understood solely by appeals to the environment of the moment. The environment of the moment directly guides behavior, but the behavior of the moment is the result of the present environment acting on a computer agent that should have been changed by an even richer history of selection. That means that the robot-perceived world is always known in terms directly related to the automatic archaeologist's current possibilities for future action.

Artificial Intelligence has been defined as "the simulation of human intelligence on a machine, so as to make the machine efficient to identify and use the right piece of 'knowledge' at a given step in solving a problem."[8] To the extent that human thinking can be regarded as information processing, it should be possible to simulate or model human thinking on other information-processing devices.

Most AI systems use some sort of model or representation of their world and task. We can take a "model" broadly to be any symbolic structure and set of computations on it that correlate sufficiently with the world about which the computations are designed to yield information. There are two classes of AI systems. One type uses *iconic* models. These involve data structures and computations that *simulate* aspects of human behavior, and the effects of human behavior on a model of the physical world. The other type of model is "feature based." This kind of model uses declarative descriptions of the environment. A set of descriptive features is generally incomplete—a virtue of feature-based representations because they can tolerate the inevitable gaps in an agent's knowledge about complex worlds.

Robotics has nothing to do with "virtual" reality. I am not speaking here of building a "virtual" engine, but a real, concrete system that:

- thinks like archaeologists
- acts like archaeologists
- thinks rationally
- acts rationally

However, in so saying, I am not arguing that artificial archaeologists (computer programs) run like human brains, nor that computer representations should be isomorphic to "mental" states. I am not pretending to simulate me, when I am doing archaeology, but to create something different. The purpose is to understand how intelligent behavior in archaeology is possible. The methodology is to design, build, and experiment with computational systems that perform tasks commonly viewed as intelligent. The goal is not to simulate intelligence, but to understand real (natural or synthetic) archaeological reasoning by synthesizing it.

### Inside an "Automatic Archaeologist"

We assume *computer actions* are abstract mathematical or logical descriptions implemented in a computer program, which takes the current percept as input from the sensors and returns it to the actuators. Therefore, the automatic archaeologist objective is to rigorously apply the scientific method to iterative archaeological object recognition and explanation. This action of machine behavior *predicts* the cause or formation process of some archaeological entity given some observed evidence of the effect of this causal process.

Prediction problems can be represented under the form of a contingent association between an input stimulus (*archaeological description*) and an output response (*explanation*). The main assumption is that the input is related causally to the action it produces. We solve the problem when we define how the input is associated to the output, that is to say, when we understand the nature of the *causal relationship*, which resides in the specific mapping between stimulus and response. As in any other inverse problem, we have access to an output associated to an input, and the task of prediction consists of inferring the mapping (linear or nonlinear) between input and the known output, in order to be able to generalize with new similar inputs. The thing that is predicted is generally termed the *target function*, and the inputs and associated outputs are treated as the arguments and values (respectively) of an unknown function.

There are two contenders for modeling the structure of the mapping between perception and epistemic action, each of which tends to emphasize different aspects of cognition. Production system places emphasis on the fact that much of our behavior, particularly problem-solving behavior, is rule-governed, or can be construed as being rule-governed, and is often sequential in nature—we think one thing after another. Each rule is a process that consists of two parts—a set of symbols acting as tests or *conditions* and a set of symbols acting as *actions*. Here, conditions are only a specific set of key features (empirical or not), and actions are explanatory concepts associated with those features, and used to reach the goal. In general, the condition part of a production rule can be any binary-valued $(0, 1)$ function of the features resulting from perception of the problem givens (initial state). The action part can be either a primitive action, a call to another production system, or a set of actions to be executed simultaneously. The conditional part enumerates those situations in which the rule is applicable. When those conditions are "true," we say that the knowledge represented in the rule consequent has been activated. The underlying logical mechanism is:

IF        FEATURE1    = true
                 (object O has Feature1)
     And    (If Feature1 then Concept X) = true
THEN      CONCEPT X = true
                 (the presence of object O allows the use of Concept X in the
                 circumstances defined by Object O)

Opposite to the use of this kind of logical rule, there is the Connectionist approach.[9] The emphasis of this architecture is on learning and categorizing, and on how we can spontaneously generalize from examples. This approach is based on the idea that functions like perception, problem solving, or memory emerge from complex interactions in highly distributed neuronal networks which, unlike conventional information-processing systems, are shaped by learning and experience-dependent plasticity. In such networks, information processing does not follow explicit rules, but is based on the self-organization of patterns of activity. In this sense, information processing is not carried out as a sequence of discrete computational steps. Just as we apply knowledge gained from previous experiences to solving new problems, an artificial Neural Network "looks" at a set of answers to previously solved examples to build a system of "connections" that makes decisions, classifications, forecasts, and finds solutions to problems in a nonlinear manner. Artificial Neural Networks seem to be the best tools to implement epistemic actions inside our "automatic" archaeologist, given that they are oriented to solve problems for which we do not have precise computational answers.

There are two modes of operation for any artificial Neural Network—the learning phase and the recall phase. During the learning phase, we would "show" the Neural Network multiple sets of historical problem data in which the outcomes are known, kind of a "case study," or sample problems and solutions created with the help of domain knowledge experts. During this process of "learning," the Neural Network adjusts the strength of its internal connections, and once the training is completed, the network should be able to classify or predict based on new inputs. The greater the range of historical data used in the learning phase, the better the resulting network at finding correlations and solving new problems. During the recall phase, the new problem's parameters are being propagated within the artificial Neural Network structure and a solution is being generated, based on what the network "knows" from its training. The recall phase takes almost no time, as only one set of parameters is being considered for the internal connections of the trained network.

In this way, using a neural network to build an automatic archaeologist, we can see problem solving as a constrain-satisfaction procedure in which a very large number of constraints act simultaneously to produce the solution. We can conceptualize a neural network as a *constraint network* in which each unit represents a hypothesis of some sort (for example, that a certain feature is present in the input), and in which each connection represents constraints among the hypotheses. Thus, for example, if feature *B* is expected to be present whenever feature *A* is present, there should be a positive connection from the unit corresponding to the hypothesis that *A* is present, to the unit representing the hypothesis that *B* is present. Similarly, if there is a constraint that whenever *A* is present *B* *is* expected *not* to be present, there should be a negative connection from *A* to *B*. If the constraints are weak, the weights should be small. If the constraints are strong, then the weights should be large. Similarly, the inputs to such a network can also be thought of as constraints. A positive input to a particular unit means that there is evidence from the outside that the relevant feature is present. A negative input to a particular unit means that there is evidence from the outside that the feature is not present. The stronger the input, the greater the evidence. If such a network is allowed to run, it will eventually *settle* into a locally optimal state in which as many as possible of the constraints are satisfied, with priority given to the strongest constraints.

### *"Visualizing" Archaeology*

One way of using the "automatic" archaeologist is for building visual models that help to understand the data.

"Visualizing" is not the same as "seeing," but an inferential process to aid the understanding of reality. The idea is not to take a "picture" of the artifact, but to decompose empirical information in terms of its location marks (shape, size, location) and retinal properties (texture, composition). That is, geometry is used as a visual language to represent a theoretical model of the pattern of contrast and luminance, which is the strict equivalent of perceptual models of sensory input in the human brain. All this means is that "visualizing" the real world is not the same as "picturing" it, because the model and the graphical means for creating and visualizing the world are distinct.

This is the task of our "automatic" archaeologist: to create a *geometric representation* of the regularity present in a data set: joining points with lines, fitting surfaces to lines, or "solidifying" connected surfaces.

The main reason for visual models is to help see what the data seems to be saying and to test what you think you see. They are used to convert input data (usually numerical) into visual concepts acting as a model for that data. That is, the mapping of abstract inputs into graphical representations (lines, surfaces, and solids) as an aid in the understanding of complex, often massive numerical inputs of scientific concepts or results. A visual model will compress a lot of data into one picture (data browsing), so it can reveal correlations between different quantities both in space and time. It can furnish new space-like structures besides those already known from previous calculations, and it opens up the possibility of viewing the data selectively and interactively in "real time."

Visual models are then the equivalent of sensory representations in the brain: a translation of empirical phenomena into a geometric language. As models, they are the result of a transformation of input data into a geometric explanation of the input, with light and texture information.

Visually based knowledge construction offers two distinct advantages. The first is support for a high degree of interaction. This means the possibility of creating an opportunity for humans and machines to work together in constructing and evaluating the objects and relationships required by the analysis, and allowing the user to explore data from several perspectives and to steer the process. The second is the rich visual environment, allowing many dimensions of data to be viewed concurrently. In both ways, visualization can play an important role in "process-pattern tracking" (visual representation that displays key aspects of a process as it unfolds) and "process steering" (interactive environments that provide controling parameters of knowledge-construction processes to shape and modify their behavior).

### Using Automatic Archaeologists

Let us see how an automatic archaeologist will use mathematics and visual models in understanding an archaeological site (figure 22.4).

Our intention here is to explore the way a computer program (an "automatic" archaeologist) can create an explanatory model to understand how the archaeological site we are excavating is the result of a sequence of social actions. Archaeologists traditionally have drawn their inferences about past behavior from dense, spatially discrete aggregations of artifacts, bones, features, debris, called *archaeological sites*. We have traditionally assumed that the main agent responsible for creating such aggregates was *only* human behavior. Even though today most archaeologists are aware of natural disturbance processes and the complexities of archaeological formation, we usually speak of human behavior as being fossilized in archaeological *sets*. Aggregates, assemblages, deposits, and accumulations seem to be, then, the only "visible" aspect of social action.

We have seen that archaeology is a kind of inverse engineering, looking for the *causes* (or formation processes) of observable effects. In our case, the automatic archaeologist will use as input data accumulations of archaeological material (effect), and it

Figure 22.4   Using a visual model to explain archaeological remains of the Tunel VII site (Tierra del Fuego, Argentina) as generated by human work inside and around a hunter-gatherer hut. Photograph provided by Jordi Estevez (Universitat Autonoma de Barcelona).

should build a visual model of that data depicting its role in the site-formation process (cause). We assume that the dispersal or differential accumulation of items may be patterned as a direct consequence of the formation process, but it does not follow that the patterning of the archaeological artifacts and the patterning of the human behavior that produced them are identical. The major problem is the *degree* to which the accumulation or deposit of archaeological items can be attributed to social action. Most post-depositional processes have the effect of disordering artifact patterning in the archaeological record, and increasing entropy. Loss, abandonment, reuse, decay, and archaeological recovery are numbered among the diverse formation processes that, in a sense, mediate between past behaviors of interest and their surviving traces. They make archaeological assemblages more amorphous, lower in artifact density, more homogeneous in their internal density, less distinct in their boundaries, and more similar (or at least skewed) in composition. Our automatic archaeologist has to deal with these problems.

The computer agent should generate an analytical decomposition of spatial relationships into geometric "units" (points, lines, areas, etc.) with the idea that if we can specify the (*spatial*) behavior of each unit, we can understand the behavior of the whole system. Spatial behavior means that the relevant properties of any entity vary from one location to another, (either temporal location or spatial location), and sometimes this variation has some appearance of continuity, which should be understood as variation between causal actions due to neighborhood relationships. The goal of the automatic archaeologist is then to analyze how the material consequences of social action (the archaeological record) vary significantly from one location to another. The idea is that temporal relationships influence the spatial position of social acts, in the same way as the spatial relationships of actions influence the temporal location (reproduction) of the same actions. In other words, the main objective should be the correlation of different social actions:

- how the spatial distribution of an action has an influence over the spatial distribution of (an)other action(s);
- how the temporal displacement of an action has an influence over the spatial distribution of (an)other action(s);
- how the temporal displacement of an action has an influence over the temporal displacement of (an)other action(s); and
- how the spatial distribution of an action has an influence over the temporal displacement of (an)other action(s).

The computer agent must decide whether what happens in one location (temporal or spatial) is the *cause* of what happens in neighboring locations, because in most occasions social actions are performed in an intrinsically better or worse location for some purpose. Obviously, the system should not limit itself to the analysis of "spatial similarity" relationships, but all effects *probabilistically* related to the spatial or temporal location of the cause. Spatial causality is a probabilistic relationship. It can be calculated from the sum of probabilities of actions performed in a given area around a location, whether or not those actions influenced the action performed on that location, or whether the sum of probabilities explain why an action was not performed in that location.

To adequately represent archaeological spatial conditions on a computer we must consider a semi-infinite continuum made up of discrete, irregular, discontinuous volumes defined by characteristics which, in turn, influence the spatial variation of an

Figure 22.5 (a,b)   Taking data in the field: spatial values (the archaeological site of Shamakush VIII, Tierra del Fuego, Argentina). Photographs by the author.

Figure 22.5 (c) (continied)

archaeological or geological feature. Those volumes represent single, differentiated accumulations, which may also be described and analyzed using global attributes:

- the average density of attributes; and
- the form of arrangement of artifacts within it—clustered, random, or systematically spaced—independent of density.

These volumes, called *archaeological phases*, are the building blocks for the model, and they should be understood as events. They are expressions of the fact that a homogenous region in a space-time continuum is delimited by an abrupt change (or *interfacial discontinuity*) in the values of some quantitative or qualitative properties. The entity we are studying is the *archaeological site*, which can be in different states, depending on the number and nature of archaeological phases identified. We want to discover whether the features defining states of that entity have changed or not, and, if so, when, how, and why (figures 22.5a,b,c).

The idea is to consider that a social action or sequence of social actions will be causally related to a *spatial change* if and only if the probability for a specific modification present at a distinct region (a *phase*) is higher in the presence of that action than in its absence. That is, changes in the visible characteristics of archaeological phases in shape are not determined univocally by the location of social actions (production, distribution, and use). But there is some probability that in some distinct locations productive, distributive, or use actions are more probable than others. A wall, an occupation floor, an activity area, a grave, a pit hole are *phases* or distinct regions of the archaeological space. Their *composition* or the specific accumulation of materials related to their boundaries depends very much on where and how they were formed. In this sense, physical modifications seem to act as classifiers associated with discrete archaeological units with distinct boundaries. Their importance to archaeological characterization lies in the fact that they frequently influence the spatial variation of differential accumulations.

By building a "visual" surrogate of the site, the automatic archaeologist is able to provide a convenient vehicle for "experimentation" with, and prediction of, the effects of causal factors, and to show how some state-change in a spatial location produces changes in other spatial locations (figure 22.6). Each phase is likely to have a unique spatial variation, and this variation is likely to be different, and therefore discontinuous, from one location to the next. It may be directionally dependent, and possibly consists of several superimposed variations.

### Beyond "Virtual" Archaeology

The word "automaton" as used in this paper conjures up a vision of a sullen android. Its "mathematical" core can also tend to repel humanists. Why do we need such an "automatic" archaeologist? There are two reasons; the first is that archaeology is too important to exist solely in the hands of humans trying to understand themselves. And the second is a mere technical aspect: computers do our job better than we do.

Of course, this is a provocative view of archaeology and cultural studies—the idea that the study of cultural matters is so easy that even a computer can do it better than we can. I have written this paper to reveal this fact.

There are still many scholars working in self-limited humanist disciplines dedicated to the unearthing of past treasures, and the static description of past ways of life. In this sense, most archaeological and cultural information seems to be *artifactual*, because it deals with the relevance of archaeological finds as self-important entities. Social action is here reduced to a mere description of objects constructed by indeterminate people.

In most cases the modern use of virtual reality in archaeology and related disciplines follows this old-fashioned and static approach. "Visual" models of past monuments

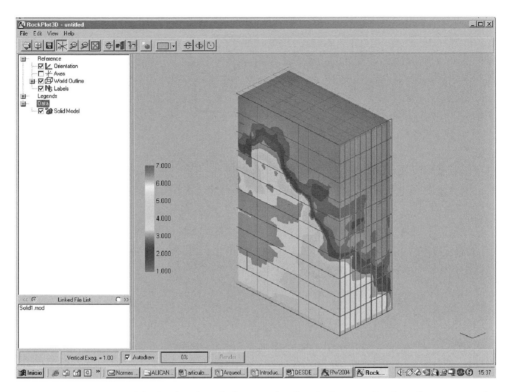

Figure 22.6   A visual model of archaeological phases at the Shamakush VIII archaeological site. (Using the Rockworks 2004 software ⟨www.rockware.com⟩).

seem more an artistic task than an inferential process. Virtual reality is the modern version of the artist that gave a "nice" reconstruction using watercolors. Computer scientists take the drawings of these "imaginative" reconstructions, and transform them into computer language. The computer gives 3D, color, and texture to the archaeologist's (or architect's) imagination. As a direct result of this uncritical acceptance, fundamental questions relating to what our expensively assembled models truly represent have been left largely unexplored. Simply put, the better and more optimized the data used in a computer model, the more faithful the visual model seems to be, and the closer it seems to come to the reality it seeks, or purports, to represent. The considerable efforts currently being expended in incorporating ever more detail into models, whether through the use of individual bricks and stones rather than simplified blocks, or the application of highly complex textures, achieve little more than the generation of an even more fastidious investigative attitude on the part of the observer.

I am suggesting a totally different approach, where archaeologists do not "paint" the past, but are involved in social science inverse engineering. Archaeology becomes a discipline dealing with the *history* of our society, that is, those processes that have *caused* our present. And we use computer technology and computer agents to *help* in this endeavor.

*Notes*

1. These general principles are presented in Barceló 1996 and 2002.

2. For a general presentation of what inverse engineering is and its methods, see Woodbury 2002 and Uhlmann 2003.

3. Definition suggested in Aleksander and Morton 1993, 97.

4. This sentence is quoted as an explicit objective by the European Commission Framework 6 Objectives for Cognitive Systems.

5. Others have used the performance of humans at a particular task to design a robot that can do the same task in the same manner (and as well): Moravec 1999, Nolfi and Floreano 2000, Murphy 2002, Santore and Shapiro 2004, King et al. 2004, and Kovács and Ueno 2005.

6. See Stein 1995 and Kirsh and Maglio 1995.

7. For a theory of human and robotic perception, see Donahue and Palmer 1994.

8. See Konar 2000, Wagman 2002, and Russell and Norvig 2003.

9. General references to this field are Seidenberg 1990, Mehrotra et al. 1997, Haykin 1999, and Principe et al. 2000.

*References*

Aleksander, Igor, and Helen Morton. 1993. *Neurons and Symbols: The Stuff That Mind Is Made Of*. London: Chapman and Hall.

Barceló, J. A. 1996. *Arqueología Automática. El uso de la Inteligencia Artificial en Arqueología*. Barcelona: Editorial Ausa.

Barceló, J. A. 2002. "Archaeological Thinking: Between Space and Time." *Archeologia e Calcolatori* 13: 237–256.

Donahue, J. W., and D. C. Palmer. 1994. *Learning and Complex Behaviour*. Boston: Allyn and Bacon.

Haykin, S. 1999. *Neural Networks: A Comprehensive Foundation*. 2d ed. Upper Saddle River, N.J.: Prentice Hall.

King, R. D., K. E. Whelan, F. M. Jones, P. G. K. Reiser, C. H. Bryant, S. H. Muggleton, D. B. Kell, and S. G. Oliver. 2004. "Functional Genomic Hypothesis Generation and Experimentation by a Robot Scientist." *Nature* 427: 247–252.

Kirsh, D., and P. Maglio. 1995. "On Distinguishing Epistemic from Pragmatic Action." *Cognitive Science* 18: 513–549.

Konar, A. 2000. *Artificial Intelligence and Soft Computing: Behavioural and Cognitive Modeling of the Human Brain*. Boca Raton: CRC Press.

Kovács, A. I., and H. Ueno. 2005. "The Case for Radical Epigenetic Robotics." *Proceedings of the 10th International Symposium on Artificial Life and Robotics (AROB 2005)*. Beppu, Oita, Japan, 4–6 February. Available at ⟨http://www.alexander-kovacs.de/kovacs05arob.pdf⟩ (accessed 20 April 2005).

Mehrotra, K., C. K. Mohan, and S. Ranka. 1997. *Elements of Artificial Neural Networks*. Cambridge, Mass.: MIT Press.

Moravec, H. 1999. *Robot: Mere Machine to Transcend Mind*. New York: Oxford University Press.

Murphy, R. R. 2002. *An Introduction to AI Robotics*. Cambridge, Mass.: MIT Press.

Nolfi, S., and D. Floreano. 2000. *Evolutionary Robotics*. Cambridge, Mass.: MIT Press.

Principe, J. C., N. R. Euliano, and W. C. Lefebvre. 2000. *Neural and Adaptive Systems: Fundamentals Through Simulations*. New York: Wiley.

Russell, S., and P. Norvig. 2003. *Artificial Intelligence: A Modern Approach*. 2d ed. Englewood Cliffs: Prentice Hall.

Santore, J. F., and S. C. Shapiro. 2004. "A Cognitive Robotics Approach to Identifying Perceptually Indistinguishable Objects." In *The Intersection of Cognitive Science and Robotics: From Interfaces to Intelligence, Papers from the 2004 AAAI Fall Symposium*, Technical Report FS-04–05, ed. A. Schultz, 47–54. Menlo Park, Calif.: AAAI Press.

Seidenberg, M. 1990. *Neural Networks in Artificial Intelligence*. Chichester: Ellis Horwood.

Stein, L. A. 1995. "Imagination and Situated Cognition." In *Android Epistemology*, ed. K. M. Ford, C. Glymour, and P. J. Hayes. Menlo Park/Cambridge, Mass.: AAAI Press/MIT Press.

Thornton, C. 2000. *Truth from Trash: How Learning Makes Sense*. Cambridge, Mass.: MIT Press.

Uhlman, G., ed. 2003. *Inside Out: Inverse Problems and Applications*. Cambridge: Cambridge University Press.

Wagman, M. 2002. *Problem-solving Processes in Humans and Computers: Theory and Research in Psychology and Artificial Intelligence*. Westport: Praeger.

Woodbury, A. 2002. *Inverse Engineering Handbook*. Boca Raton: CRC Press.

# Contributors

*Nadia Arbach*, E-Learning Curator, Tate, London, United Kingdom.

*Juan A. Barceló*, PhD, Professor of Prehistory, Universitat Autonoma De Barcelona, Departament De Prehistòria, Facultat De Lletres. Edifici B (B9-119), Bellaterra, Spain.

*Deidre Brown*, PhD, Senior Lecturer, School of Architecture, National Institute of Creative Arts and Industries, University of Auckland, New Zealand.

*Fiona Cameron*, PhD, Research Fellow, Museum and Cultural Heritage Studies, Centre for Cultural Research, University of Western Sydney, Sydney, Australia.

*Erik Champion*, PhD, Lecturer, Interaction Design, Information Environments Program, ITEE, University of Queensland, Australia.

*Sarah Cook*, PhD, Postdoctoral researcher and curator, School of Arts, Design, Media and Culture, University of Sunderland, Sunderland, UK, in collaboration with BALTIC Centre for Contemporary Art, Gateshead, United Kingdom.

*Jim Cooley*, Director, Technology Incubation and Strategy, Microsoft Corporation, One Microsoft Way, Redmond, WA, USA.

*Bharat Dave*, PhD, Associate Professor of Information Technology, Associate Dean (Research), Faculty of Architecture, Building and Planning, University of Melbourne, Australia.

*Suhas Deshpande*, Technology Assessment Analyst, Canadian Heritage Information Network.

*Bernadette Flynn*, Lecturer Screen Production, Griffith Film School, Queensland College of Arts, Griffith University, Australia.

*Maurizio Forte*, PhD, Primo Ricercatore, Senior Scientist, Archaeologist CNR-ITABC, Istituto per le Tecnologie Applicate ai Beni Culturali, Rome, Italy; Vice-President Virtual Heritage Network.

*Kati Geber*, Manager of Research and Business Intelligence, Canadian Heritage Information Network, Canada.

*Beryl Graham*, PhD, Professor of New Media Art, School of Arts, Design, Media, and Culture, University of Sunderland, Sunderland, United Kingdom.

*Susan Hazan*, PhD, Curator of New Media, New Media Unit, Computer and Information Systems, The Israel Museum, Jerusalem, Israel.

*Sarah Kenderdine*, Director of Special Projects, Museum Victoria; Project Manager, The Virtual Room, Melbourne Museum, Victoria, Australia; Director Virtual Systems and Multimedia; Director Virtual Heritage Network.

*José Ripper Kós*, coordinator of the Laboratory of Urban Analysis and Digital Representation, and professor at the Postgraduate Program in Urbanism, Faculty of Architecture and Urbanism, Federal University of Rio de Janeiro, Brazil.

*Harald Kraemer*, PhD, Project Director of Artcampus, Institute of Art History, University of Berne, and Producer, Creative Director of TRANSFUSIONEN, Zurich, Switzerland.

*Ingrid Mason*, Epublications Librarian, Alexander Turnbull Library, National Library of New Zealand, Te Puna Mātauranga o Aotearoa, Wellington, New Zealand—Aotearoa.

*Gavan McCarthy*, Director, Australian Science and Technology Heritage Centre, The University of Melbourne, Australia.

*Slavko Milekic*, MD, PhD, Department of Art Education and Art Therapy, University of the Arts, Philadelphia, USA.

*Rodrigo Paraizo*, member of the Laboratory of Urban Analysis and Digital Representation, Postgraduate Program in Urbanism, Federal University of Rio de Janeiro and professor at the Architecture Graduation course of the Catholic University of Rio de Janeiro, Brazil.

*Ross Parry*, PhD, Lecturer in Museums and New Media, Department of Museum Studies, University of Leicester, United Kingdom.

*Scot Thrane Refsland*, PhD, Managing Director of RedClay Studios, USA; Executive Officer, Virtual Systems and Multimedia, Secretariat Virtual Heritage Network.

*Helena Robinson*, Curator and Collection Manager, Harold Nagley Moriah Heritage Centre, Moriah College, Sydney, Australia.

*Angelina Russo*, PhD, Senior Research Associate, Creative Industries Research and Applications Centre, Queensland University of Technology, Australia.

*Corey Timpson*, Visual and Interaction Designer, Canadian Heritage Information Network, Canada.

*Marc Tuters*, co-founder of Locative Media Lab; Principal Investigator, Mobile Digital Commons Network, Montreal, Canada.

*Peter Walsh*, Art and Architecture Critic, WBUR.Arts/National Public Radio; Contributing Writer, Museums Magazines; Consultant, Dartmouth College, United Kingdom.

*Jerry Watkins*, Senior Research Associate, Creative Industries Research and Applications Centre, Queensland University of Technology, Australia.

*Andrea Witcomb*, PhD, Associate Professor, Citizenship and Globalsation, Faculty of Arts, Deakin University, Melbourne, Australia.

# Index